Building Construction: Materials, Strength and Properties

Building Construction: Materials, Strength and Properties

Edited by **Seth Royal**

Willford Press
New York

Published by Willford Press,
118-35 Queens Blvd., Suite 400,
Forest Hills, NY 11375, USA
www.willfordpress.com

Building Construction: Materials, Strength and Properties
Edited by Seth Royal

International Standard Book Number: 978-1-68285-243-9 (Hardback)

Printed in China.

Contents

Preface

Building construction has become a prominent field of study and research with the rising need for settling the ever growing human population. Properties of building materials are studied under various disciplines of engineering like civil, geological and material sciences. This book is a compilation of chapters that discuss the most vital concepts of building material and technology including corrosion technology, concrete reinforcements, bricks, ceramics and cement. Researches and case-studies by eminent industry experts and academicians are also incorporated that discuss the strength and various properties of materials that are currently used in construction industry. It provides an insight into construction of various buildings and their maintenance. This book will prove to be immensely beneficial to students, researchers and professionals engaged in construction industry.

This book unites the global concepts and researches in an organized manner for a comprehensive understanding of the subject. It is a ripe text for all researchers, students, scientists or anyone else who is interested in acquiring a better knowledge of this dynamic field.

I extend my sincere thanks to the contributors for such eloquent research chapters. Finally, I thank my family for being a source of support and help.

Editor

Experimental determination of damping ratios at higher modes for use in modal superposition

V. Ramamurti[1]*, S. Mithun[2], N. Prabhakar[2] and T. Sukumar[2]

[1]Indian Institute of Technology Madras (IIT Madras), Chennai, India.
[2]WABCO India Limited, Plot No. 3 (SP), III Main Road, Ambattur Industrial Estate, Chennai, Tamil Nadu 600058, India.

While carrying out the transient analysis, the major challenge is to provide accurate damping ratio data for higher modes as this is not reported. Hence an attempt is made to find out the damping ratios experimentally in this paper. Two approaches of experimental evaluation of damping ratios are discussed, for free and forced vibration, with four different materials namely: mild steel, aluminum, cast iron and plastics. For the first set, eleven cantilevers of uniform cross section with varying lengths are taken for evaluation. For the second set, fourteen components made of these materials used in air brake system of automobiles are also examined. For the free vibration, impact test is used to find the fundamental natural frequency and damping ratio. In forced vibration study, electro dynamic shaker is used to find the first three natural frequencies and damping ratios. Besides the above methods, natural frequencies are calculated analytically using finite element analysis (FEA). Based on the exhaustive experimental work carried out design guidelines on damping ratios for industrial problems are presented.

Key words: Finite element method, experimental determination of damping ratios at higher modes for different materials.

INTRODUCTION

One of the major problems encountered in transient analysis is the data needed on damping ratios corresponding to higher modes, when finite element method is used. It is well known that this depends on the material subjected to transient response and the natural frequency of the corresponding mode of vibration. This is the only method available to compute the damping matrix [C], even when one analyses transient vibration problems using time integration scheme (Ramamurti, 2012). Basic requirement for the transient analysis is to get the damping ratios corresponding to the first few modes if the concept of Rayleigh damping is used in modal superposition technique.

Since damping matrix is expressed as a combination of stiffness and mass matrices, two damping ratios are needed, one corresponding to the fundamental mode and the other one corresponding to a higher mode. These two are obtained experimentally in this article to determine the multipliers α and β.

These transient vibrations are experienced by marine structures subjected to severe impact due to floating ice bergs, rocket launchers during firing, machine tools like forging hammers and guillotine shears, inlet and delivery valves of compressors, brake pads of heavy duty vehicles during braking and space vehicles during launching.

Key researchers in the early era around 1950 like Ungar, Kerwin and Lazan addressed the damping problem. Companies like Bruel and Kjaer and LMS International collected considerable data, but they are not

*Corresponding author. E-mail: ramamurti@hotmail.com.

Abbreviations: ([C], [K], [M]), Damping, stiffness, mass matrices of size (n*n); **(B, L, T),** width, length, thickness of the plate; **E,** Young's modulus; **[X],** eigen vector; **(f_1, f_2, f_3),** natural frequencies Hz; **x,** amplitude; **t,** time; **(α, β),** mass and stiffness multipliers; ρ, Density; **(η^1, η^2),** frequency ratios on two sides of resonance; **(ω_1, ω_2, ω_3),** natural frequencies rad/s; (ζ_1, ζ_2, ζ_3), damping ratios; **(S_1, S_2, S_3),** steel samples; **(A_1, A_2, A_3),** aluminum samples; **(C_1, C_2, C_3),** cast Iron samples; **(P_1, P_2, P_3),** plastic samples; **REV,** relay valve; **QRV,** quick release valve; **MTS,** modular door system; **BC,** brake chamber.

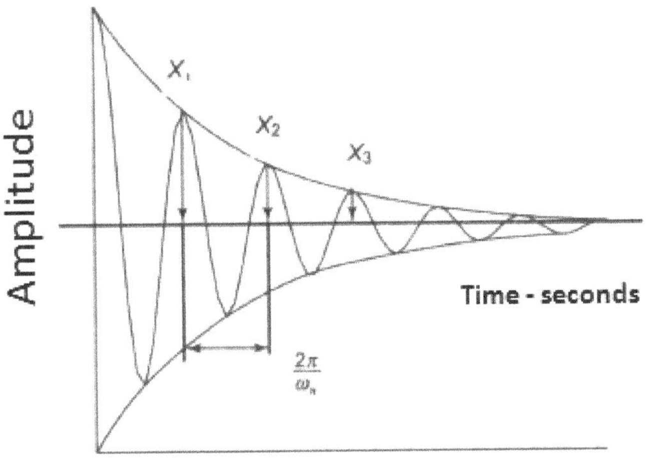

Figure 1. Free vibration response.

available in public domain. Besides most of them are not readily adaptable to modal super position technique. Standard packages like NASTRAN, ANSYS, ABAQUS and ADINA demonstrated Rayleigh damping concept through examples but did not spell out the procedure for computing the stiffness (β) and mass matrix multipliers (α). It is understandable since user manuals of these packages are not expected to provide design guidelines. They did not explain the basis of arriving at these coefficients. Obviously, these are not constants like acceleration due to gravity, but variables which depend on the material with which the vibrating member is fabricated and its first few natural frequencies.

There are not many papers on damping ratios of the first few modes of vibration reported in the last few years. Most experimental investigations are on vibration response of linear and non-linear systems (Bellette et al., 2008; Chow et al., 2005; Huang, 2005; Kuljamic et al., 2008; Mann et al., 2008; Issa et al., 2008; Li hen et al., 2008; Xiao et al., 2005; Zhang et al., 2004). There are a few experimental results on bodies surrounded by water (Grace et al., 2008; Li et al., 2010; Ramajayathilagam et al., 2004)

COMPUTING DAMPING RATIOS

There are two ways of experimentally figuring out the damping present in a system, one by the free vibration record and another by the forced vibration method (√2 method). By the first, one can compute the damping ratio for the fundamental mode only while by the second method it is possible to obtain the damping ratios for the first few modes.

Free vibration response when disturbed

It is well known, that any complicated system, when disturbed, vibrates, response gradually decreases and, towards the last phase of decay, exhibits the response corresponding to the fundamental mode.

The response of a single degree freedom can be written as

$$x = \exp(-\zeta_1 \omega_1 t)(c_1 \cos \omega_1 t + c_2 \sin \omega_1 t) \quad (1)$$

This can be represented by the curve shown in Figure 1. In this figure three successive amplitudes x_1, x_2, x_3 are separated by time difference

$$t = 2\pi/\omega_1 \quad (2)$$

Ratio of successive amplitudes is given by

$$\frac{x_1}{x_2} = e^{2\pi\zeta_1} \quad (3)$$

hence,

$$\log\left(\frac{x_1}{x_2}\right) = 2\pi\zeta \quad (4)$$

Logarithmic ratio of successive displacements is equal to 2πζ. By making use of the time period and successive amplitudes towards the dying phase of the multi degree freedom system one can compute ω_1 and ζ_1 of practical vibrating systems. Since these systems have very low damping ratios Equations (3) and (4) used in finite element methods are valid.

Determination of ζ by √2 Method

In the forced vibration method, the system is excited harmonically, at a chosen frequency and the system has the maximum response corresponding to a given natural frequency falling steeply on either side of this natural frequency. Shown in Figure 2, is a resonance zone of the frequency response curve of forced vibration. For a given natural frequency ω_i, let the maximum displacement be x_{max} corresponding to a given natural frequency. Let the frequency ratios for $x_{max}/\sqrt{2}$ values on either side of this resonance be ω_* and ω_{**}. If the natural frequency under consideration is ω_i, the frequency ratios $\eta_1$and η_2 be ω_* / ω_i and ω_{**} / ω_i with the ratio η_m =1at the peak frequency. Let the damping ratio present in the system be ζ.

From the expression for forced vibration response [18], one can write,

$$\frac{1}{2}\left(\frac{1}{2\zeta}\right)^2 = \frac{1}{(1-\eta^2)^2 + (2\xi\eta)^2} \quad (5)$$

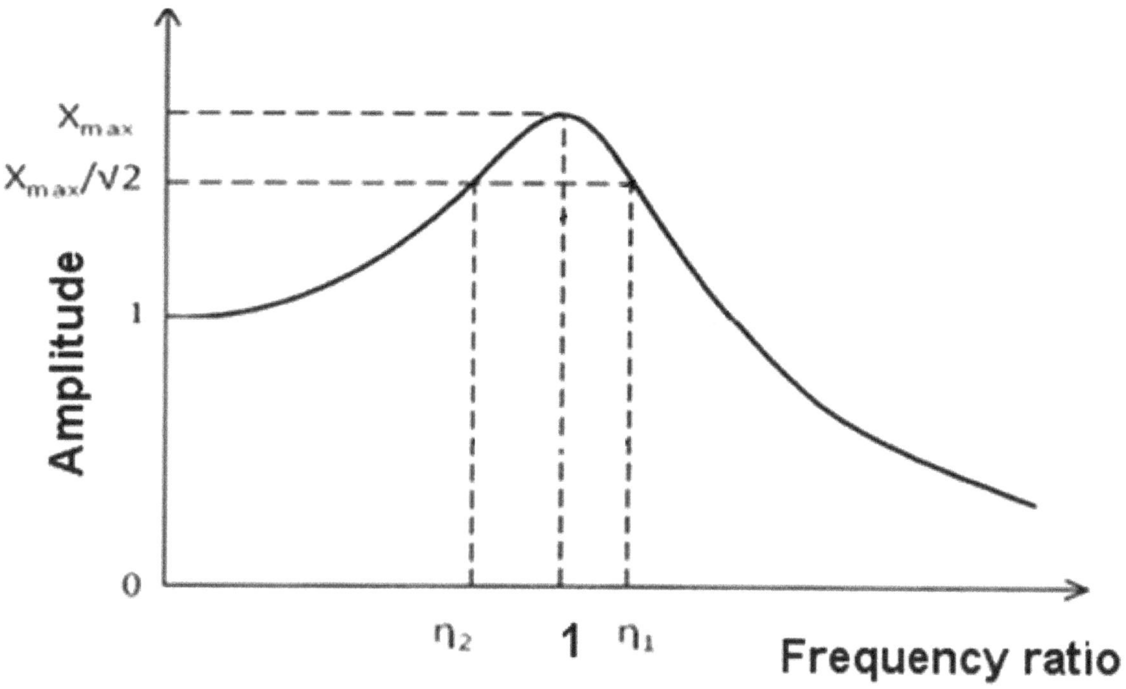

Figure 2. Forced vibration response.

Since ζ, for practical systems is <<1.It can be proved that

$$\zeta = \frac{(\eta_1 - \eta_2)}{2} \tag{6}$$

where η_1 and η_2 being the two frequency ratios as shown in Figure 2.

This method is known as the √2 method. This forced vibration test is a very useful tool for computing the damping ratios of the first few modes even for multi degree freedom systems.

RAYLEIGH DAMPING

For the multi degree freedom system, the determination of damping matrix [C] is not easy. Since only [M] and [K] are available Rayleigh damping assumes that [C] can be expressed as a linear combination of [M] and [K]. It is expressed as

$$[C]=\alpha[M]+\beta[K] \tag{7}$$

α and β are matrix multipliers to be determined. Since K is expressed in N/mm and M in Ns^2/mm it implies from Equation (7) α must be in s^{-1} and β in s. If Equation (8) is pre multiplied by eigen vectors [X], and post multiplied by $[X]^T$, one gets

$$[X]^T[C][X]=\alpha[X]^T[M][X]+\beta[X]^T[K][X] \tag{8}$$

From the basic properties of eigen pairs (Ramamurti, 2012)

$$[X]^T[M][X]=\begin{bmatrix}1 & 0 & 0\\ 0 & 1 & 0\\ 0 & 0 & 1\end{bmatrix}, \quad [X]^T[K][X]=[\Omega^2]=\begin{bmatrix}\omega_1^2 & 0 & 0\\ 0 & \omega_2^2 & 0\\ 0 & 0 & \omega_3^2\end{bmatrix} \tag{9}$$

If $[X]^T[C][X]$ = is expressed as [2 $\zeta_i\omega_i$] to make it diagonal and expanding Equation (9) one gets

$$\begin{aligned}2\xi_1\omega_1 &= \alpha + \omega_1^{\,2}\beta\\ 2\xi_2\omega_2 &= \alpha + \omega_2^{\,2}\beta\\ 2\xi_m\omega_m &= \alpha + \omega_m^{\,2}\beta\end{aligned} \tag{10}$$

Equation (10) indicates α and β are not constants, but matrix multipliers which depend on the first few damping ratios and the natural frequencies of the member analyzed. In this paper, forced vibration response is used for experimental determination of the first three natural frequencies and damping ratios. This approach became popular after the finite element method was successfully used to solve problems in the area of dynamics with computational efficiency in mind.

RELEVANCE OF INVESTIGATION

It may appear, as if, "Computing Damping Ratios" and "Rayleigh Damping" are well known and are reported in most books. If one looks at the industrial scenario, the following may be observed. A marine steel structure experiences disturbance due to floating iceberg, probably once an hour, a multi barrel rocket launcher impact once in four seconds, a guillotine shear made of a steel housing every two seconds (in continuous mode of shearing of thick plates) and exhaust and inlet valves of a reciprocating compressor every 0.02 s (when the compressor runs at 3000 RPM). A typical case of two gear pairs in mesh provides two extremities of time duration, the first, a high speed pinion running at 3000 RPM every 0.02 s and the second, a girth gear running at 2 to 3 RPM every 30 s. All these disturbances are transients which completely die down before successive hits. With duration of disturbance varying very widely, the two parameters ζ and ω play a crucial role in the behavior of the vibrating member. They depend on the material, geometry and boundary conditions of the vibrating member.

If the damping is underestimated, the member receives hits before the influence of the previous hit disappears. If overestimated, the predictions negate what is being experienced. It is easy to compute the first few natural frequencies accurately analytically or experimentally, but damping ratios can be determined only experimentally. It is for this purpose this investigation is taken up for different materials commonly used in mechanical engineering.

PROCEDURE FOR INVESTIGATION

Impulse hammer

The hammer used excites the test structure with constant force over the frequency range of interest and produces an analog voltage pulse which is an exact representation of the input impulse (F.t) both in amplitude and phase. It uses an accelerometer of Low Impedance Voltage Mode (LIVM) which is designed to measure shock and vibration phenomena over a wide frequency range. It contains integral IC electronics that converts the high impedance signal generated by the piezo crystals to a low impedance voltage that can drive long cables with excellent noise immunity. These accelerometers utilize quartz and piezo ceramic crystals in compression and shear mode. The accelerometer is mounted on the test object at the position defined by the customer and the impact is given using Impact Hammer at the specified point and in specified direction and the data is collected using Focus Dynamic Signal Analyzer. Figure 4 shows the arrangement for this test and Figure 3 the response.

Forced vibration test

One of the major problems encountered in forced vibration test is the range of excitation available in the exciter. Mechanical double mass eccentric exciter produces pure sinusoidal excitations, but the range is limited up to 100 Hz. While heavy engineering units are having lower natural frequencies (in the range 5 to 100 Hz), components in automobiles like inlet and exhaust valves have natural frequencies as high as 18000 Hz. It is impossible to cover this wide range experimentally. In this investigation, a heavy duty electrodynamic exciter with a range up to 3500 Hz is used to collect meaningful data (Figures 5 and 6).

The vibration testing unit consists of a signal generator, a power amplifier and a vibration shaker. In addition to this, instrumentations like accelerometers, signal conditioners, charge amplifiers, Fast Fourier Transform (FFT) are also used.

Dynamic tests are performed with sinusoidal signal sweeping between the defined frequency range. The output of the signal generator is amplified in the power amplifier to a level sufficient to drive the vibration shaker. This vibration shaker converts this amplified signal to mechanical motion at the work table which transfers the same to the test samples. The mechanical motion at the table surface is sensed through the accelerometer. The electrical signal can be used for monitoring and feedback purposes to make the operation as a close loop. The acceleration level imposed on the specimen is measured and continuously compared with the program by the Digital signal controller. The difference between the command and the input signal is used to provide a continuous correction signal to the shaker, which causes this difference to minimum. Specifications of the experimental set up are shown in Table 1.

Digital amplifier

Model DSA 5K15, class D, Technology PWM, Output power 15 KVA (Suitable to drive shaker), Input power 415 V, 50 Hz, 3 phase, Frequency range DC to 3500 Hz.

System control unit

Sinusoidal signal generator routes the path of signal towards the power amplifier. It can use its own signal or signal from the external source and thereafter to the shaker after amplifications.

DUT: Steel Plate S1. **Condition: Impact in Z direction Response in Z direction**

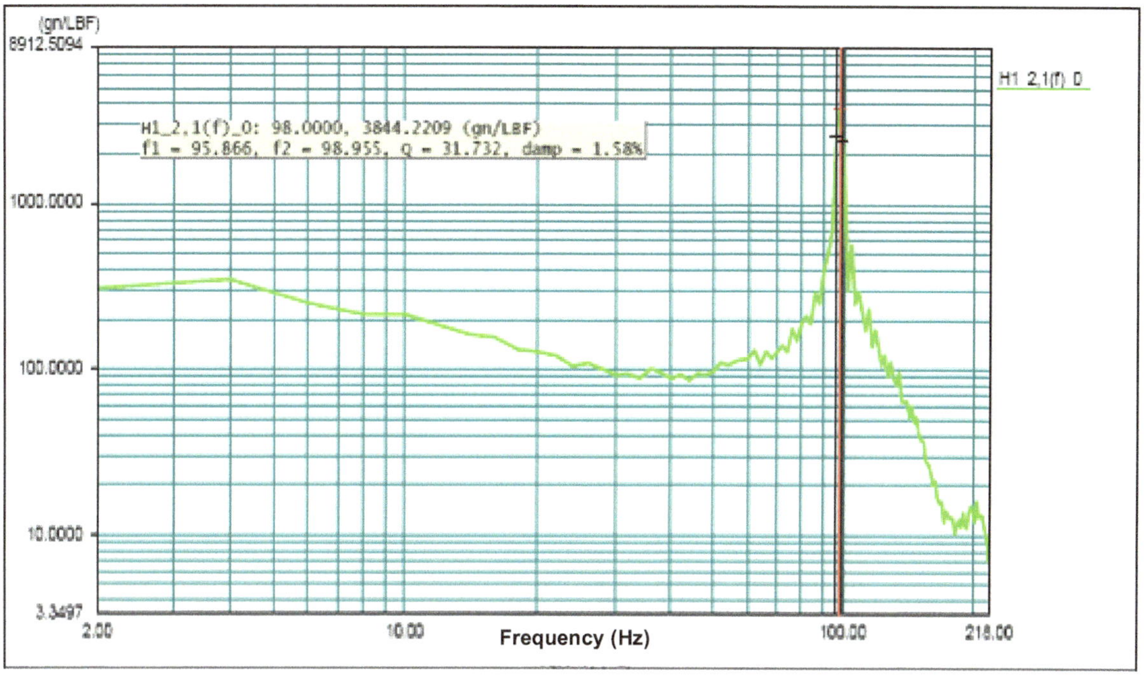

Figure 3. Frequency Vs Amplitude response –Impulse hammer (S_1).

DUT: Steel Plate S1. **Condition: Impact in Z direction Response in Z direction**

Figure 4. Impact Hammer-Decoupled excitation.

Figure 5. Vibration shaker.

Figure 6. Experimental setup-forced vibration for the first set.

Accelerometers

Accelerometer of Model 2224C is a high precision Piezo electrical pickup over a wide frequency range. Nominal charge sensitivity 10 pC/g, Resonant frequency > 32000 Hz, mechanical shock limits 2000 'g', Sensing element shear ceramic , Transverse sensitivity < 3%, Mounting thread 10 to 32 coax type, Weight 15 g, Cable Microdot to Microdot 5 m length.

TESTS CONDUCTED

Twenty five specimens have been chosen for the analysis. The first set of 11 specimens are cantilevers of three varying lengths (Table 2) and four different materials (Mild Steel, Cast Iron, Aluminium and Plastics) of uniform cross section (Figure 4). It is well known that their first three natural frequencies in bending mode are decoupled. This program is expected to give controlled values of damping ratios. The other set of fourteen components constituting four materials (Mild Steel, Cast Iron, Aluminum and Plastics) (Ramamurti et al., 2010) are used in air brake system of automobiles and heavy vehicles. These details are shown in Table 3. Test specimens are shown in Figure 7. The first few mode of vibrations are coupled. These are also studied to find out how the complex systems behave.

Assumed ratio of values of E/ρ (taken from Hand books) are the same for steel and aluminium and hence computed natural frequencies of these two will be the same for geometrically identical plates.

All the 25 specimens listed in Tables 2 and 3 are modeled using finite element method (ANSYS) and the first three natural frequencies are computed. Figure 8 shows the first three modes for S_1 using finite element method. The specimens are subjected to free vibration test using impulse hammer. Figures 9, 10, 11 and12 show the free vibration records for steel, aluminum, cast iron and plastics. Forced vibration on electro dynamic exciter is also performed. The results on damping ratios are shown in Figures 13 to 16. Firm lines indicate the response for plates of uniform rectangular cross section (uncoupled modes) whereas dots represents products of manufacture of complicated shapes (coupled modes).

RESULTS AND DISCUSSION

From the Tables 2 and 3, and Figures 13 to 16, the following observations have been summarized and are shown in Table 4

The plots of ω Vs ζ for the first and third modes are made available for the four materials investigated (Figures 13 to 16). ζ obtained by two approaches were compared. Damping ratios ζ_1 for all the specimens computed by using logarithmic decrement approach and Tables 2 and 3 by using the √2 method with electro dynamic exciter agree well.

(i) ζ values of these materials are in the following order aluminum < mild steel < cast iron < plastics.
(ii) Plastics exhibit a maximum value of ζ_1 4.1% whereas for mild steel it does not exceed 2.9%.
(iii) Values of ζ_3 are less than the corresponding values of ζ_1, for most of the specimens tested (Tables 1 and 2).

INFLUENCE ON DYNAMIC STRESS

As per international standards, for continuous operation, the threshold of vibration disturbance is of the order of 2 mm/s. It is presumed that very severe vibration due to

Table 1. Specifications of shaker and subsystem

Subsystem	Model	Shaker	SEV 150
Electro dynamic shaker	SEV 150	Rated force	1500 Kgf (Peak SINE)
Digital power amplifier	DSA 5K15	Operating frequency range	1 Hz to 3500 Hz
Digital system interface	SCU 011C	Stroke	40 mm
Piezo electric accelerometers	2224 C (2 nos)	Acceleration bare table	80 'g'
'SINE" S/W package	SVC 2205-FW	Maximum velocity	1200 m/s
		Moving Platform diameter	240 mm

Table 2. Details of standard plates investigated (set-1)

Plates	L mm	B mm	T mm	Finite element method		Free vibration method (Experimental)		Forced vibration method (Experimental)			
				ω_1	ω_3	ω_1	ζ_1	ω_1	ω_3	ζ_1	ζ_3
S_1	420	50	4	132	2262	138	0.0145	128	2284	0.0150	0.0030
S_2	220	50	4	522	9117	540	0.0115	548	9192	0.0130	0.0064
S_3	220	50	8	1043	18234	993	0.0104	1087	19578	0.0110	0.0100
A_1	420	50	4	132	2262	138	0.0085	130	2314	0.0096	0.0031
A_2	220	50	4	522	9117	528	0.0200	547	9079	0.0097	0.0090
A_3	220	50	8	1043	18234	1005	0.0180	1030	18414	0.0088	0.0090
C_1	420	50	8	201	3569	214	0.0300	196	3625	0.0208	0.0031
C_2	220	50	8	817	11520	754	0.0360	804	11451	0.0144	0.0080
P_1	375	50	4	50	880	63	0.0320	63	921	0.0300	0.0260
P_2	185	50	4	239	4191	264	0.0360	283	4659	0.0330	0.0450
P_3	220	50	8	371	5623	415	0.0230	425	5329	0.0177	0.0180

Table 3. Details of components tested (Set-2)

Components	Material	Finite Element method		Free vibration method (Experimental)		Forced vibration method (Experimental)			
		ω_1	ω_3	ω_1	ζ_1	ω_1	ω_3	ζ_1	ζ_3
RE 6	Steel	477	1903	483	0.0270	444	1717	0.0290	0.0110
QRV	Steel	1149	4090	852	0.0200	784	3596	0.0220	0.0160
BC type 12	Steel	1746	6729	2418	0.0290	2500	4481	0.0290	0.0080
MTS	Steel	3623	8664	3443	0.0270	3237	7546	0.0270	0.0130
Fixture	Steel	3631	9710	3321	0.0140	3257	9555	0.0120	0.0110
RE6	Al	477	4803	483	0.0140	440	4506	0.0140	0.0100
QRV	Al	5400	11828	6063	0.0100	6195	10687	0.0100	0.0040
MTS	Al	3623	8664	3443	0.0190	3670	7516	0.0150	0.0120
Flange (compressor)	Al	5510	10499	4423	0.0180	4668	10675	0.0130	0.0080
Compressor (over hang)	CI	2022	3700	2150	0.0240	2023	4668	0.0200	0.0087
Compressor	CI	5015	7320	4724	0.0370	4324	6014	0.0330	0.0046
Compressor	CI	5510	10499	4461	0.0290	4706	10710	0.0280	0.0100
QRV	Plastics	823	4012	911	0.0440	924	3522	0.0410	0.0035
MTS	Plastics	3623	8664	3443	0.0380	3675	7053	0.0400	0.0030

transients acting for a very short period of time will not exceed 300 mm/s. The ratio of fall in amplitude for transient vibration is of the order 150 (300/2). Fall in amplitude for every oscillation for viscous damping is $e^{2\pi\zeta}$.

Figure 7. Test specimens.

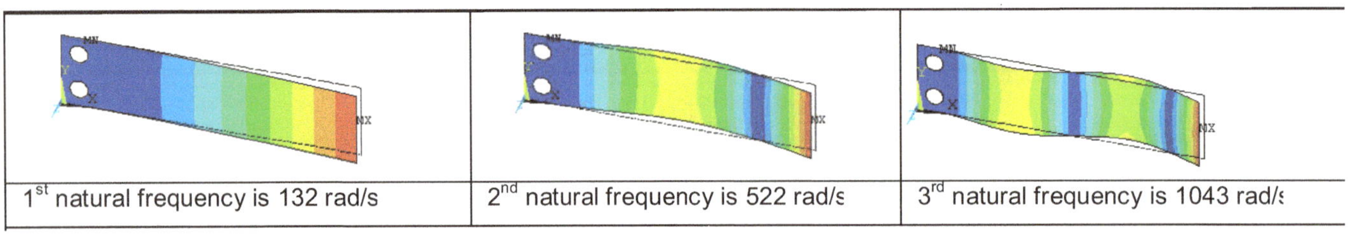

Figure 8. Finite element analysis of S_1 along with its mode shape.

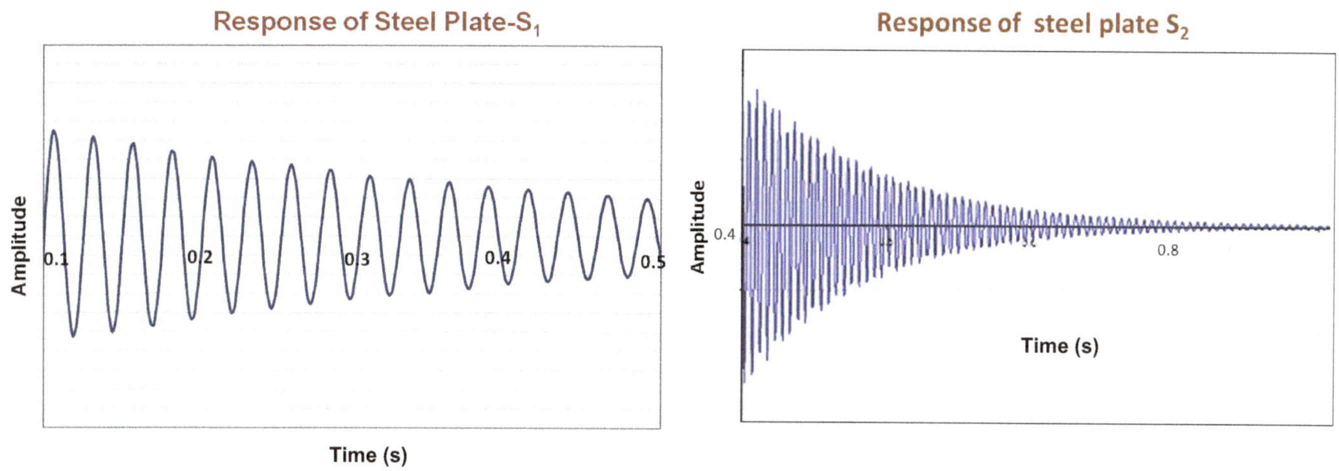

Figure 9. Free vibration response of steel S_1, S_2.

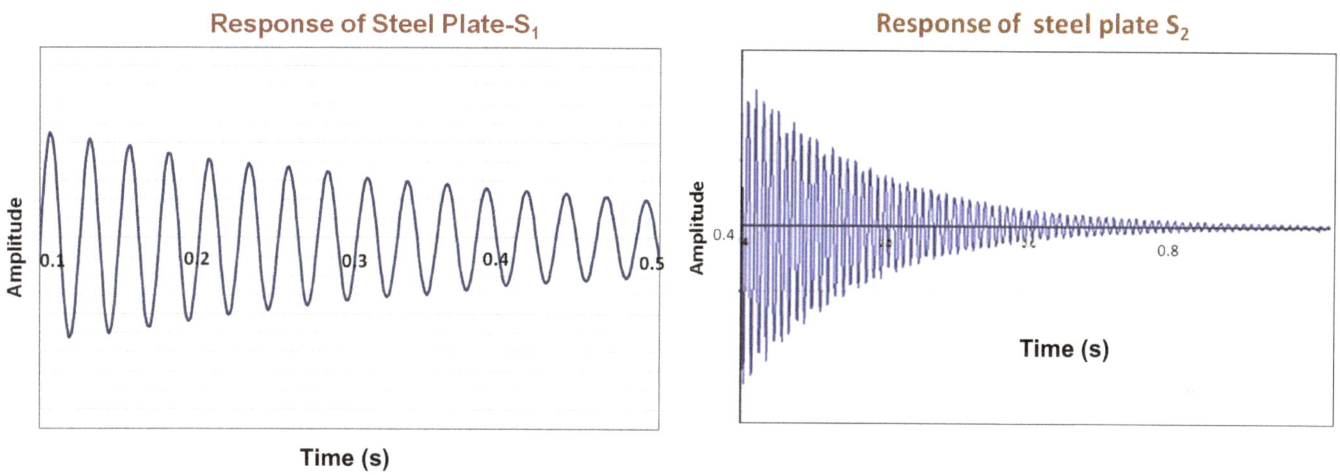

Figure 10. Free vibration response of steel A_1, A_2.

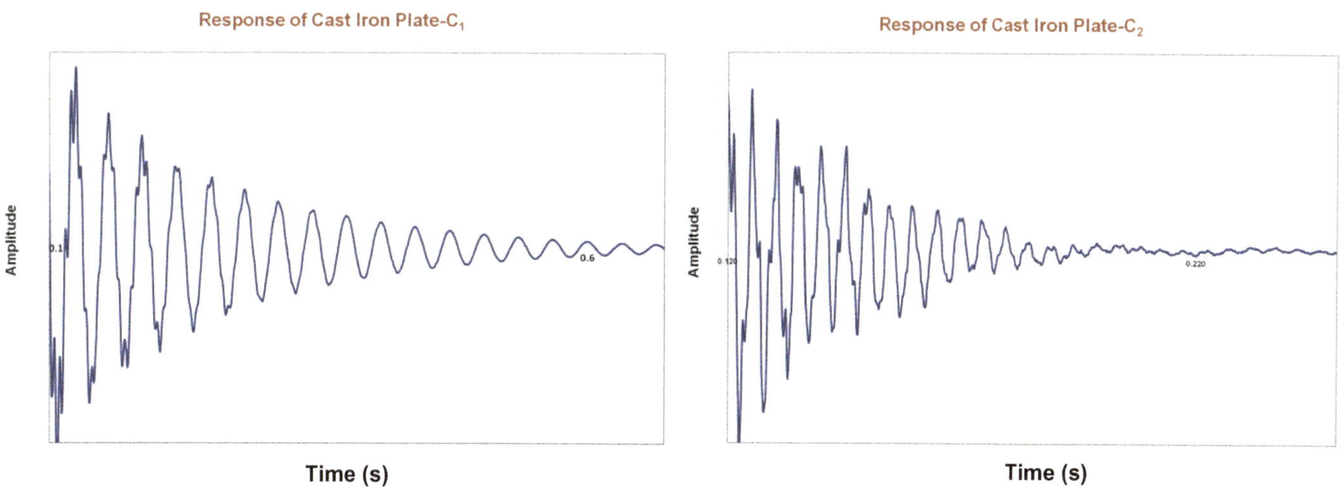

Figure 11. Free vibration response of steel C_1, C_2.

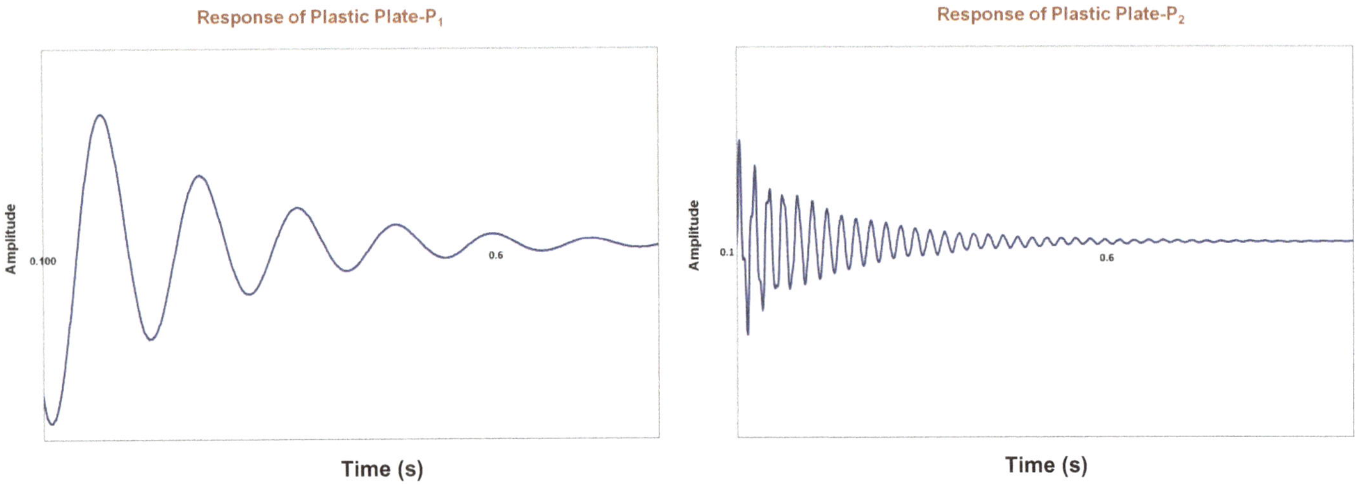

Figure 12. Free vibration response of plastics P_1, P_2.

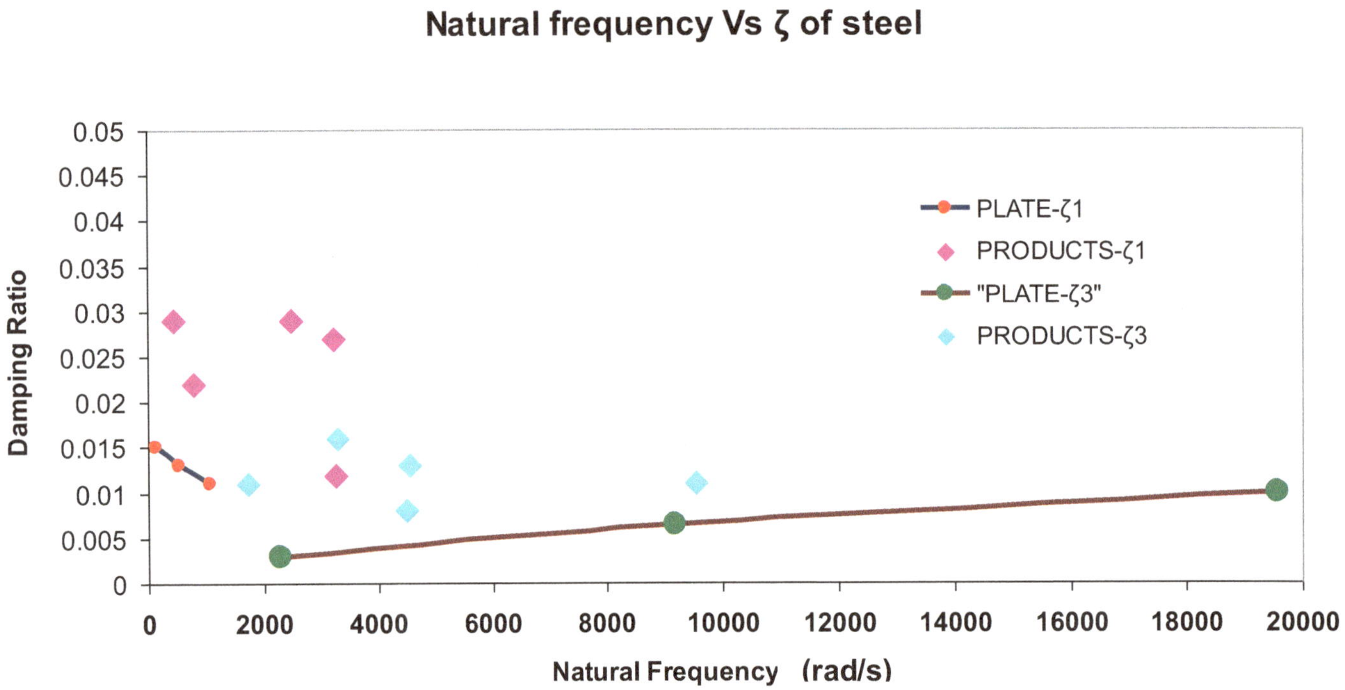

Figure 13. ω Vs ζ for steel.

Hence the ratio of fall in amplitude in one second $(e^{2\pi\zeta})^{\omega/2\pi} = e^{\zeta\omega}$. If $e^{\zeta\omega}$ is to have a maximum value of 150, then $\zeta\omega$ should be of the order of 5 ($e^5 \sim 150$).

To study the effect of damping ratio, as an example delivery valve stopper used in reciprocating air compressor is taken for investigation. This is very light with very high fundamental natural frequency of 18000 Hz.

Figure 17 shows the delivery valve stopper sub assembly with a shoulder held by helical spring. Figure 18 shows the stopper alone for the delivery valve (Ramamurti et al., 2010). This experiences, dynamic stress due to movement of ring valve which discharges air from the compressor. Since this valve stopper is subjected to repetitive shock pulses when the force acts for a very short period of time Figure 19, transient dynamic analysis is carried out using mode superposition technique.

In order to know the influence of damping ratio on the dynamic response of the valve stopper, four cases are

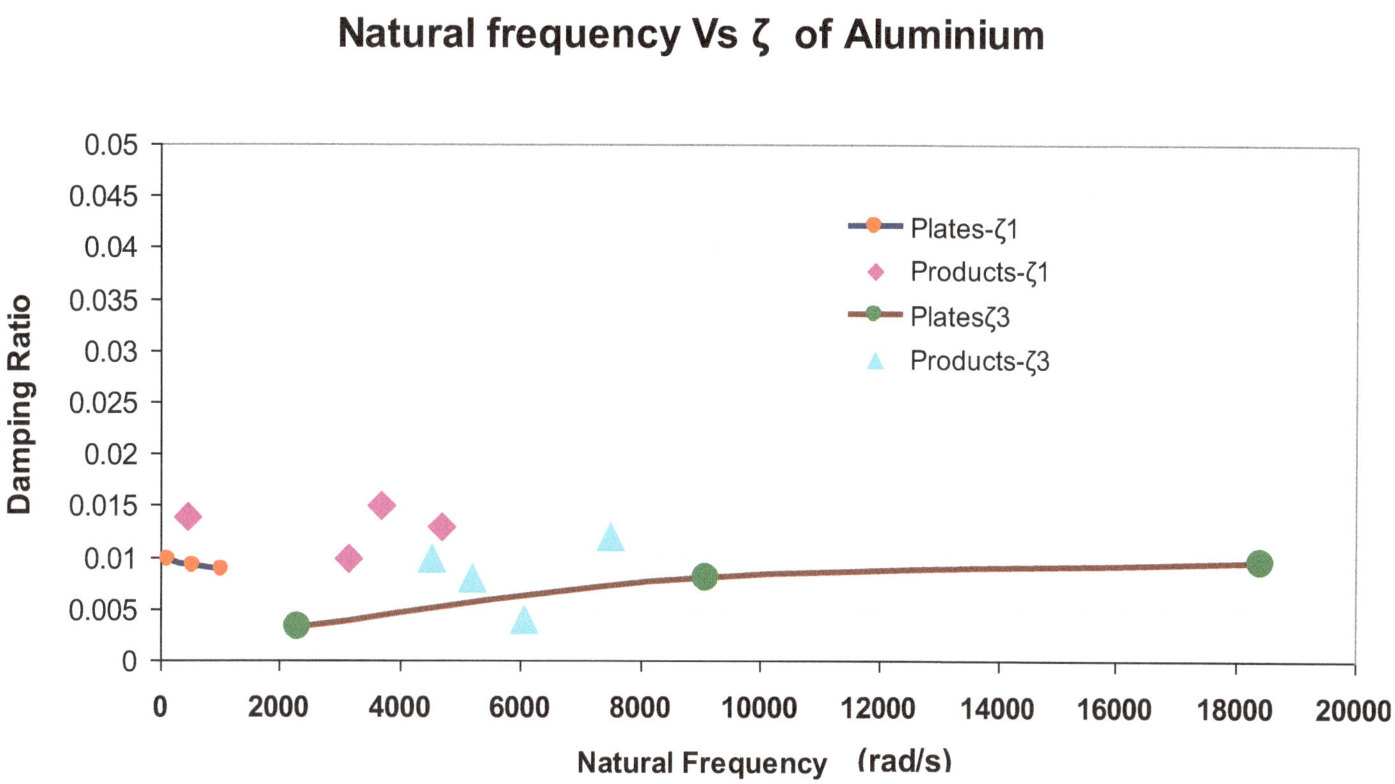

Figure 14. ω Vs ζ for aluminum.

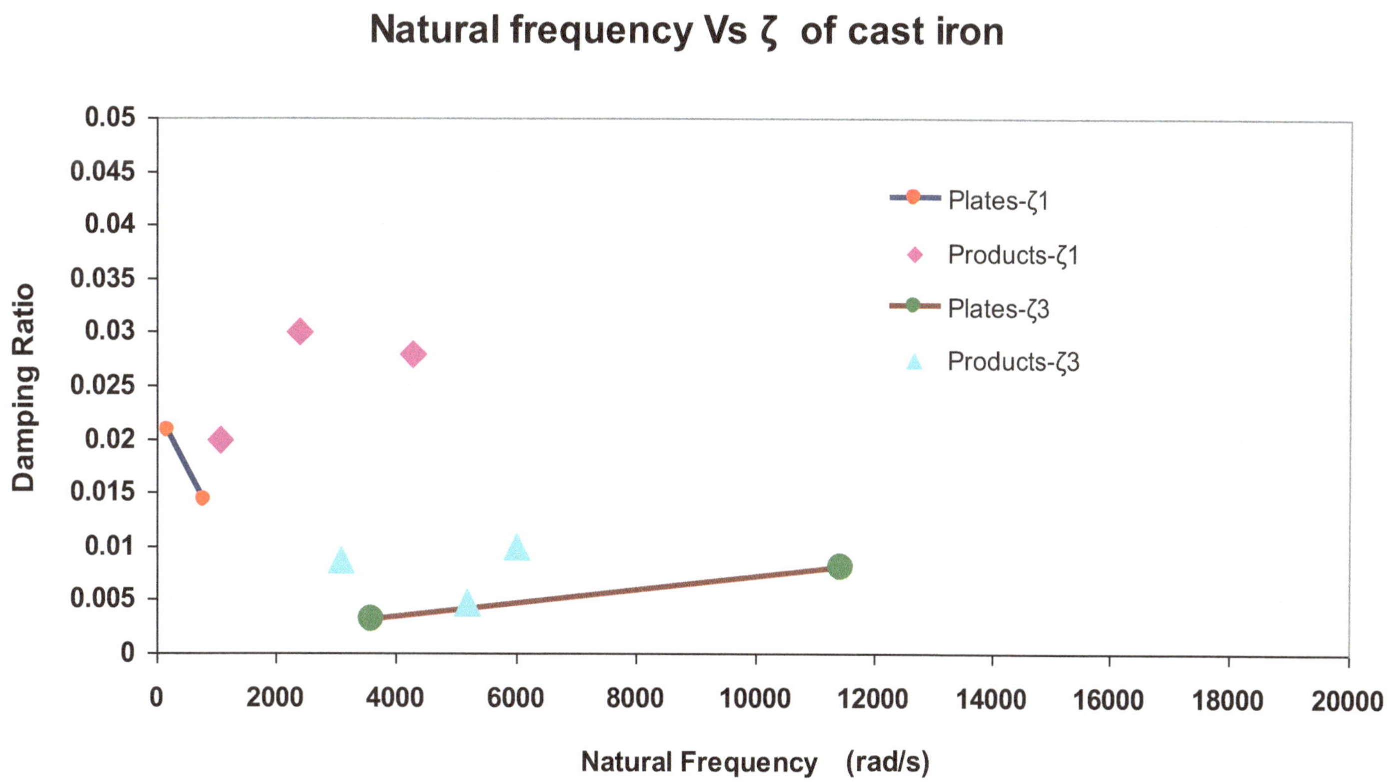

Figure 15. ω Vs ζ for cast iron.

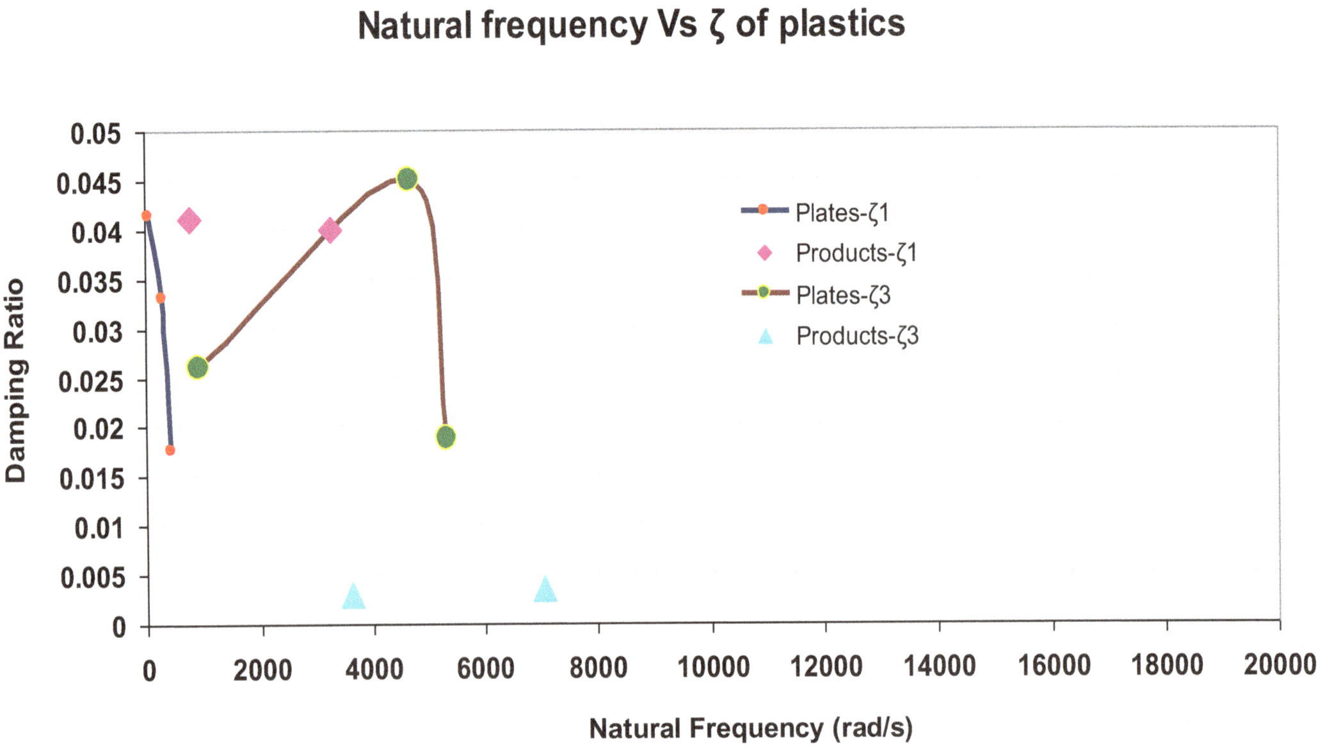

Figure 16. ω Vs ζ for plastics.

Table 4. ζ and f of different materials.

Material	f_1 (Hz)	ζ_1	f_3 (Hz)	ζ_3
Steel	21 to 518	0.011 to 0.029	273 to 1521	0.003 to 0.016
Aluminum	21 to 985	0.01 to 0.015	368 to 2930	0.003 to 0.012
Cast Iron	31 to 748	0.014 to 0.033	576 to 1822	0.003 to 0.01
Plastics	10 to 584	0.017 to 0.041	146 to 1122	0.003 to 0.045

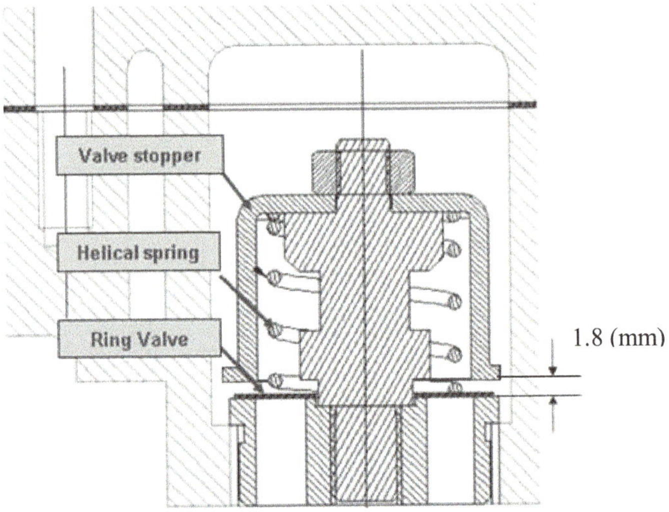

Figure 17. Delivery valve assembly in the cylinder head.

Figure 18. Valve stopper.

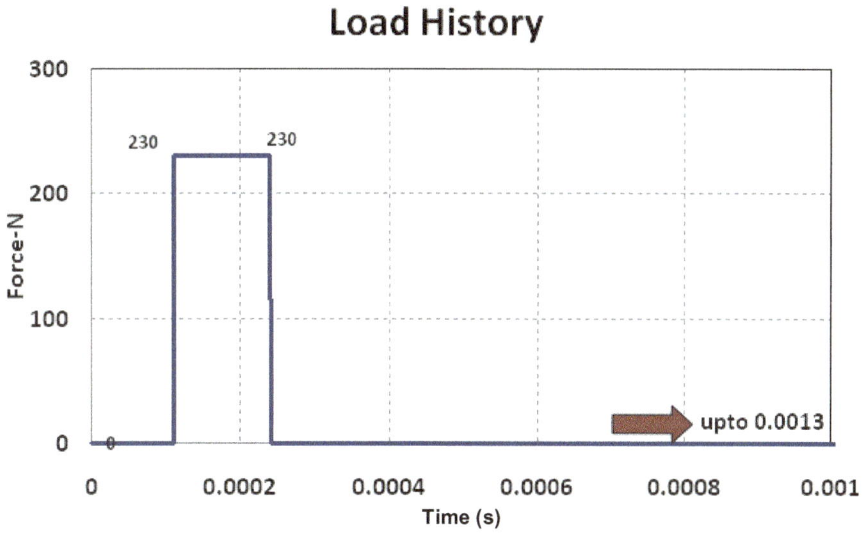

Figure 19. Shock pulse as input to delivery valve.

Table 5. Dynamic response of valve stopper.

Case	ζ_1	ζ_3	Maximum dynamic stress (MPa)	Time of decay(s)
1	0.015	0.01	40.70	1.5 e-3
2	0.015	0.02	40.50	1.4 e-3
3	0.010	0.01	40.83	2.04 e-3
4	0.010	0.02	40.60	2.01 e-3

taken for the analysis By varying ζ_1 and ζ_3 obtained from the experimental findings transient dynamic analysis is carried out. For the first two cases, ζ_1 is assumed to be 0.015, ζ_3 is varied with 0.01 and 0.02. For the second set ζ_1 is assumed to be 0.01 and ζ_3 is varied with value 0.01 and 0.02.The dynamic stress obtained in these cases is shown in Figure 20 (i) to (iv).

When ζ_1 is kept constant, the effect of ζ_3 on dynamic stress is negligibly small and the time of decay (fall in ratio of amplitude around 150) is also small. When ζ_3 is kept constant at 0.01 if ζ_1 is altered from 0.01 to 0.015, the maximum value of dynamic stress varies from 40.83 to 40.7 MPa. When ζ_3 is 0.02 the dynamic stress varies from 40.6 to 40.5. With ζ_1 0.01 the dynamic stresses dies down at $2.01e^{-3}$ s whereas it takes $1.4e^{-3}$ s for ζ_1 of 0.015 (Table 5).

Conclusion

Experimental evaluation has been carried out to find out the damping ratios for free and forced vibration using four different materials. Besides the above, the natural frequencies have been calculated analytically using finite element analysis (FEA). The natural frequency values obtained using the analytical method, free and forced vibration methods have a fairly good agreement with each other for both standard plates investigated and components tested. It was observed that for all the four materials covered in this paper Figures (13 to 16), ζ_1 has never exceeded 4.5% (mostly less than 2%) Table 4. The value of ζ_1 obtained by free vibration method and √2 method agree well. Since the natural frequencies cover a wide range of values in industrial parlance, the damping ratios ζ_1 and ζ_3 obtained give a very good database for a designer using steel, aluminium, cast iron and plastics. As an example, in "Influence on Dynamic Stress", axis symmetric element is used for modeling a valve stopper transient vibration study. Figures 13 to 16 provide good data base for computing ζ_1 and ζ_3. After computing the first few natural frequencies of a member which is to be analyzed (eigen pairs) these damping ratio values are to be used in modal super position technique. "Influence on Dynamic Stress" also indicates that the effect of damping ratio ζ_3 on dynamic stress is negligible whereas ζ_1 has some influence. However, it is premature to conclude that the change in damping ratio has marginal influence on the magnitude of dynamic stress. Since Equation 8 is the basic relationship connecting the damping matrix [C] with stiffness and mass matrices [K] and [M], in our opinion, the picture is likely to change if other types of finite elements are used for modeling.

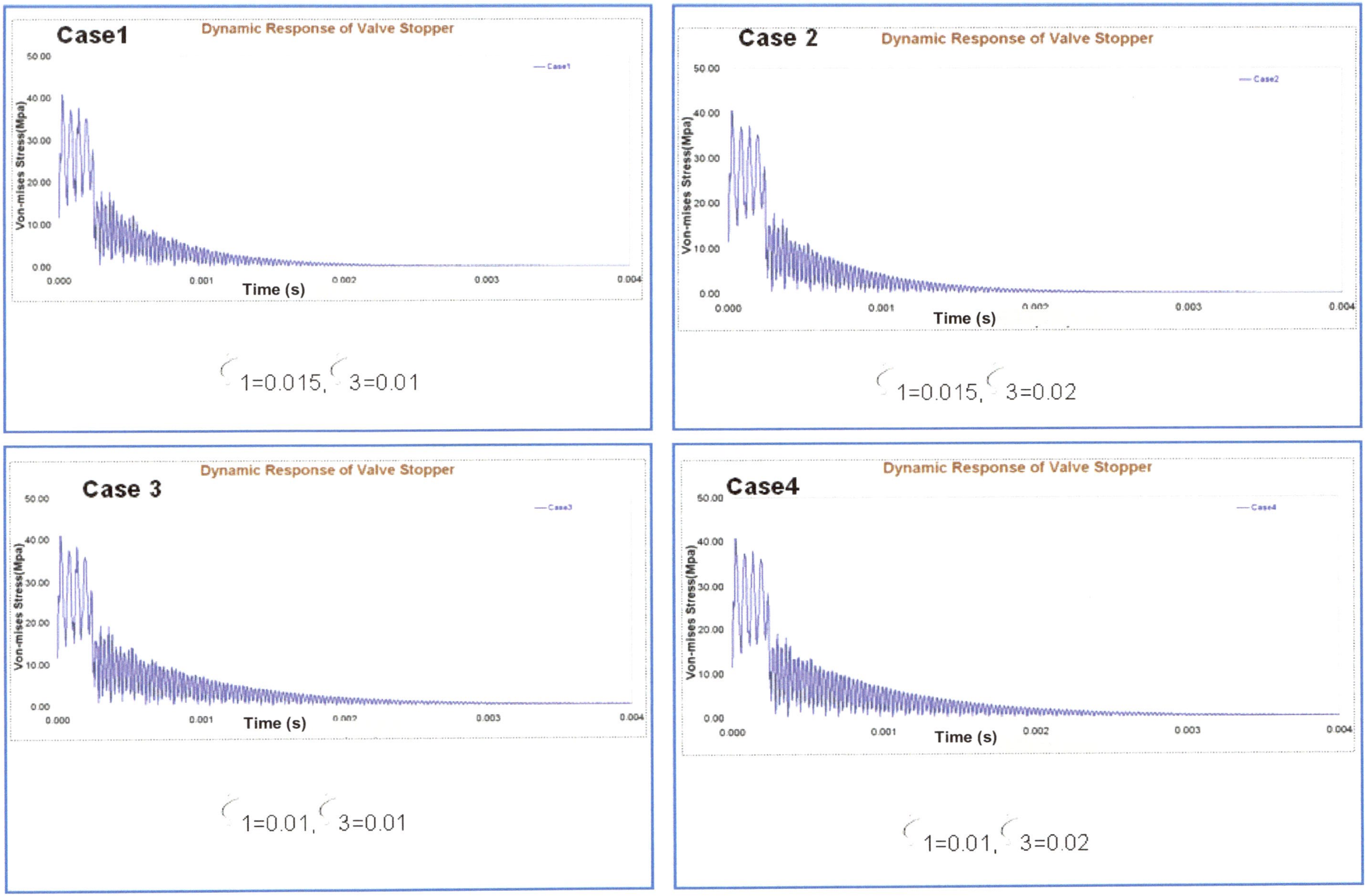

Figure 20. Dynamic stress plot.

ACKNOWLEDGEMENT

The authors gratefully acknowledge the support given by Messer's WABCO India Limited, Chennai in carrying out this investigation and permitting publication of this article in a journal.

REFERENCES

Bellette PA, Meehan PA, Daniel WJK (2008). Effect of variable pass speed on wear type corrugation growth. J. Sound Vib. 314(3-5):616-634

Chow F, Lu W (2005). Experimental observation of nonlinear vibration in a rub impact rotor system. J. Sound Vib. 283(3-5):621-643.

Grace IF, Ibrahim RA (2008). Modelling and analysis of ship roll oscillation intensity with stationary ice berg. Proc. IMechE, Part C:J. Mech. Eng. Sci. 222(C10):1873-1884

Huang CH (2005).Transverse vibration analysis and measurement for the Piezo ceramic annular plate with different boundary conditions. J. Sound Vib. 283:655-683.

Issa J, Mukherjee R, Diaz AR, Shaw SW (2008). Modal disparity and its experimental verification. J. Sound Vib. 311:465-1475.

Kuljamic E, Sorting M, Totoos G (2008). Multisensor approaches for chatter detection in milling. J. Sound Vib. 312(4-5):622-693.

Li hen Z, Zhengjia He, Wang ZY, Xue Y (2008). Customized wavelet devices using intra and inter scale dependency for bearing fault detection. J. Sound Vib. 313:342-359

Li LJ, Jiang WK, Ai YH (2010). Experimental study of deformation and shock damage of cylindrical shell strut subjected to under water explosion. Proc. IMechE, Part C: J. Mech. Eng. Sci. 224(C11):2505-2514.

Mann BP, Liu J, Hazra SS (2008). Measurement of nonlinearity interpreted as material behavior in dynamic nano indentation. J. Sound Vib. 311(3-5):1358-1370.

Ramajayathilagam K, Vendhan CP (2004). Deformation and rupture of thin wave rectangular plate subjected to under water shock. J. Impact Eng. 30:699-719.

Ramamurti V (2012). Finite Element Method in Machine Design. Alpha Sciences International, London.

Ramamurti V, Sridhar S, Mithun S, Kumaravel B (2010). Transient dynamic analysis of valve stoppers used in reciprocating air compressor. Proc. INSA 76(3):119-128.

Xiao L, Cox TX, Aris UR (2005). Active diffuser measurement, some prototypes and 2D measurements. J. Sound Vib. 285:321-339.

Zhang D, Engeda A, Hardin JR, Aungier RH (2004). Experimental study of steam turbine control valves. Proc. IMechE, Part C: J. Mech. Eng. Sci. 218(C2):493-508.

2

A simple mechanical device for the measurement of discharge in a tubewell

Samanpreet Kaur[1]*, Rajan Aggarwal[1], Satvinder Singh[2] and Harjit Singh Gulati[1]

[1]Department of Soil and Water Engineering PAU, Ludhiana – 141004, India.
[2]Zonal Research Station for Kandi Area, Ballowal Saunkhri, Distt Nawanshaher, India.

About 85% of the total water resources are utilized in agriculture. Agriculture is more dependent upon groundwater resources. It is important to know the accurate discharge of the tubewell so that required amount of water may be applied to the field. Presently, Co-ordinate method is the most common method used to measure tubewell discharge, but this is not very accurate. Therefore, to overcome the limitations, an orifice bucket was fabricated and calibrated. It was found that orifices of diameter 11.3, 10, 7.5 and 5 cm measured discharges ranging from 15 -20, 10 -15, 5 -10 and 1 - 5 lps, respectively.

Key words: Tube well discharge, water measurement, orifice bucket

INTRODUCTION

India is having 2% of the world geographical area and 4% fresh water resources. Whereas total population of India is approximately 17% and cattle population is more than 20%. In developing countries more than 85% of water resources are used for agricultural purpose. So considering these facts India needs to use water judiciously especially in the agricultural sector. The state of Punjab with 1.5% area of the country, has been contributing 35 - 40% rice and 50 - 60% wheat to central pool since the last three decades. Intensive agriculture has adversely affected the state's water resources as it has increased the irrigation water requirement (Aggarwal et al., 2009). However, the surface water resource is not sufficient to meet the total irrigation needs. The percentage irrigated area from canals and tube wells in the state is 30 and 70% respectively. The number of tube well in the state has increased from 0.192 million in 1970 to 1.246 million in 2008 (Anonymous, 1970, 2008). Thus, the pressure on our water resources is continuously increasing. Therefore, to maximize production per unit of water resources, it is emphasized to utilize every drop of water available judiciously and carefully.

The efficiency of water utilization on the farmers' fields has been poor due to water losses in conveyance and during application at the field level. The situation arises due to lack of awareness among the farmers about the consequences of inefficient water application and lack of appropriate tools and instruments for uniform application of water. In the next decade, water is going to become a major limiting factor for sustained production. Appropriate instrumentation for water measurement, land levelling, water flow and soil moisture measurement and irrigation scheduling are needed in addition to a well-maintained conveyance network to have effective irrigation with a view to realizing high water use efficiencies. Not many tools and instruments are available for the farmers to undertake these tasks. In the absence of appropriate devices, farmers adopt obsolete and inefficient alternatives (Hamdy et al., 2003). Thus, under the present scenario, flow regulation on the farmers' fields through appropriate structural interventions are practically non-existent or in disuse. Lack of scientific water management causes loss of valuable water resources and the concept of application of a measured quantity of water in accordance with the crop water requirement is hardly practiced (Rajput and Patel, 2002). With the number of tubewells increasing in the state, it is important to know the accurate discharge of the tubewell so that required amount of water may be applied to the field. About 3.5 lps of discharge is required for 8 h to irrigate one hectare of land in case of non- paddy crop whereas for paddy about 4 l/sec discharge is required for same duration to irrigate same area. The method, which is frequently used at present, is the coordinate method, but this is not very

*Corresponding author. E-mail: samanpreet@rediffmail.com.

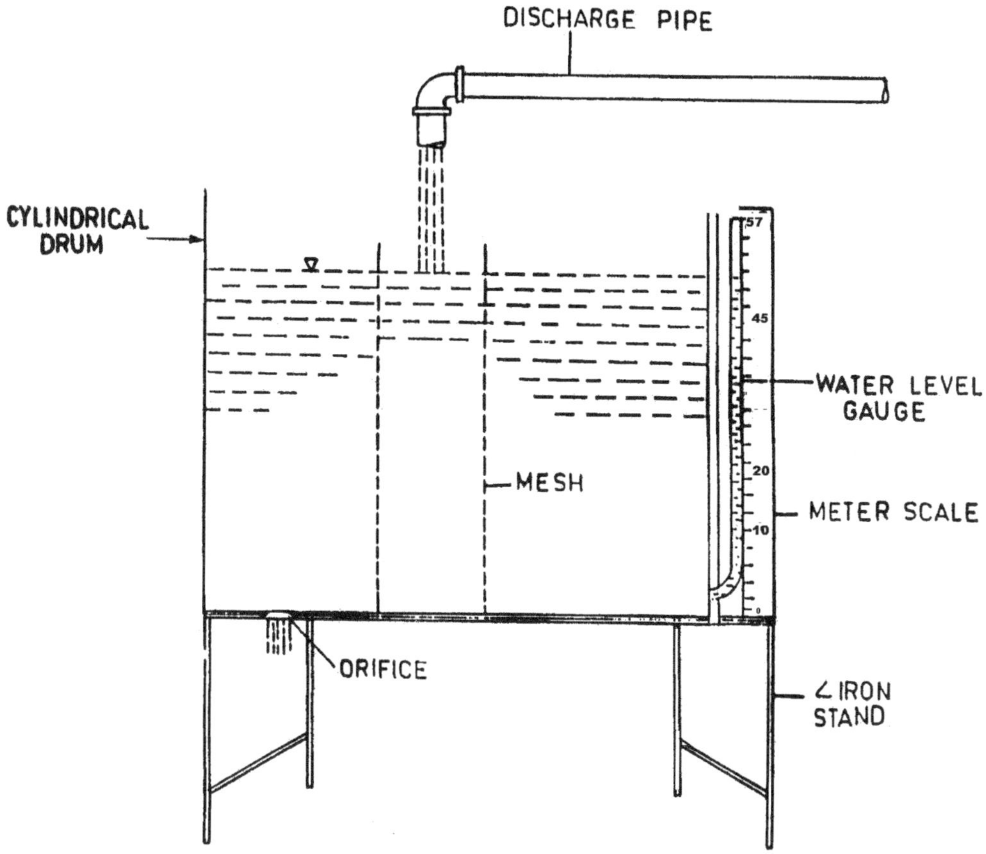

Figure 1. Schematic sketch of orifice bucket (not to scale).

accurate. Certain direct and indirect methods have only limited application in the fields as they need specific conditions to be met which may not be available. So, keeping in view the above difficulties and drawbacks efforts have been made to fabricate and calibrate an orifice bucket that can give accurate measurement and is easy to transport and operate. The water to be measured flows into the tank and discharges through the orifice provided at the bottom of the tank. The container fills with water to a level where the pressure head causes the outflow through the orifice to just equal the inflow from the pump based on Bernoulli's theorem. A suitable vertical scale is fastened to the outside so that the level of water in the container can be read accurately. The device is calibrated and the calibration curve shows the rate of discharge through an orifice of a given size for various values of pressure head.

EXPERIMENTATION

Construction of orifice bucket

It consists of a cylindrical drum of 74 cm diameter and 57 cm height fabricated from mild steel sheet in the Research Hall of Department of Soil and Water Engineering, PAU, Ludhiana. A large diameter hole was cut through the bottom of the drum and a leak proof arrangement was made on the backside of the bottom sheet for attachment different orifice plates (of 11.3, 10, 7.5 and 5 cm diameter). Near the base of the drum an elbow of 2.5 cm diameter was welded. A simple glass tube was connected to the elbow with the help of a rubber cork. A wooden scale (meter rod) was fastened to the outside of the drum to read the water level in the drum accurately. To reduce the turbulences produced by falling water from the delivery pipe, a 20 cm diameter perforated metallic screen cylinder was installed in the center of the drum and the water from the tubewell pipe was allowed to fall in the cylinder. Figure 1 shows the details of construction of an orifice bucket. Plate 1 and Plate 2 shows a fabricated orifice bucket.

Calibration of orifice bucket

The orifice bucket was calibrated on the Pump Testing Rig which is as per the BIS standards and is installed in the research hall of the college. Figure 2 shows the schematic sketch of the calibration circuit. The water was allowed to fall in the bucket and pass through the water-measuring tank fitted with a sharp edge 90° V-notch

Plate 1. Fabricated orifice bucket (Top view).

Plate 2. Fabricated orifice bucket (Side view).

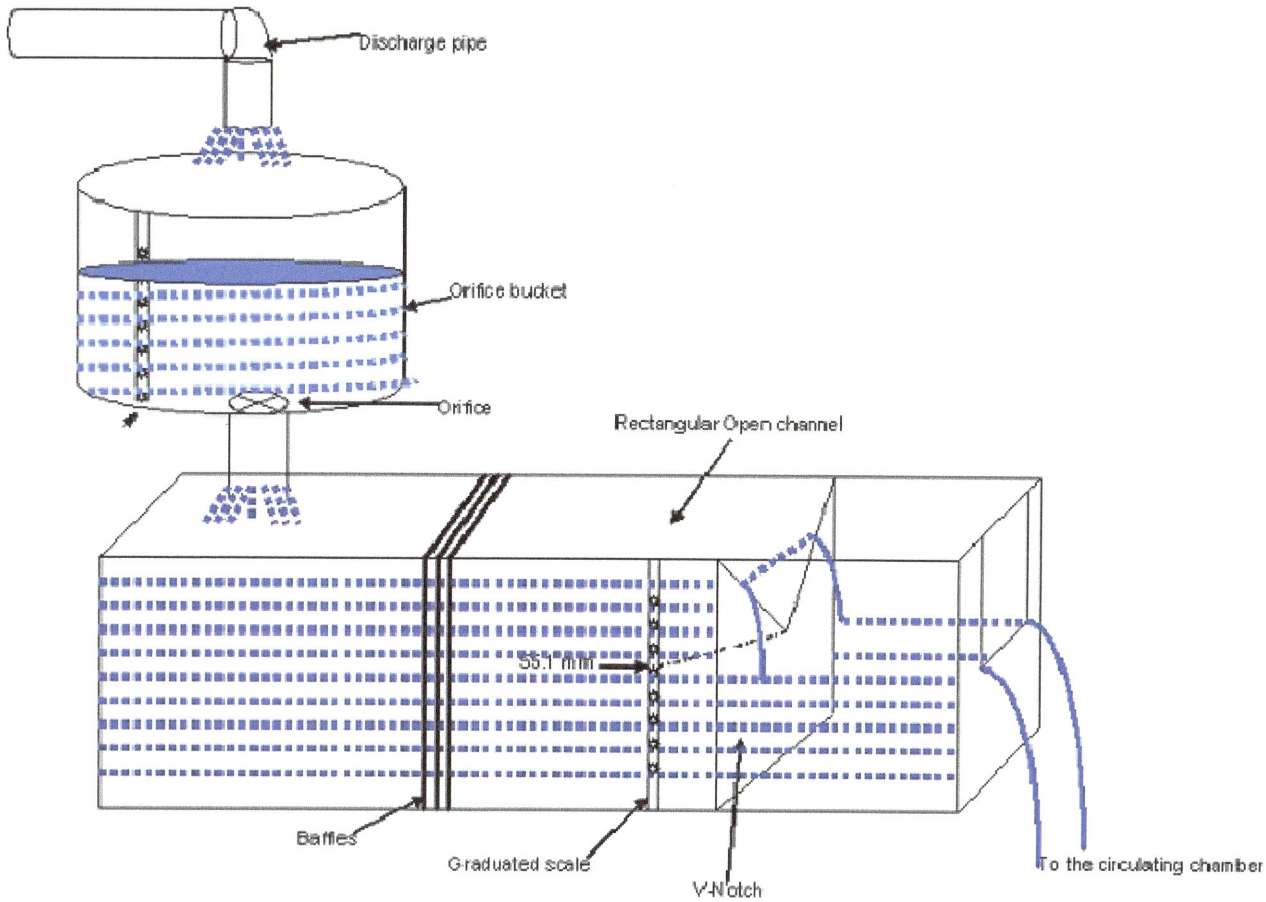

Figure 2. Schematic sketch of the calibration circuit.

made from brass sheet. The V-notch weir is a triangular channel section, used to measure small discharge values. The upper edge of the section is always above the water level, and so the channel is always triangular simplifying calculation of the cross-sectional area. V-notch weirs are preferred for low discharges as the head above the weir crest is more sensitive to changes in flow compared to rectangular weirs. After the steady state condition is achieved, the head on the upstream side of the V-notch was observed and the discharge through V-notch was computed by using the following formula (Henderson 1966).

$$Q = \frac{H^{2.48}}{21900}$$

Where Q = discharge, lps (litre per second)
H = head over the crest of the notch, mm

Modifications

During the calibration studies, it was observed that there were wide fluctuations in the water level of manometer so a second filter was also placed around the first filter. The flexible pipe was attached with the delivery pipe of the tubewell to put the discharge at the bottom of the bucket, to further reduce turbulences in water level.

Observations

The water level reading in the orifice bucket corresponding to a given discharge was noted from the manometer. The discharge through the delivery pipe was varied with the help of a gate valve. Using different settings of the gate valve calibration curve was prepared showing the discharge through orifice of a given size for various values of pressure head. Table 1 gives the value of head above orifice with respect to datum versus discharge through V-notch for orifices of 11.3 cm diameter. It was observed that it was possible to measure discharges ranging between 15 - 20 lps accurately by using 11.3 cm diameter orifice, so for discharges less than 15 lps replaceable orifices of 10, 7.5 and 5 cm diameter were fabricated and calibrated and are given in Table 2, 3 and 4 respectively. The calibration curve showing the rate of discharge through a given orifice for

Table 1. Observations of discharge with orifice diameter 11.3 cm.

Initial V-Notch reading corresponding to no discharge = 55.1 mm				
Sr. No.	**Head above orifice w.r.t. datum (mm)**	**V-Notch reading (mm)**	**Head above V-Notch (h-mm)**	**Discharge (l/s)**
1	87.5	176.5	121.4	6.74
2	110.0	187.0	131.9	8.27
3	123.0	191.0	135.9	8.91
4	177.0	205.0	149.9	11.16
5	195.0	207.8	152.7	11.90
6	250.0	215.7	160.6	13.48
7	265.0	218.0	162.9	13.97
8	295.0	220.5	165.4	14.51
9	337.5.0	225.5	170.4	15.62
10	365.0	228.4	173.3	16.28
11	390.0	230.5	175.4	16.78
12	405.0	232.8	177.7	17.33
13	417.5	233.0	177.9	17.38
14	435.0	235.0	179.9	17.87
15	458.0	238.0	182.9	18.50
16	490.0	239.7	184.6	19.05
17	515.0	242.0	186.9	19.64
18	545.0	243.7	188.6	20.09

Table 2. Observations of discharge with orifice diameter 10 cm.

Initial V-Notch reading = 55.1 mm				
Sr. No.	**Head above orifice w.r.t. datum (mm)**	**V-Notch reading (mm)**	**Head above V-Notch (h-mm)**	**Discharge (l/s)**
1	120	179.1	124.0	7.10
2	160	185.5	130.4	8.04
3	250	200.2	145.1	10.48
4	315	206.5	151.4	11.65
5	355	209.5	154.4	12.23
6	390	212.7	157.6	12.87
7	415	215.2	160.1	13.38
8	465	221.6	166.5	14.75
9	492	223.4	168.3	15.15
10	545	223.7	168.6	15.21

Table 3. Observations of discharge with orifice diameter 7.5 cm.

Initial V-Notch reading = 55.1 mm				
Sr. No.	**Head above orifice w.r.t. datum (mm)**	**V-Notch reading (mm)**	**Head above V-Notch (h-mm)**	**Discharge (l/s)**
1	117	159.5	104.4	4.63
2	150	166.3	111.2	5.42
3	200	172.0	116.9	6.13
4	280	178.8	123.7	7.06
5	355	184.3	129.2	7.86
6	395	186.8	131.7	8.24
7	460	190.3	135.2	8.80
8	505	193.5	138.4	9.32
9	545	195.0	139.9	9.58

Table 4. Observations of discharge with orifice diameter 5 cm.

Initial V-Notch reading = 55.1 mm				
Sr. No.	**Head above orifice w.r.t. datum (mm)**	**V-Notch reading (mm)**	**Head above V-Notch (h-mm)**	**Discharge (l/s)**
1	165	127.7	72.6	1.88
2	225	136.2	81.1	2.48
3	300	148.8	93.7	3.54
4	345	151.2	96.1	3.77
5	404	156.9	101.9	4.36
6	440	161.2	106.1	4.83
7	475	162.7	107.6	5.00
8	541	166.4	111.2	5.43

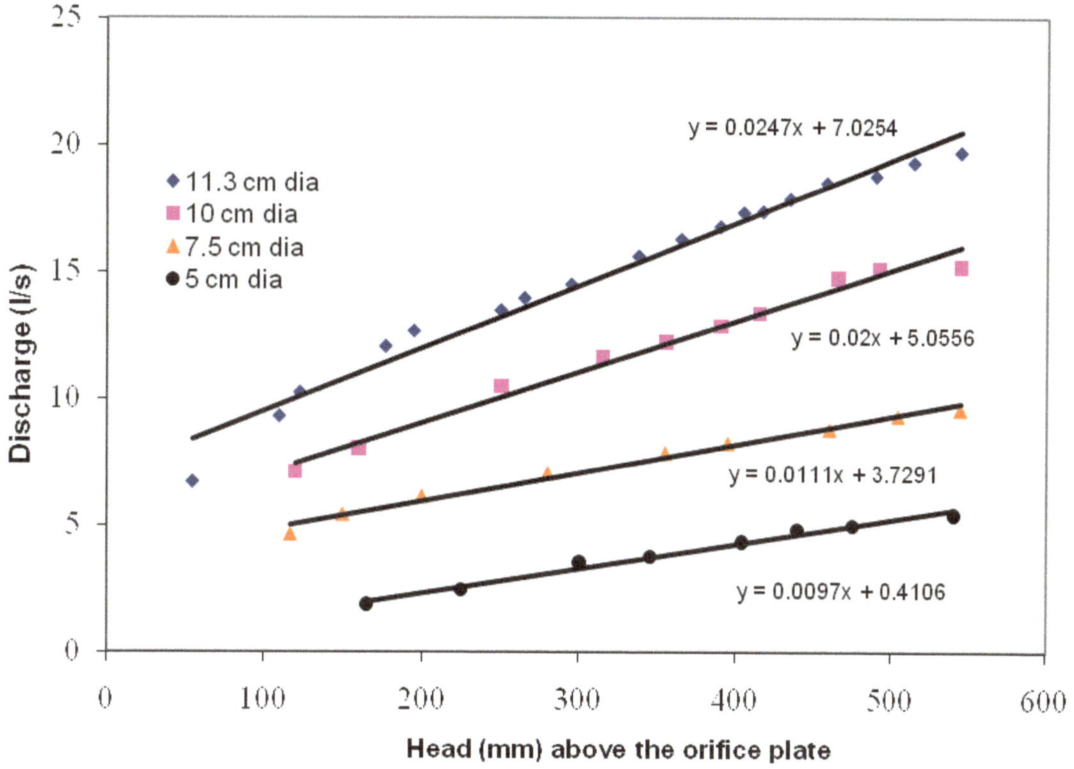

Figure 3. Rating curve for orifice bucket.

various values of pressure head is plotted in Figure 3. It was found that orifices of diameter 10cm, 7.5 cm and 5 cm accurately measured discharges ranging from 10 -15, 5 -10 and 1- 5 lps respectively.

Conclusions

An orifice bucket was designed, fabricated and calibrated and it was found that orifices of diameter 11.3, 10, 7.5 and 5 cm measured discharges ranging from 15 - 20, 10 - 15, 5 -10 and 1 - 5 lps respectively. The fabricated orifice bucket was found to be suitable for measuring discharges up to 20 lps at farmer's field. However for tubewell discharging greater than 20 lps larger diameter orifice bucket has to be fabricated.

REFERENCES

Aggarwal R, Kaushal M, Kaur S, Farmaha B (2009). Water resource management for sustainable agriculture in Punjab, India. Water Sci. Technol. 60(11): 2905 -2911.

Anonymous (1970, 2008). Statistical Abstract of Punjab - The Economic Advisor to Govt. of Punjab, Chandigarh, India.

Hamdy A, Ragab R, Scarascia-Mugnozza E (2003). Coping with water scarcity: water saving and increasing water productivity. Irrigation and Drainage, ICID 52(1): 3–20.

Henderson FM (1966). Open Channel Flow. MacMillan Company, New York, USA.
Rajput TBS, Patel N (2002). Determination of status of use of irrigation equipments on farmers' field through participatory rural appraisal techniques. In Proceedings of 2nd International Agronomy Congress, Nov. 26–30, 2002, New Delhi, 2: 1470–1471.

3

A combined high and low cycle fatigue model to estimate life of steel bridges

Kamal Karunananda[1]*, Mitao Ohga[1], Ranjith Dissanayake[2] and Sudath Siriwardane[1]

[1]Department of Civil and Environmental Engineering, Ehime University, Bunkyo-cho 3, Matsuyama 790-8577, Japan.
[2]Department of Civil Engineering, University of Peradeniya, Peradeniya 20400, Sri Lanka.

This paper proposes a new model to estimate life of structures for combined damage of high and low cycle fatigue. The model mainly consists of a new damage indicator and strain-life fatigue curve. The model predictions are verified by comparing with fatigue test results of some materials. The proposed model is applied to estimate the fatigue life of a bridge member for combined damage of high cycle and low cycle fatigue caused by usual traffic and earthquake loadings. Finally, the importance and applicability of the proposed model is confirmed.

Key words: High cycle fatigue, low cycle fatigue, combined damage, railway, bridge.

INTRODUCTION

Bridges are generally subjected to high cycle fatigue (HCF) caused by usual traffic loading (Chen, 1987). Recently, many fatigue failures of bridges have been reported in different parts of the world" or to give similar idea. Some of these failures cannot be explained by HCF phenomenon alone. Studies of these failures reveal that overloading caused by earthquakes is one of the reasons for these incidents. When a bridge is subjected to an earthquake, some members may undergo stresses of plastic range. These plastic stresses may cause low cycle fatigue (LCF) damage during the earthquake while subjecting to HCF in service conditions. This combined damage of HCF and LCF may be a reason for much reduced life than that predicted by HCF alone (Kondo and Okuya, 2007). Therefore, it is important to study life estimation methods based on combined damage of HCF and LCF to prevent unexpected failures of such bridges in the future.

Most of available literature on fatigue life estimation of steel bridges is focused on HCF caused by usual traffic loading (Mori et al., 2007; Righiniotis et al., 2008; Caglayan et al., 2009; Pipinato et al., 2009). To the knowledge of authors, there exists a considerable lack of information regarding the combined damage of HCF and LCF of bridges. However, in fields such as aircraft and mechanical engineering, combined damage of HCF and LCF has considerably been studied (Oakley and Nowell, 2007; Hou et al., 2009; Colin and Fatami, 2010). There, Coffin-Manson strain-life curve (total strain-amplitude versus number of reversals to failure) is generally used with Miner's rule (Miner, 1945) as the method to estimate the life for combined damage of HCF and LCF (Suresh, 1998) since it can be used easily to estimate life of components. This method gives better predictions in situations where HCF damage is lower than LCF damage. Recently, a more accurate method (Kim et al., 2009) has been proposed to estimate combined damage of HCF and LCF where HCF damage is higher than LCF damage. This approach is based on a modified Coffin-Manson curve and the Miner's rule.

The modification of Coffin-Manson curve is made by changing the slope of the original curve in the HCF regime to take account of damage interaction effect and damage below the fatigue limit. The factors which are used to modify the slope of the curve depend on the particular materials and these are not readily available for majority of materials. Therefore, obtaining the modified Coffin-Manson curve is difficult for majority of materials.

Miner's rule is the simplest and most widely used fatigue life prediction technique. One of its interesting features is that life calculation is simple and reliable when the detailed loading history is unknown. However under much variable amplitude loading conditions, Miner's rule based on life predictions have been found to be unreliable since it cannot capture loading sequence effect

*Corresponding author. E-mail: kamal@cee.ehime-u.ac.jp.

(Mesmacque et al., 2005; Siriwardane et al., 2007, 2008).

Since bridges are subjected to variable amplitude loading, the use of Miner's rule may give inaccurate life predictions. Therefore, these reasons restrict the application of this method (Kim et al., 2009) to life estimation for combined damage of HCF and LCF in bridges which are generally subjected to variable amplitude loading and higher HCF damage.

The studies (Mesmacque et al., 2005; Siriwardane et al., 2007, 2008) were conducted with a new damage indicator that can capture loading sequence effect in HCF regime. Especially, the studies of (Siriwardane et al., 2007, 2008) explain the results of an existing bridge in HCF regime. However, its direct application for combined damage is inaccurate since combined damage is more complex and different from the situation of only HCF presence (Lanning et al., 2001).

Therefore, it is necessary to have a different model, which is based on commonly available material properties, to estimate more accurately the life for combined damage of HCF and LCF due to variable amplitude loadings.

The objective of this paper is to propose a new model to estimate the fatigue life for combined damage of HCF and LCF due to variable amplitude loading conditions. The model consists of a new damage indicator and strain-life curve and these are original points of this paper. The new damage indicator can accurately capture the loading sequence effect. The strain-life curve caters with new damage indicator for better estimation of combined damage of HCF and LCF. The model can be applied to a wide range of materials since it depends on commonly available material parameters.

First, the paper proposes the new fatigue model. Verification of the model is then performed by comparing it with experimental fatigue lives and theoretical predictions. Finally, the model is applied to estimate the fatigue life of a bridge member for combined damage of HCF and LCF caused by usual traffic and earthquake loadings.

PROPOSED FATIGUE MODEL

This section proposes the new fatigue model to estimate life of steel structures for combined damage of HCF and LCF due to variable amplitude loading. Total strain is considered as the damage variable in the proposed model. Initially, the details relevant to proposed strain-life fatigue curve are discussed. Then, the section clearly describes the proposed damage indicator.

Strain-life fatigue curve

Summation of independent damages of HCF and LCF does not directly represent real damage behaviour for combined damage of HCF and LCF (Constantinescu et al., 2003). To take into account the damage dependent effect in combined HCF and LCF, it is necessary to modify the strain-life fatigue curve in HCF regime. This, together with the requirement of hypothetical ultimate strain in HCF regime for the proposed damage indicator (damage indicator) leads to construct of a full range of different strain-life curve for HCF regime in addition to the strain-life curve used in LCF regime. In this paper, the damage transfer technique proposed by Mesmacque et al. (2005) is used for combined damage of HCF and LCF. However, strain is used as the damage variable for both HCF and LCF regimes.

The proposed curve consists of two parts as shown in Figure 1. The first part of the curve describes fatigue life of plastic strain cycles ($\varepsilon \geq \varepsilon_y$) which usually affect LCF and the second part for elastic strain cycles ($\varepsilon < \varepsilon_y$) which usually affect HCF. To describe the first part, Coffin-Manson strain-life curve is utilized as shown.

$$\varepsilon = \frac{\sigma_f'}{E}(2N)^b + \varepsilon_f'(2N)^c \tag{1}$$

Where ε is the applied strain amplitude, N is the number of cycles to failure, σ_f' is the fatigue strength coefficient, b is the fatigue strength exponent, ε_f' is the fatigue ductility coefficient, c is the fatigue ductility exponent and E is the elastic modulus of the material.

The ultimate strain of low cycle fatigue $(\varepsilon)_{ULCF}$ which is the total strain amplitude corresponding to failure in half reversal (a quarter of a cycle) is obtained from equation (1) as,

$$(\varepsilon)_{ULCF} = \varepsilon_f' \tag{2}$$

These parameters are generally known for most materials and can be found in the literature. Therefore, the first part of the curve can be obtained for most materials easily without conducting additional material test.

The second part of the curve, as mentioned earlier, describes the fatigue life of elastic strain cycles ($\varepsilon < \varepsilon_y$) which usually affects HCF. This part of the curve represents hypothetical fully known curve. The shape of the curve is obtained by directly transforming the previous fully known stress-life curve (Siriwardane et al., 2007, 2008) to elastic strain-life curve as shown below.

$$\varepsilon = \varepsilon_e \left(\frac{N + N_u}{N + N_e} \right)^{b'} \tag{3}$$

Where ε_e is the strain amplitude of the fatigue limit, N_e is the number of cycles to failure at strains of ε_e. The ε_y and N_y are the yield strain and the corresponding number of cycles to failure. The b' is the slope of the finite life region of the curve and it can be determined from the coordinates of points; (ε_y, N_y) and (ε_e, N_e).

The $(\varepsilon)_{UHCF}$ is the ultimate strain of high cycle fatigue which is the elastic strain amplitude corresponding to half reversal (a quarter of a cycle) is expressed as,

$$(\varepsilon)_{UHCF} = \left(\frac{\sigma_u}{E} \right) \tag{4}$$

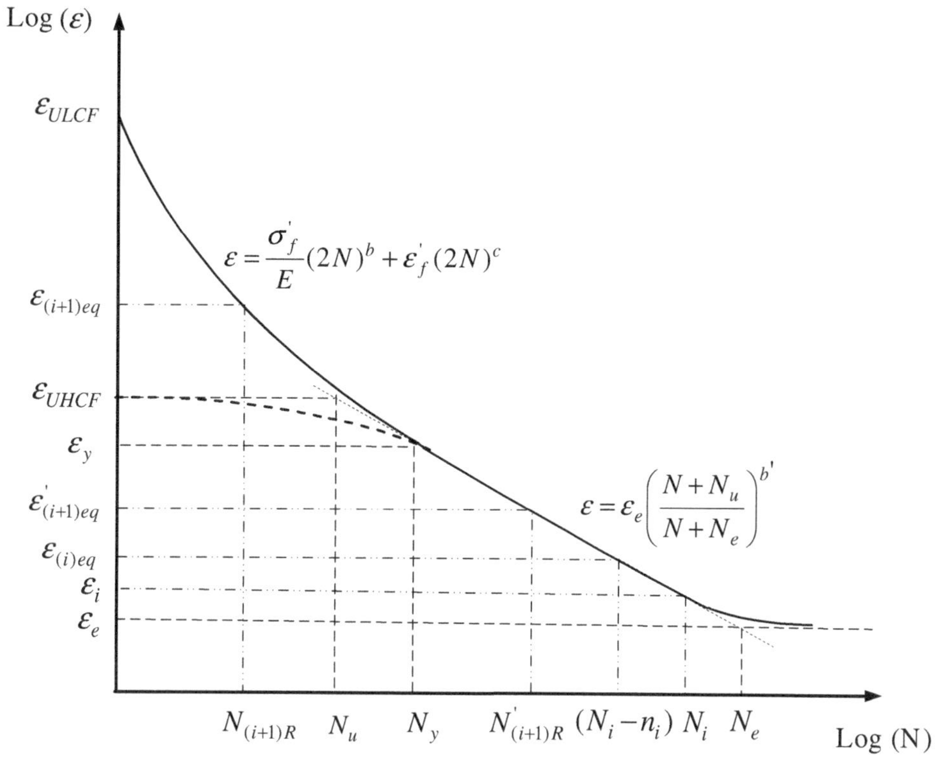

Figure 1. Schematic representation of the proposed strain-life fatigue curve.

where σ_u is the ultimate tensile strength of material. The N_u is the number cycles corresponding to the intersection of the tangent line of the finite life region and the horizontal asymptote of the ultimate elastic strain amplitude $(\varepsilon)_{UHCF}$ as shown in Figure 1.

Like the first part of the curve, the second part of the curve can be obtained using commonly available parameters without conducting additional material test. Therefore, the proposed strain-life curve can be obtained using commonly available parameters of materials.

Damage indicator

The proposed damage indicator considers combined damage of HCF and LCF due to variable amplitude loading. The hypothesis behind this fatigue law is that if the physical state of damage is the same, then fatigue life depends only on the loading condition. Suppose a component is subjected to a strain history; a strain amplitude $(\varepsilon)_i$ of n_i number of cycles at load level i and a strain amplitude $(\varepsilon)_i$ of n_i number of cycles at load level *i*, N_i is the fatigue life (number of cycles to failure) corresponding to $(\varepsilon)_{(i)eq}$ (Figure 1). Therefore, the reduced life at the load level i is obtained as (N_i–n_i). The equivalent strain $(\varepsilon)_{(i)eq}$ (Figure 1), which corresponds to the failure life (Ni–ni) is defined as ith level damage equivalent strain. Hence, the new damage indicator, Di is stated as,

$$D_i = \frac{(\varepsilon)_{(i)eq} - (\varepsilon)_i}{(\varepsilon)_u - (\varepsilon)_i} \tag{5}$$

where the $(\varepsilon)_u$ is

$$(\varepsilon)_u = \begin{cases} \varepsilon_{ULCF} & (\varepsilon)_i \geq \varepsilon_y \\ \varepsilon_{UHCF} & (\varepsilon)_i < \varepsilon_y \end{cases}$$

As shown in Equation (6), the appropriate value for $(\varepsilon)_u$ is selected based on the fatigue regime (HCF or LCF) in which strain amplitude $(\varepsilon)_i$ lies.

This damage Di has to be transformed to the next loading level i+1. It is assumed that at the end of ith loading level, damage Di has been accumulated (occurred) due to the effect of $(\varepsilon)_{i+1}$ loading cycles as follows;

$$D_i = \frac{(\varepsilon)'_{(i+1)eq} - (\varepsilon)_{i+1}}{(\varepsilon)_u - (\varepsilon)_{i+1}} \tag{7}$$

and $(\varepsilon)_u$ is expressed as

$$(\varepsilon)_u = \begin{cases} \varepsilon_{ULCF} & (\varepsilon)_{i+1} \geq \varepsilon_y \\ \varepsilon_{UHCF} & (\varepsilon)_{i+1} < \varepsilon_y \end{cases} \quad (8)$$

Then, $(\varepsilon)'_{(i+1)eq}$ is the damage equivalent strain at loading level i+1 and it is calculated as,

$$(\varepsilon)'_{(i+1)eq} = D_i[(\varepsilon)_u - (\varepsilon)_{i+1}] + (\varepsilon)_{i+1} \quad (9)$$

Thus the corresponding equivalent number of cycles to failure $N'_{(i+1)R}$ is obtained from the strain-life curve as shown in Figure 1. The $(\varepsilon)_{i+1}$ is the strain at the level i+1 and supposing that it is subjected to $n_{(i+1)}$ number of cycles, then the corresponding residual life at load level i+1, $N_{(i+1)R}$ is calculated as,

$$N_{(i+1)R} = N'_{(i+1)R} - n_{(i+1)} \quad (10)$$

Therefore, strain $(\varepsilon)_{(i+1)eq}$, which corresponds to $(\varepsilon)_{(i+1)eq}$ at load level i+1, is obtained from the strain-life curve as shown in Figure 1. Then the cumulative damage at the end of load level i+1 is defined as,

$$D_{(i+1)} = \frac{(\varepsilon)_{(i+1)eq} - (\varepsilon)_{i+1}}{(\varepsilon)_u - (\varepsilon)_{i+1}} \quad (11)$$

where $(\varepsilon)_u$ is expressed as

$$(\varepsilon)_u = \begin{cases} \varepsilon_{ULCF} & (\varepsilon)_{i+1} \geq \varepsilon_y \\ \varepsilon_{UHCF} & (\varepsilon)_{i+1} < \varepsilon_y \end{cases} \quad (12)$$

Similarly, the damage indicator, D_i, can be calculated for any given strain history. At the first cycle the equivalent strain $(\varepsilon)_{(i)eq}$ is equal to $(\varepsilon)_i$ and the corresponding damage indicator becomes D_i=0. Similarly at the last cycle, the damage indicator becomes D_i=1 when $(\varepsilon)_{(i)eq}$ is equal to $(\varepsilon)_u$.

Therefore, the damage indicator is normalized to one (D_i=1) at the fatigue failure of the material. Hence, the above procedure is followed until D_i=1. The proposed damage indicator-based algorithm is more clearly summarized in the flow chart shown in Figure 2. Here, the defined fatigue failure is the time taken for initiation of crack at the location of maximum stress of the structural component.

VERIFICATIONS OF PROPOSED FATIGUE MODEL

In this section, predicted fatigue lives of the proposed model are compared with combined HCF and LCF test results of two materials and predicted fatigue lives of a precise LCF model. The following subsections describe these comparisons in detail.

Comparison with experimental lives of combined HCF and LCF tests

The fatigue test results of two materials are compared with the predicted fatigue lives of the proposed model. Two materials are P355NL1 steel (Pereira et al., 2008) and Inconel 718 nickel base super alloy (Cook, 1982). The first material is a low carbon steel which is recommended for manufacturing of pressure equipments. The second material is used in manufacturing of turbines. The tests were performed both in the HCF and LCF regimes under different kinds of variable amplitude loading nature such as increasing or decreasing step loadings, variable amplitude repeating block loadings.

Six fatigue tests of P355NL1 steel were carried out with two steps of strain ranges with values of 1 and 0.5% (Pereira et al., 2008). The specimens were tested under two loading sequences which are increasing (Pattern A) and decreasing (Pattern B) as shown in Figure 3. Three similar types of specimens were tested under each loading sequence. The tests were carried out in such a way that first block is applied for a specified number of cycles, not causing material failure. Then, the second block of loading is applied until the failure is observed. Further, all tests were strain controlled and conducted in null strain ratio. A detailed summary of the tests is given in Table 1.

Failure number of cycles of these tests was predicted by the proposed model. In addition, Miner's rule employed previous model was used to predict the number of cycles to failure. The obtained results are given in Table 1 and plotted in Figure 4. Both models give a close relation of fatigue life for increasing loadings (Pattern A). But predicted lives by previous model differ significantly from experimental lives for decreasing loading (Pattern B). Correlation coefficients of the proposed model predictions and previous model predictions with experimental test results were estimated as 0.97 and 0.64, respectively. The illustrations of Figure 4 and the difference in correlation coefficients convinced us that the proposed method has better correlation with experimental results than Miner's rule employed previous model.

Six fatigue tests (Cook, 1982) of Inconel 718 nickel base super alloy were carried out both under increasing type step loading (Pattern A) and variable amplitude repeating block loading (Pattern B) as shown in Figure 5. The four specimens were tested under increasing type loading with different number steps of strain ranges and two specimens were tested under variable amplitude repeating block loadings (Cook, 1982). A detailed summary of the tests is given in Tables 2 and 3 for increasing type step loading and variable amplitude repeating

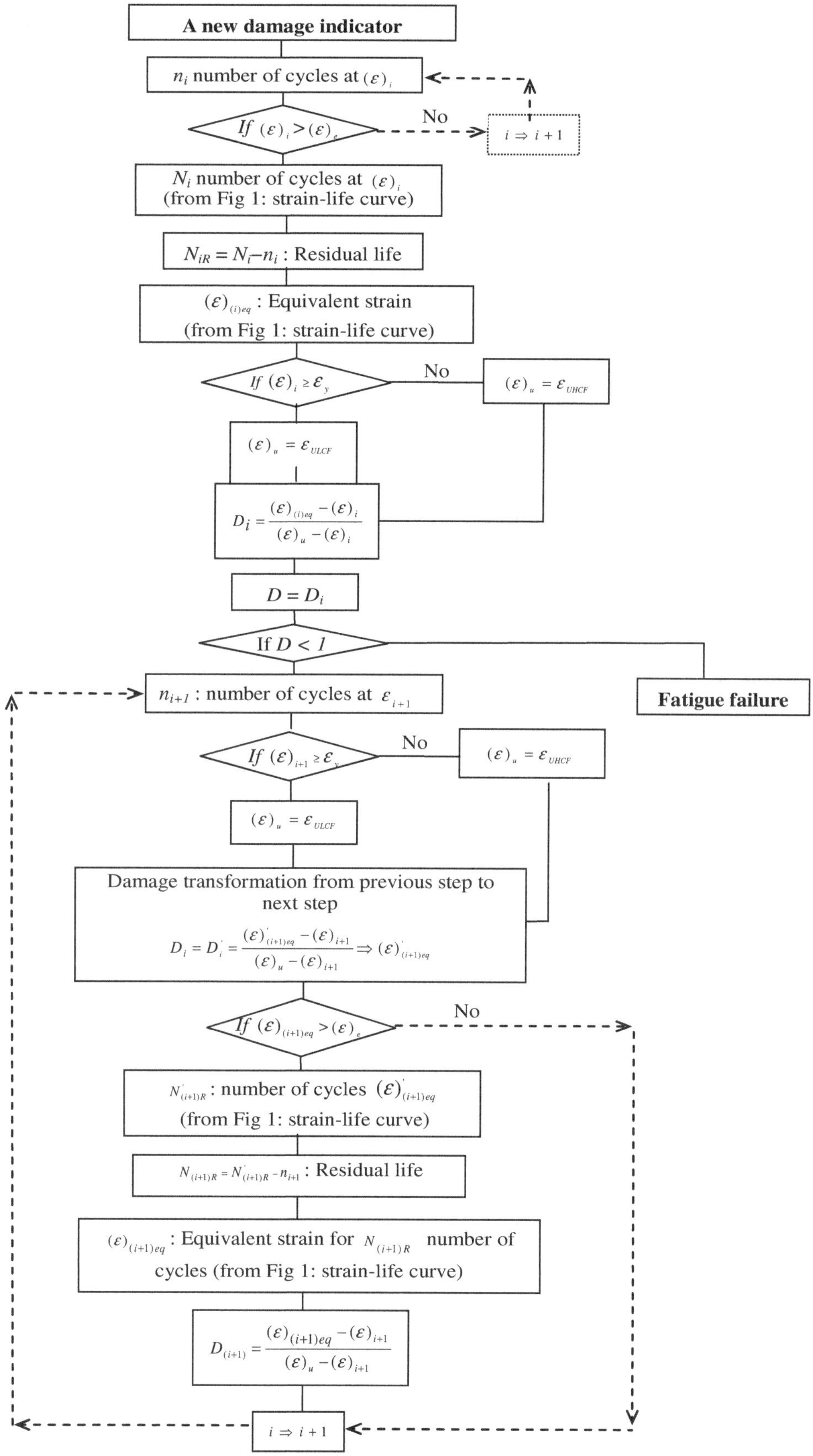

Figure 2. Flow chart of the proposed damage indicator.

Table 1. Comparison of predicted lives with experimental lives for P355NL1 steel.

Specimen	First strain step		Second strain step		Experimental fatigue life (cycles)		Predicted fatigue life (cycles)	
	Strain range (%)	Number of cycles (n_1)	Strain range (%)	Number of cycles (n_2)	Test result (n_1+n_2)	Average life	Previous model*	Proposed model
D201	1.0	337	0.5	2525	2862			
D202	1.0	337	0.5	3600	3937			
						3399	9024	4577
D203	1.0	673	0.5	977	1650			
D204	1.0	673	0.5	2727	3400			
						2525	5837	2007
D205	1.0	1010	0.5	1790	2800			
D206	1.0	1010	0.5	1404	2414			
						2607	2641	1805
D301	0.5	3620	1.0	922	4542			
D302	0.5	3620	1.0	1092	4712			
						4627	4440	4607
D303	0.5	7240	1.0	1507	8747			
D304	0.5	7240	1.0	1044	8284			
						8515	7715	7972
D305	0.5	9377	1.0	1528	10905			
D306	0.5	9377	1.0	1693	11070			
						10987	9648	9832

*: Coffin-Manson curve with Miner's rule.

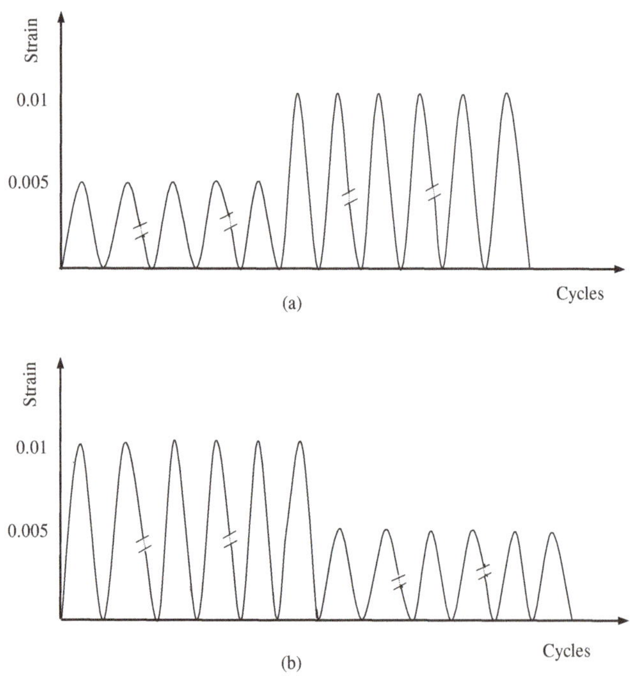

Figure 3. Loading patterns used for P355NL1 steel: (a) Pattern A (increasing loading) (b) Pattern B (decreasing loading).

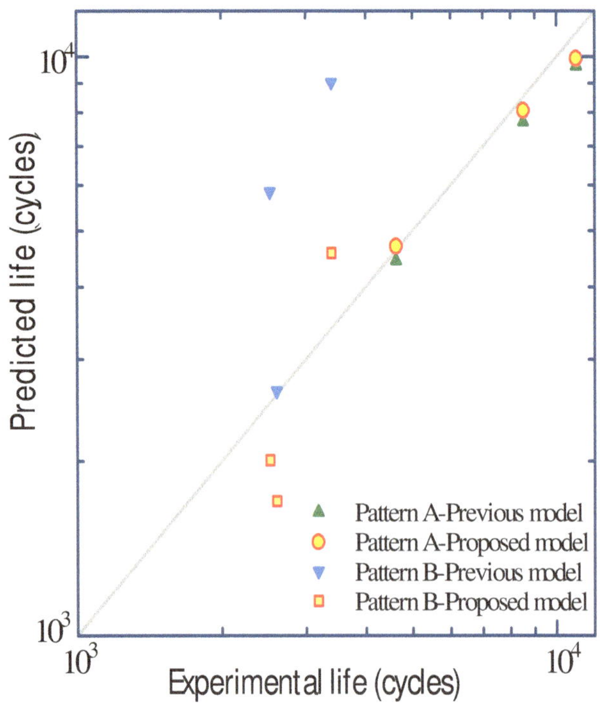

Figure 4. Comparison of predicted lives versus experimental lives for P355NL1 steel.

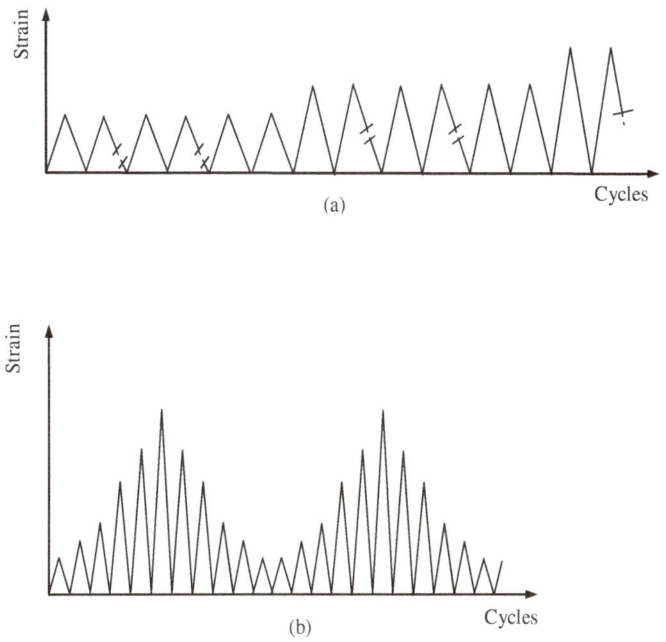

Figure 5. Loading patterns used for Inconel 718: (a) Pattern A (increasing type step loading) (b) Pattern B (variable amplitude repeating block loading).

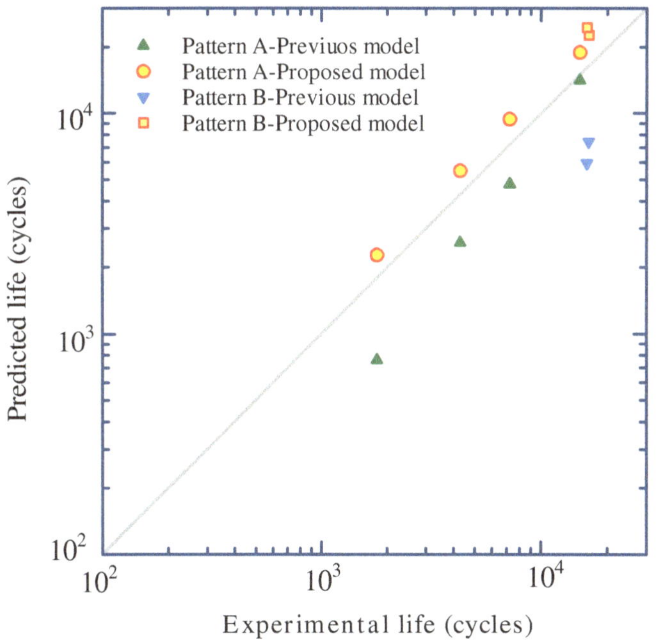

Figure 6. Comparison of predicted lives versus experimental lives for Inconel 718.

repeating block loading, respectively. The maximum strain amplitude of each block was subjected to increase in four stages during the strain history.

Failure number of cycles of these tests was predicted by the proposed model. In addition, Miner's rule employed previous model was used to predict the number of cycles to failure. The obtained results are given in Tables 2 and 3, and plotted in Figure 6. Correlation coefficients of the proposed model predictions and previous model predictions with experimental test results were estimated as 0.99 and 0.76, respectively.

The illustrations of Figure 6 and the difference in correlation coefficients convinced us that the proposed method has better correlation with experimental results than Miner's rule employed previous model.

Comparison with predicted lives of a precise LCF model

Liu et al. (2005) have recently proposed and verified an LCF model for accurate life prediction of A36 steel under variable amplitude loading. The model consists of a new damage indicator which was obtained by modifying the Miner's rule and considering three multiplication factors such as amplitude change factor, power coefficient for relative strain and partial cycle factor. Further this model describes a new cycle counting method. However, application of this model to other materials was found to be less since model parameter determination procedures are lengthy and difficult. On the other hand, corresponding cycle counting technique for random loading is totally different from well known cycle counting technique such as rain flow or reservoir counting, etc. In this section, this model (Liu et al., 2005) was used to compare with the proposed model prediction in LCF regime. The considered material is A36 steel and four repeating block loading patterns were considered as shown in Figure 7. For each pattern, six different strain history blocks were obtained by varying the amplitude of strain (that is, changing the value "a" of Figure 7 as shown in Table 4). For each history, failure number of cycles was predicted using the proposed model, Liu et al. (2005) model, as well as Miner's rule employed previous model. The comparisons are given in Table 4 and Figure 8. These comparisons indicate that the proposed model has better correlation with Liu et al. (2005) model for LCF regime than previous model predictions.

CASE STUDY: FATIGUE LIFE ESTIMATION OF A BRIDGE MEMBER

Fatigue life estimation of a bridge member is discussed in this section. The purpose of the case study is only to explain the application of the proposed model to an existing bridge member. The evaluations are especially based on secondary stresses and strains, which are generated around the riveted connection of the member due to stress concentration effect of primary stresses caused by usual traffic and earthquake loadings.

Table 2. Comparison of predicted lives with experimental lives of Inconel 718 for increasing type step loading.

Specimen	Stress range (GPa)	Strain range (%)	N_i (cycles)	Experimental fatigue life (cycles)	Predicted fatigue life (cycles)	
					Previous model*	Proposed model
N-7	1.02	0.5	7600			
	1.22	0.6	3400			
	1.44	0.7	1680			
	1.63	0.8	1000			
	1.78	0.9	640			
	1.89	1.0	460			
	1.94	1.1	260			
				15040	14102	18959
L-6	1.91	1.1	510			
	1.94	1.2	360			
	1.97	1.3	290			
	1.98	1.4	220			
	2.00	1.5	170			
	2.03	1.6	140			
	2.02	1.7	110			
				1800	759	2276
N-8	1.63	0.8	2360			
	1.86	1.0	1065			
	1.95	1.2	560			
	1.98	1.4	325			
				4310	2588	5501
T-7	1.41	0.7	4000			
	1.71	0.9	1830			
	1.84	1.1	800			
	1.85	1.2	600			
				7230	4779	9443

*: Coffin-Manson curve with Miner's rule.

Table 3. Comparison of predicted lives with experimental lives of Inconel 718 for variable amplitude repeating block loading.

Specimen	Stress range (GPa)	Strain range	N blocks	Experimental fatigue life (cycles)	Predicted fatigue life (cycles)	
					Previous model*	Proposed model
M-6	1.70	0.9	217			
	1.88	1.0	150			
	1.98	1.2	100			
	1.97	1.4	400			
				16473	7441	22639
X-7	1.62	0.8	150			
	1.93	1.1	150			
	1.89	1.2	150			
	1.89	1.3	400			
				16150	5956	24559

*: Coffin-Manson curve with Miner's rule

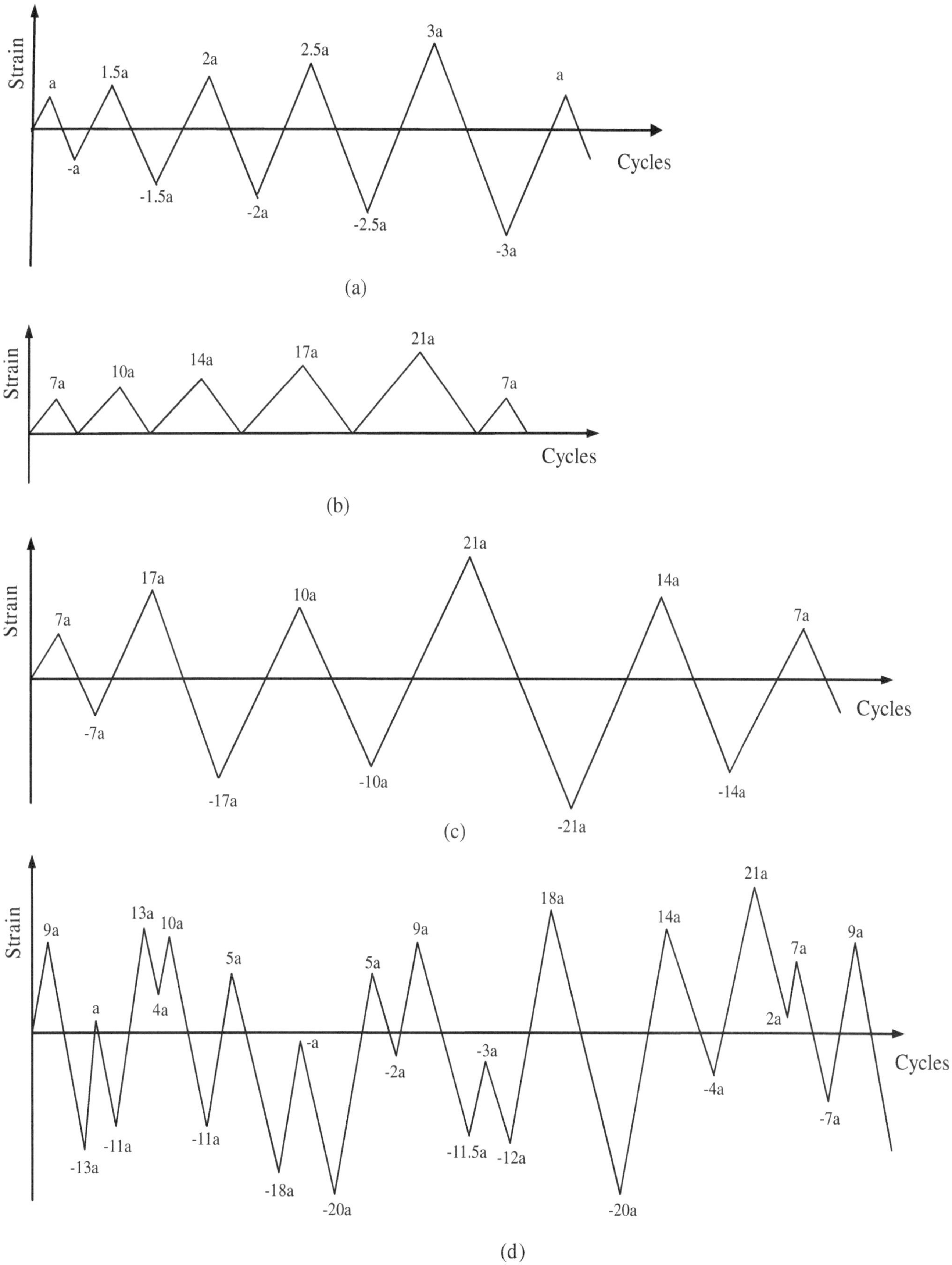

Figure 7. Loading patterns for A36 steel: (a) Pattern I (b) Pattern II (c) Pattern III (d) Pattern IV.

Table 4. Comparison of predicted lives by previous and proposed models with Liu's life (Liu et al., 2005) for A36 steel.

Pattern	Amplitude magnification factor	Predicted fatigue life (cycles)		
		Liu et al. model	Previous model*	Proposed model
I	0.03	40.5	49	44
	0.0275	54.5	64	58
	0.025	75.5	84	79
	0.0225	108.0	119	111
	0.02	159.0	173	162
	0.015	399.0	427	403
II	0.0045	33.0	43	38
	0.004	49.5	61	56
	0.0035	80.0	93	86
	0.003	136.5	153	142
	0.0025	251.0	271	254
	0.002	523.0	548	517
III	0.0055	31.5	51	43
	0.005	43.0	74	59
	0.0045	63.0	101	82
	0.004	92.5	140	119
	0.0035	141.5	209	182
	0.003	232.5	342	298
	0.0025	414.5	608	532
IV	0.0055	35.5	70	58
	0.005	52.5	89	76
	0.004	113.0	178	154
	0.0035	179.0	268	236
	0.003	299.5	434	384
	0.0025	542.5	764	684

*: Coffin-Manson curve with Miner's rule.

The considered bridge is one of the longest and busiest railway bridges in Sri Lanka situated near Colombo and carries the main railway line of the country. It was constructed in 1885 and since then it has been in operation without any major structural alterations except some strengthening conducted in 1939. A view of the bridge is shown in Figure 9a.

The bridge is semi-through having 6 double-system Warren girders supported on cylindrical piers. Its length, height and width are 154.8, 13.5 and 8.85 m, respectively and consisting of 6 spans. The super structure is made of wrought iron and the substructure is cast iron casings within filled concrete. The present service load varies from 28.86 to 112.8 tons.

Considered member

A structural appraisal was carried recently by University of Peradeniya (Ranaweera et al., 2002) in response to the request made by Sri Lankan Railway to find out current serviceable condition of the bridge. The visual, experimental and analytical investigation of the study led us to perceive the present condition of the bridge and identify critical members of the bridge. It was understood that a particular bracing member is highly subjective for fatigue failure. In history, it has been noticed that some of these types of members have failed but the reason is not yet known. Further, previous time-history analysis of the global structure of the bridge revealed that this member is also the one of highly stressed members for earthquake loadings. Therefore, that member was selected for fatigue life estimation to explain the application of the proposed model. A view of the selected member is shown in Figure 9b.

Stress-strain analysis

The combined damage of HCF and LCF is evaluated based on the state of strain due to release of contact (tightness) of rivets while all the riveted locations have no clamping force. The clamping force is generally defined as the compressive force in the plates which is induced

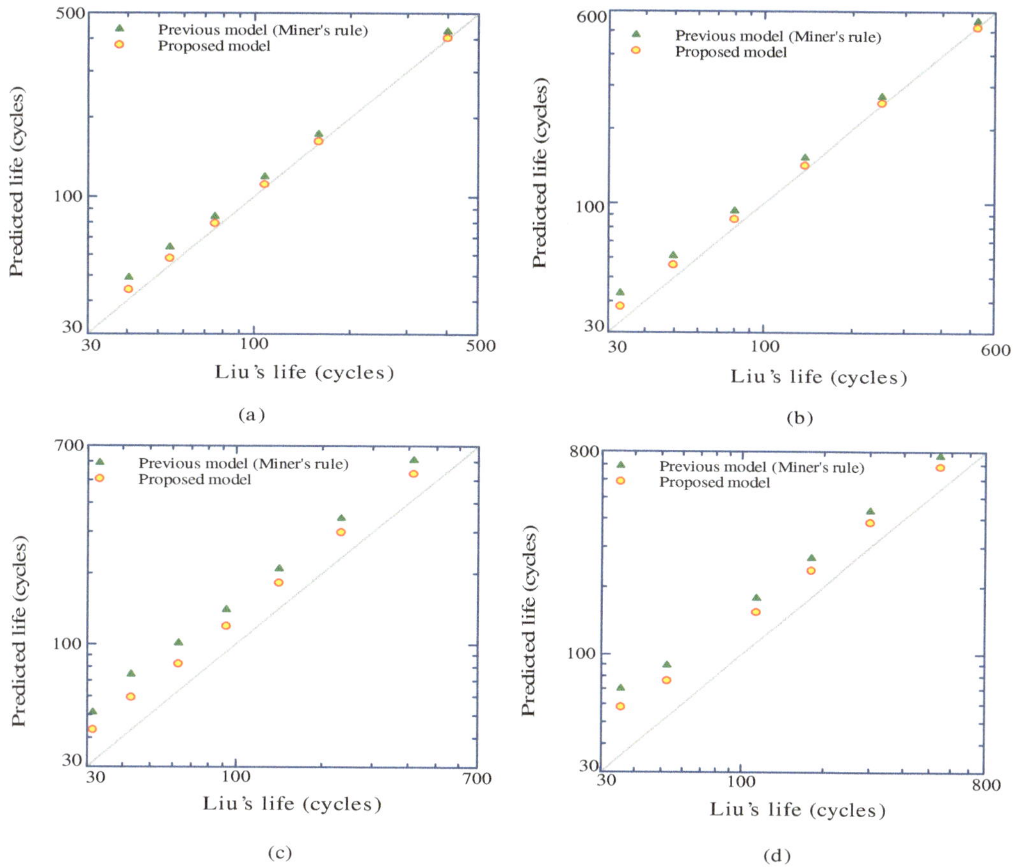

Figure 8. Comparison of predicted lives versus Liu's lives (Liu et al. 2005) for: (a) Loading pattern I (b) Loading pattern II (c) Loading pattern III (d) Loading pattern IV.

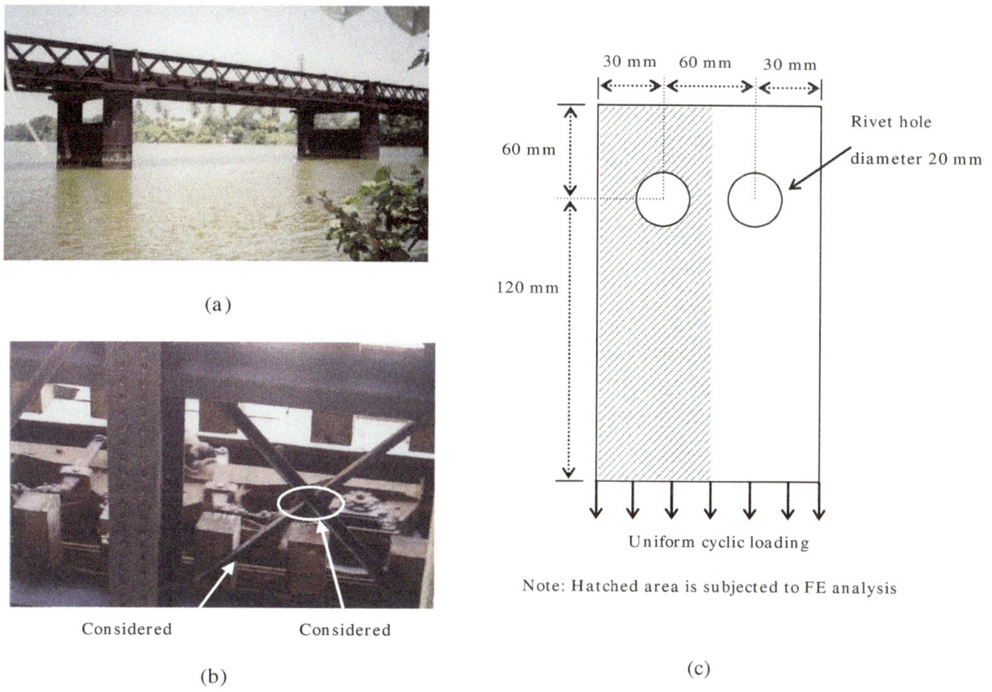

Figure 9. Considered member: (a) General view of the bridge, (b) Close view of considered connection and member, (c) Geometric details of connection.

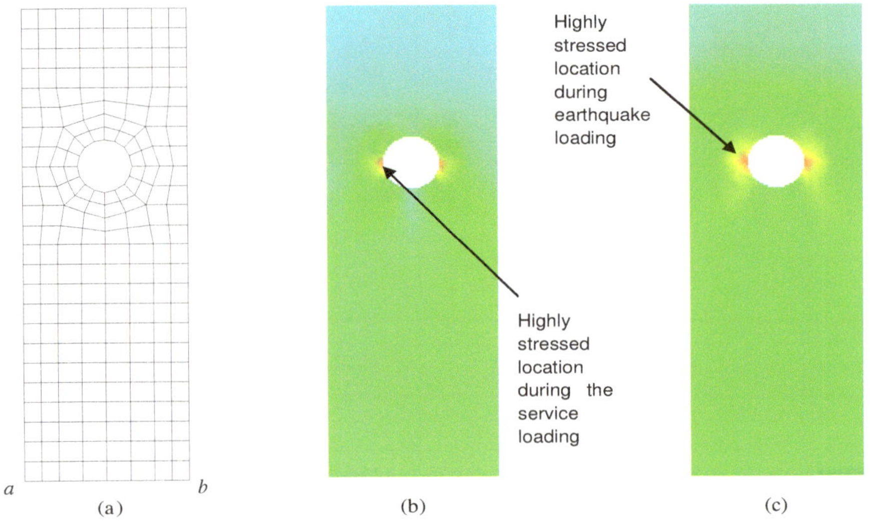

Figure 10. Stress analysis: (a) FEM mesh (b) Maximum von Misses stress distribution during service loading (c) Maximum von Misses stress distribution during earthquake loading.

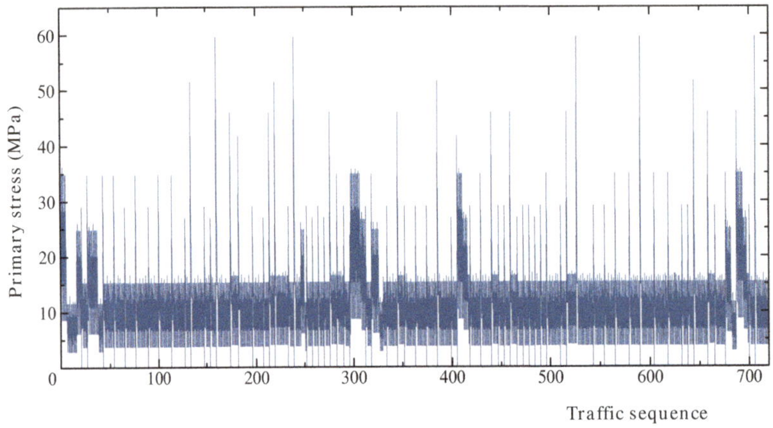

Figure 11. Primary stress variation of the member with traffic sequence per single day.

by the residual tensile force in the rivet. The residual force in the rivets occurs when the rivet get shortened in length due to cooling after a hot rivet is inserted into the hole of plates in order to connect them, and a second head is formed from the protruding shank. Finally clamping force generates a triaxial state of stress in the connected plate (Siriwardane et al., 2007; 2008). Since this study assumes that the riveted locations have no clamping force (value of clamping force is zero), the connected members are considered to subject to the biaxial stress state. Therefore, a critical member without rivets can be considered to analyze the biaxial state of stress. Considering symmetry, one half (hatched area) of the member was considered for FE analysis shown in Figure 9c. A FEM program with a FORTRAN source code which was developed in Ehime University was used in FE analysis. It can accurately model elastic and elasto–plastic behavior of materials. The nine node isoperimetric shell elements were used for the finite element analysis as shown in Figure 10a. The actual air gap restraint conditions were considered in the model to represent unilateral contact between the rivet and the plate. To make the continuity of the stress field between the global structure and the sub structure, it is required to use any interface between the two structures at every iterative step. Therefore, in this model, primary tensile stress history (Figure 11) of the member, which has been obtained from measured strain histories at the mid span of the bridge due to usual traffic loadings (Figure 11), is applied to the bottom face as uniform pressure. The position of

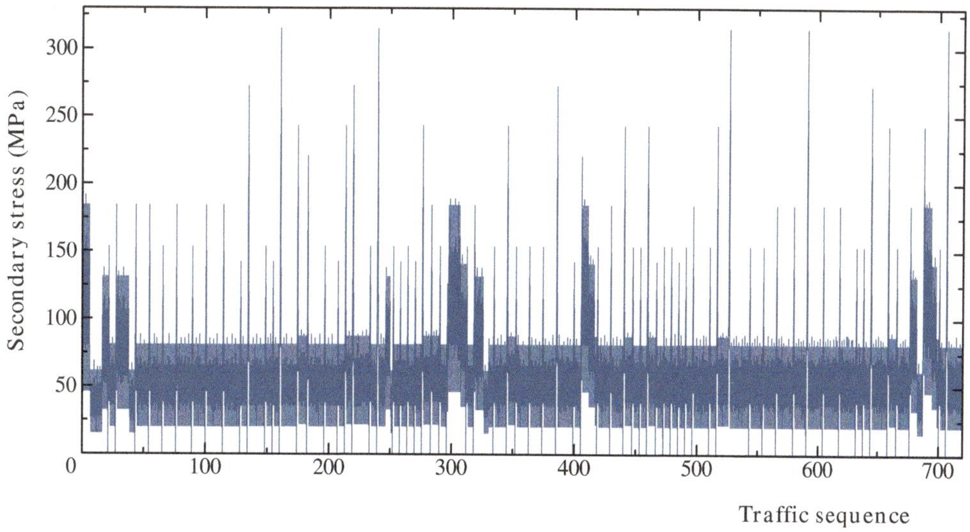

Figure 12. Secondary stress variation of the critical location with traffic sequence per single day.

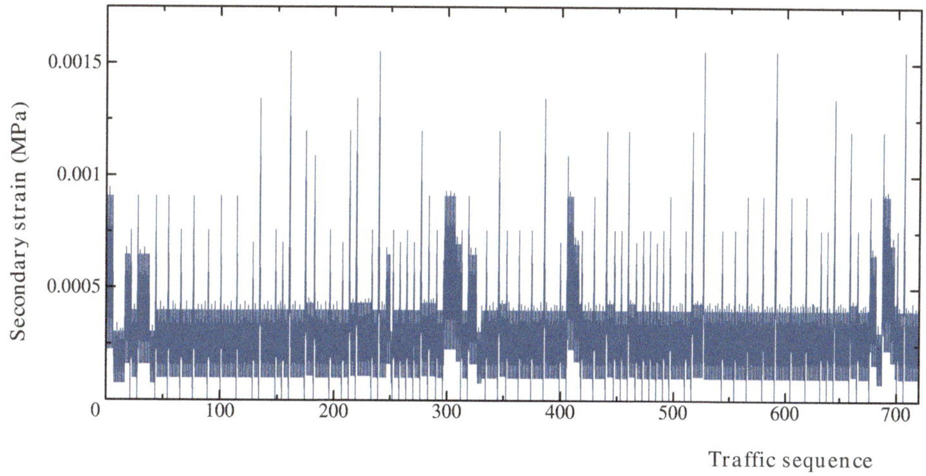

Figure 13. Secondary strain variation of the critical location with traffic sequence per single day.

position of the ab boundary of the sub model was determined considering the distribution of the far field primary stress of the member. The obtained maximum stress contours are shown in Figure 10b. This shows that stresses are operating well below the yield limit of the material (therefore HCF regime) and the highly stressed locations are subjected to uniaxial stress state. Assuming that a single day traffic sequence is repeating every day, one day time history is considered as the loading block in this study. The obtained secondary stress and strain histories for normal traffic loadings are shown in Figures 12 and 13 respectively.

In this case study, fatigue life was calculated for combined damage of usual traffic and earthquake loadings. Therefore, the same geometry of the member is subjected to further analysis assuming it is subjected to an earthquake loading. Then the elasto-plastic analysis was conducted by applying the primary stress history (Figure 14) which was obtained by time history analysis of global structure under earthquake loading was applied to the ab interface as same as above. The obtained maximum stress contours are shown in Figure 10c and it was decided that state of stress is uniaxial. The obtained secondary stress and strain histories for earthquake loadings are shown in Figures 15 and 16, respectively.

The obtained secondary strain variations (Figures 13 and 16) are complex and also of irregular shape. These strains should be reduced to a series of equivalent strain cycles at zero mean strain. In order to achieve this objective, initially the famous rainflow cycle counting technique (Dowling and Socie, 1982) is used to identify the strain ranges and sequences of closed strain cycles.

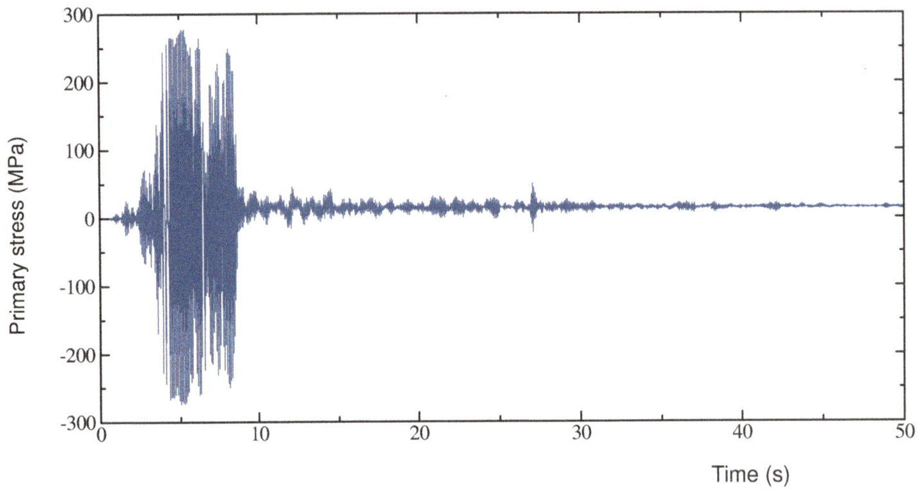

Figure 14. Primary stress history of the member during the earthquake.

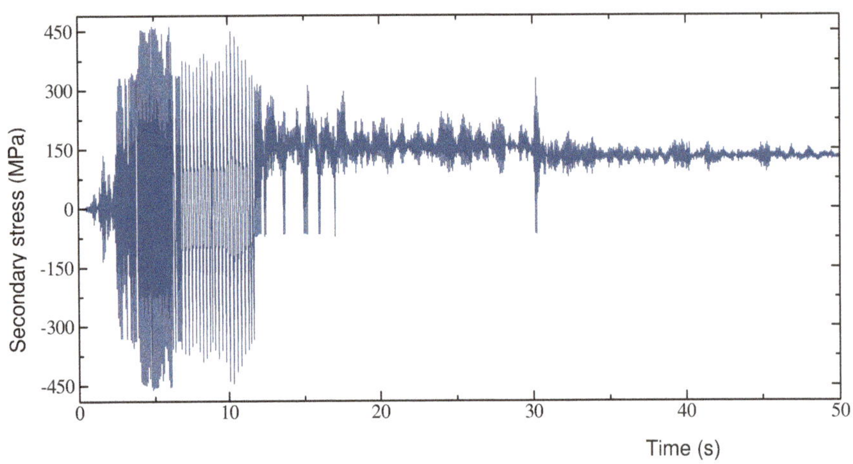

Figure 15. Secondary stress history of the critical location during the earthquake.

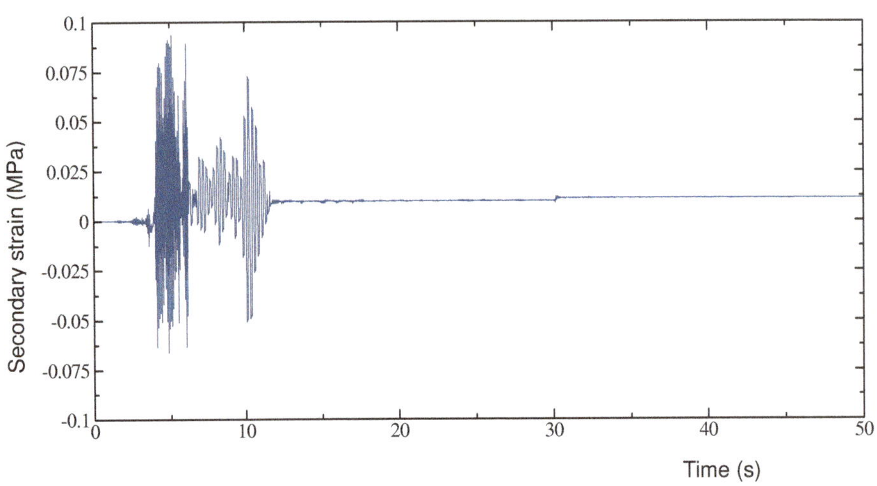

Figure 16. Secondary strain history of the critical location during the earthquake.

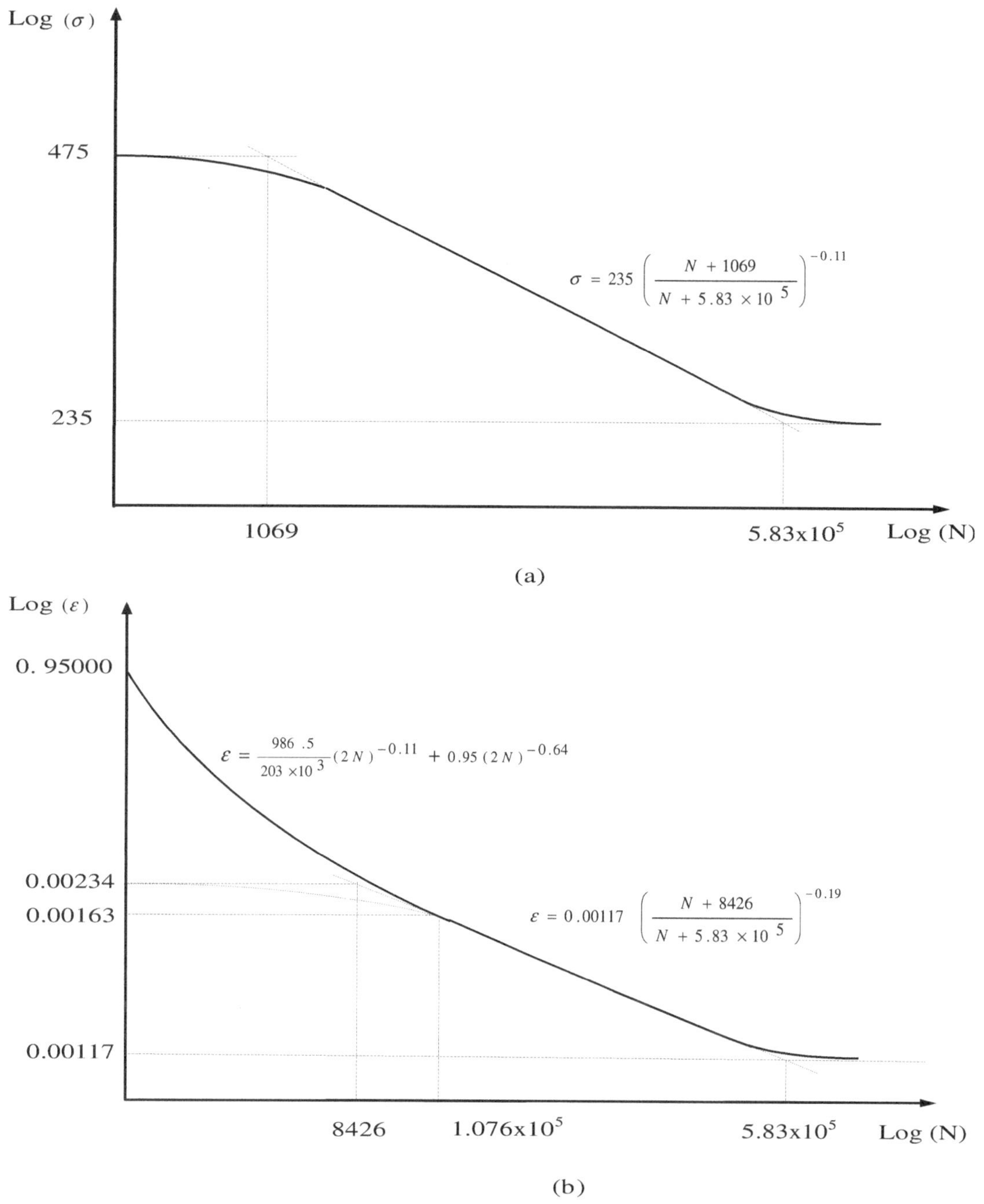

Figure 17. Fatigue curves for considered bridge material for use (a) Before the earthquake (b) During and after the earthquake.

Then modified Goodman relation is used to transfer these counted cycles to mean strain zero stabilized cycles.

Fatigue curves

Fatigue properties, $\sigma_f^{'}$, b, $\varepsilon_f^{'}$, c, σ_y, σ_u, σ_e and E of the considered member are 986.5 MPa, -0.11, 0.95, -0.64, 330 MPa, 475 MPa, 235 MPa and 203 GPa, respectively.

From these properties, ε_{ULCF}, ε_{UHCF}, ε_y, ε_e, N_y and N_e are estimated as 0.95, 0.00234, 0.00163, 0.00117, 1.076 × 10^5 and 5.83 × 10^5 respectively as explained in the strain-life fatigue curve. Then, $b^{'}$ and N_u were estimated as -0.19 and 8426, respectively (strain-life fatigue curve). From these parameters, the corresponding stress-life curve for HCF and the strain-life curve for combined HCF and LCF of the considered bridge material are constructed as shown in Figure 17.

Table 5. Comparison of calculated fatigue lives of the riveted connection.

Time of earthquake* (years)	Previous model (Miner's rule)		Proposed curve with Miner's rule		Proposed model	
	Fatigue life (years)	Percentage reduction of life (%)	Fatigue life (years)	Percentage reduction of life (%)	Fatigue life (years)	Percentage reduction of life (%)
5	26.5	35.4	34.0	17.0	22.0	63.3
10	26.5	35.4	34.0	17.0	24.0	60.0
20	26.5	35.4	34.0	17.0	29.0	51.7
30	30.0	26.8	34.0	17.0	34.0	43.3
40	40.0	2.4	-	-	41.0	31.7
50	-	-	-	-	50.5	15.8
54	-	-	-	-	54.0	10.0
Without earthquake	41.0	-	41.0	-	60.0	-

*: After construction.

Fatigue life estimation

Until the earthquake occurrence, the damage due to usual traffic loads (HCF) was evaluated using a previously proposed HCF model (Siriwardane et al., 2008) with the stress-life curve shown in Figure 17a. During and after the earthquake, fatigue damage was evaluated using the proposed combined HCF and LCF model (Proposed fatigue model). The updated damage (damage accumulation) was calculated by following the flow chart given in Figure 2. The fatigue life was estimated when the damage indicator reaches to the unity.

Earthquake was considered to occur at different times of the bridge life such as 5, 10, 20, 30, 40, 50 and 54 years as shown in Table 5. It is assumed that usual traffic load is followed after the earthquake until the member failure. The fatigue life of the member was estimated using three approaches: (1) proposed model; (2) previous model; (3) The Miner's rule with the proposed strain-life curve. The purpose of using approach (3) was to estimate the effect of the proposed strain-life curve on life estimation instead of Coffin Manson strain relationship.

The obtained results are given in Table 5 for three approaches. The results indicate that LCF damage by earthquake loading causes an appreciable reduction of bridge life. For the proposed model, percentage reduction of life is higher when the earthquake occurs at the beginning of the bridge life compared to those occurring in later times. For example, when the earthquake occurs 5 years after construction, the previous model predicts the fatigue life as 26.5 years causing 35.4% of life reduction. However, the proposed model gives a fatigue life as 22.0 years causing 63% of life reduction. When the earthquake occurs later in the bridge life, percentage reduction of the service life gets lower for the proposed model. However, Miner rule gives a constant percentage reduction of service life since it cannot capture the loading sequence effect. The amount of fatigue damage predicted by the Miner's rule is constant irrespective of its occurrence time. The results obtained using Miner's rule with the proposed curve (approach 3) lies between the previous and the proposed methods. Therefore, it indicates that the proposed curve more accurately represent the combined damage of HCF and LCF than the Coffin Manson strain relationship. The differences in case study results confirm the importance of accurate combined HCF and LCF model to estimate the fatigue life of existing connections.

Conclusion

A new model for combined damage of HCF and LCF was proposed to estimate the fatigue life of steel structures such as bridges. A verification of the model was conducted by comparing the predicted lives with experimental lives due to variable amplitude loading for two materials available in the literature. Further, the model was verified with a previously proposed model only in LCF regime. It was shown that the proposed fatigue model gives a realistic fatigue life for the combined damage of HCF and LCF in variable amplitude loading situations where detailed stress histories are known. The proposed fatigue model was used to estimate the fatigue life of a bridge member of an old wrought iron railway bridge. Case study realized the importance of consideration of the earthquake induced LCF damage in addition to HCF damage due to usual traffic loading in steel bridges. The importance of accurate prediction of combined damage of HCF and LCF was also confirmed. Further verifications of proposed model in more complex loading situations are currently under way. Since this model only describes the uniaxial stress state, it is recommended to extend these models for multiaxial stress state in the future.

ACKNOWLEDGEMENTS

The authors wish to express their sincere gratitude to Professor M.P. Ranaweera and the team of experts who worked in the Sri Lankan Railway Bridge project, for their great advice in carrying out this research. The kind support given by the Sri Lanka Railways (SLR) is also appreciated.

REFERENCES

Caglayan BO, Ozakgul K, Tezer O (2009). Fatigue life estimation of a through –girder steel railway bridge, Eng. Failure Anal., 16(3): 765-774.

Chen WF, Lui EM (1987). Handbook of Structural Engineering. CRC Press, New York.

Colin J, Fatami A (2010). Variable amplitude cyclic deformation and fatigue behavior of stainless steel 304L including step, periodic, and random loadings, Fatigue Fracture Eng. Materials Structures. 33(4): 205-220.

Constantinescu A, Van KD, Maitournam MH (2003). A unified approach for high and low cycle fatigue based on shakedown concepts. Fatigue and Fracture of Eng. Materials Structures, 26(6): 561-568.

Cook TS (1982). Stress-strain behavior of Inconel 718 during low cycle fatigue. J. Eng. Materials Technol., 104(3): 186-191.

Dowling SD, Socie DF (1982). Simple rainflow counting algorithms, Int. J. Fatigue, 4(1): 31-40.

Hou NX, Wen ZX, Yu QM, Yue ZF (2009). Application of a combined high and low cycle fatigue life model on life prediction of SC blades, Int. J. Fatigue, 31(4): 616-619.

Kim YH, Song JH, Park JH (2009). Expert system for fatigue life prediction under variable loading, Expert Syst. Appl., 36(3): 4996-5008.

Kondo Y, Okuya K (2007). The effect of seismic loading on the fatigue strength of welded joints, Material Sci. Eng. A., 468-470: 223-229.

Lanning D, Haritos GK, Nicholas T, Maxwell DC (2001). Low cycle fatigue/high cycle fatigue interactions in notched Ti-6Al-4V, Fatigue Fracture Eng. Materials Structures, 24(12): 565-577.

Liu W, Liang Z, Lee G (2005). Low cycle bending fatigue of steel bars under random excitation. Part II: design considerations. J. Structural Eng., 131(6): 919-923.

Mesmacque G, Garcia S, Amrouche A, Rubio-Gonzalez C (2005). Sequential law in multiaxial fatigue, a new damage indicator, Int. J. Fatigue, 27(4): 461-467.

Miner MA (1945). Cumulative damage in fatigue. J. Appl. Mechanics, 12: 159-64.

Mori T, Lee H, Kyung K (2007). Fatigue life estimation parameter for short and medium span steel highway girder bridges, Eng. Structures, 29(10): 2762-2774.

Oakley SY, Nowell D (2007). Prediction of the combined high- and low cycle fatigue performance of gas turbine blades after foreign object damage, Int. J. Fatigue, 29(1): 69-80.

Pereira HFSG, de Jesus AMP, Fernandes AA, Ribeiro AS (2008). Analysis of fatigue damage under block loading in a low carbon steel, Strain, 44(6): 429-439.

Pipinato A, Pllegrino C, Bursi OS, Modena C (2009). High cycle fatigue behavior of riveted connections for railway metal bridges, J. Constructional Steel Res., 65(12): 2167-2175.

Ranaweera MP, Aberuwan H, Mauroof ALM, Herath KRB, Dissanayake PBR, Siriwardane, SASC, Adasooriya AMND (2002). Structural appraisal of Railway Bridge at Colombo over Kelani River, Engineering Design Centre, University of Peradeniya.

Righiniotis TD, Imam BM, Chryssanthopoulos MK (2008). Fatigue analysis of riveted railway bridge connections using theory of critical distances. Eng. Structures, 30(10): 2707-2715.

Siriwardane S, Ohga M, Dissanayake R, Taniwaki K. (2008). Application of new damage indicator-based sequential law for remaining fatigue life estimation of railway bridges, J. Constructional Steel Res., 64(2): 228-237.

Siriwardane SC, Ohga M, Dissanayake R, Taniwaki K (2007). Different approaches for remaining fatigue life estimation of critical members in railway bridges, Int. J. Steel Structures, 7(4): 263-276.

Suresh S (1998). Fatigue of materials. UK: Cambridge University Press, Cambridge.

Development of a multipurpose, efficient and inexpensive bearing test rig

S. P. S. Matharu[1]*, S. Sanyal[1] and D. S. Bal[2]

[1]Department of Mechanical Engineering, National Institute of Technology Raipur (C.G.) 492010, India.
[2]Directorate of Technical Education, Government of Chattisgarh, Raipur (C.G.) 492001, India.

The basic objective of this research is to develop an inexpensive, multipurpose and efficient test rig for condition monitoring of rolling element bearings. The fundamental philosophy of measurement of lubricant film thickness for monitoring the condition of rolling element bearings has been exploited and transformed in the form of a test rig. The developed test rig is potent enough for performance evaluation and condition monitoring of rolling element bearings which is vital for its prolonged life and enhanced performance. A significantly correct prediction of life of bearing can also be made which in turn gives an idea about possible mode and time of failure. Certain machineries like aircraft, battle tanks, rocket launchers, ships, etc. need impeccable working of the system. Failure of such systems is not acceptable at any cost. In the heart of such system, bearings play a central and key role. Still there is no such standard mechanism available for ascertainment of performance and testing of rolling element bearings with high degree of reliability. In the continuing process of improvement and modification in the existing test rig, a simple, versatile, and inexpensive test rig has been developed and presented through this paper.

Key words: Condition monitoring, low cost, lubricant film thickness, performance evaluation, rolling element bearing, test rig.

INTRODUCTION

Rolling element bearing is a basic component of almost all machinery transmitting power. It finds its applications in land automobiles, aircrafts, ships, space vehicles, large and small industrial installations, defense machinery, etc. The extent of damage or loss can be thought off when the bearing suddenly fails of an automobile overtaking a vehicle, of an aircraft after attaining its maximum height, of a ship in deep sea, of a space vehicle when placed in orbit, of an industrial installation during its peak production run, of a defense vehicle when encircled by enemies, etc.

The testing of rolling element bearings involves various parameters to be considered such as the range of speed and load, type of loading such as radial or axial or both, type of lubricant, etc. A good test rig essentially consists of a prime mover with speed control options, a supporting shaft assembly for test bearing with options to incorporate multiple diameter bearings, arrangement for application of load such as hydraulic or mechanical, measuring instruments related to parameters to be measured, etc. The test rigs (Prashad, 1987, 2006) along with measuring instruments used by researchers mostly are very costly. This limits the bulk of research in the field. Keeping in view the limitation of cost and to promote research in the field, a low cost experimental setup is developed with essential instruments in the Dynamics of Machines Laboratory of National Institute of Technology, Raipur.

The developed test rig have tapping of two electrical points, one connected to inner race of the rolling element bearing and other connected to the outer race. A simple electrical circuit, discussed in methodology, is connected between the tapped points to measure the overall electrical resistance between the inner and outer race of the bearing due to the presence of lubricant between rolling elements and races. The voltage drop across

*Corresponding author. E-mail: spmatharu123@gmail.com.

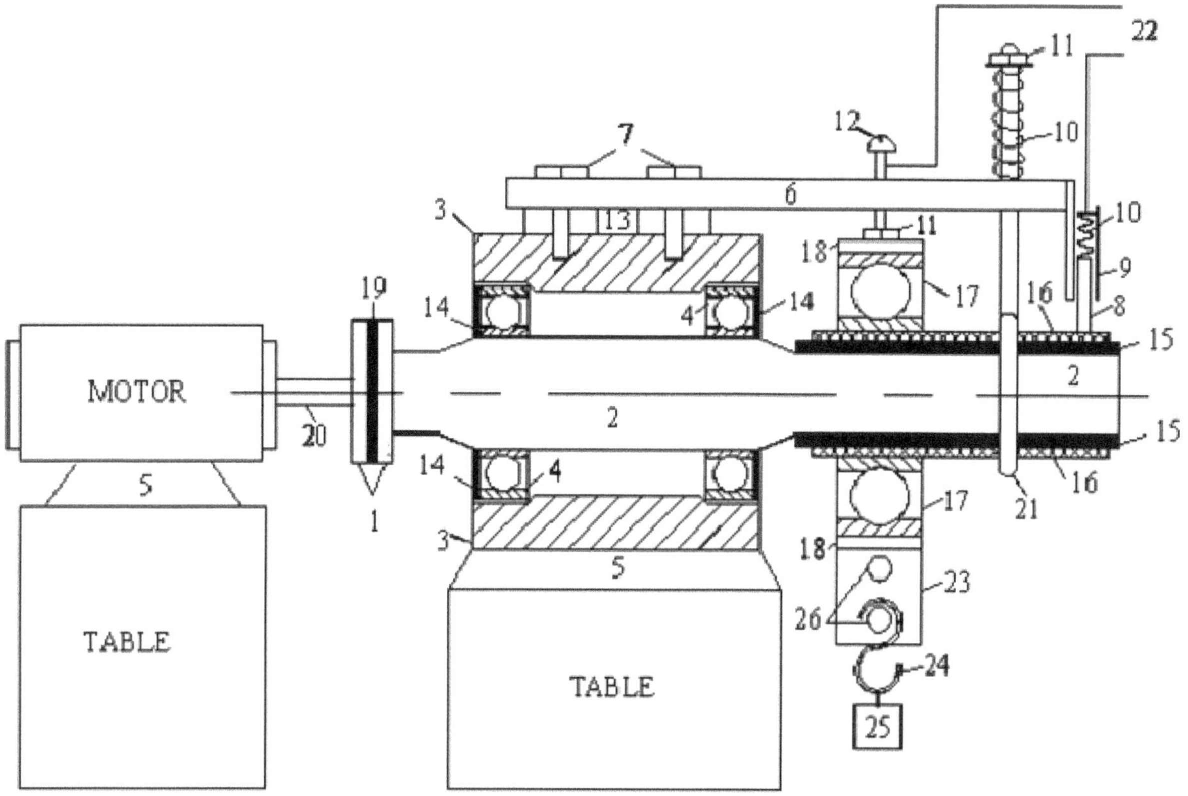

Figure 1. Schematic diagram of rolling element bearing test rig; 1, Coupling; 2, Test rig shaft; 3, Thick mild steel tube; 4, Supporting ball bearings; 5, Supporting stand; 6, Cantilever wooden plate; 7, Bolts; 8, Carbon brush; 9, Brush guide; 10, Helical spring; 11, Nuts; 12, Screw; 13, Spacer washer; 14, Bearing seal; 15, Insulating sleeve; 16, Mild steel sleeve; 17, Test ball bearing; 18, Bearing outer copper sleeve; 19, Rubber pad; 20, Motor shaft; 21, Brass ring; 22, Wires to electrical circuit; 23, Projection on copper sleeve; 24, Steel hook; 25, Suspended weight; 26, Holes for bolts.

bearing inner and outer race is measured by using an Oscilloscope. By using suitable formula, electrical resistance between inner and outer race of bearing is calculated. The value of resistance measured with varying speed and load will directly depict the amount of lubricant present between rolling elements and races. This will lead to convenient, reliable and inexpensive online condition monitoring of rolling element bearings.

EXPERIMENTAL SETUP

Experimental setup has been shown through Figures 1 and 2. Figure 1 depicts the schematics of the setup whereas Figure 2 is the photograph of the setup developed and used in the present study. The test rig essentially consists of a shaft supported at one end by two well placed bearings (press fitted in a steel housing at either ends) and overhung at the other end that supports the test bearing. One end of the shaft is coupled to a DC motor and the other end carries the test bearing. An AC variable transformer is used to vary the input voltage to the motor through the rectifier.

The steel shaft is fitted upon an insulating sleeve on which a steel sleeve of appreciable length is tightly fitted. The outer diameter of steel sleeve is equal to the inner diameter of bearing. Similarly inner diameter of steel sleeve is equal to outer diameter of the insulating sleeve. This steel sleeve accommodates the test bearing and the carbon brush arrangement.

Insulation is needed to eliminate the influence of support bearings in the measurement of resistance and capacitance of the test bearing. The outer diameter of the steel sleeve varied from one end just below 35 mm and at the other end just above 35 mm after considering different tolerances. The steel sleeve was made tapered to facilitate easy fitting and removal of the different test bearing with inner diameter of 35 mm. The test bearings are deep groove ball bearings (6207and 6307) whose inner race is press fitted on one end of the steel sleeve and the carbon brush/brass ring arrangement is accommodated on the other side (free end of sleeve).

A split copper outer sleeve, tightly fitted on the outer race of the ball bearing, acts as the bearing housing. The lower end of the copper sleeve is split end which is tightened with each other by using a nut and bolt. One

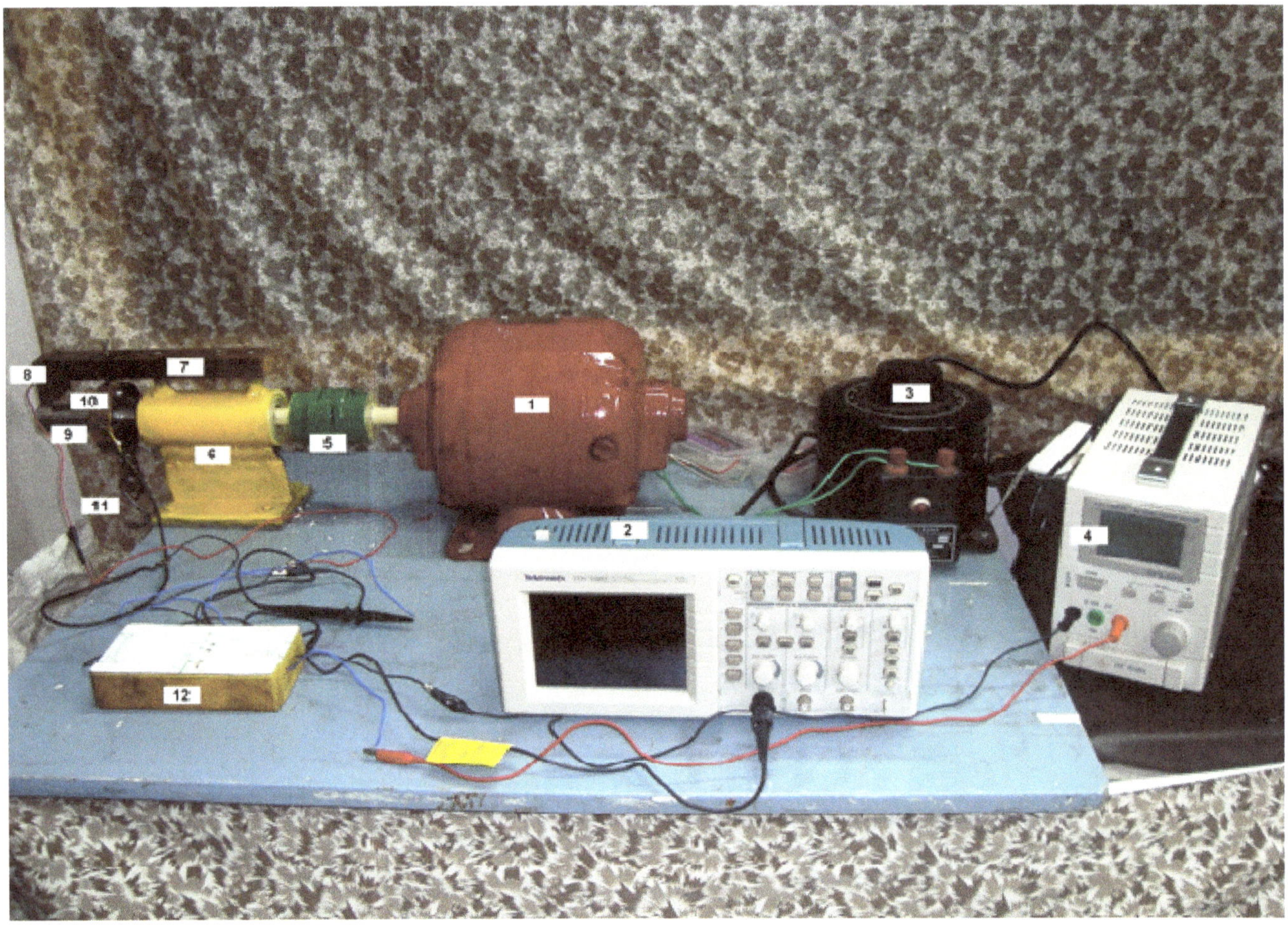

Figure 2. Photograph of rolling element bearing test rig; 1, DC motor; 2, Storage oscilloscope; 3, Variable AC transformer; 4, Regulated DC supply; 5, Flexible coupling; 6, Supporting shaft assembly; 7, Wooden cantilever; 8, Carbon brush assembly; 9, Stainless steel sleeve; 10, Copper sleeve over test bearing; 11, Hook for radial loads; 12, Electrical circuit.

more nut and bolt is used just below the previously fit nut and bolt which incorporates hook for application of radial load. A nut on the top is welded on the copper outer sleeve before fitting. Inside the sleeve, some spots of copper are made by using gas welding to ensure good electrical contact with outer race of the bearing when tightened appreciably. The first electrical connection is tapped from this outer copper sleeve.

A longitudinal slit is made in the wooden cantilever beam through which passes a brass rod having a ring at one end, encircling the projected steel sleeve of test bearing, and the other end have threads used to fit a nut with washer after inserting a spring on the cantilever above. The nut is tightened on the washer so as to compress the spring slightly such that the ring always bears against the steel sleeve with some force. The size of the spring is such that its diameter is more than the width of slit in cantilever and the length such that force exerted is just enough to make a good electrical contact of brass ring with the steel sleeve. The second electrical connection is tapped from the threaded portion of the rod with the ring.

As an alternative to brass ring arrangement (which is found later to generate heat due to friction and thus affected the observations taken over a time) the carbon brush arrangement is used to make the electrical circuit. One carbon brush is fitted on the wooden cantilever at the end such that the carbon brush always makes contact with the steel sleeve which is in contact with the inner race of the test bearing. The carbon remains in contact with the steel sleeve by the spring force. The second electrical connection is tapped from the carbon brush in contact with inner race.

Thus two electrical connections are tapped, one in contact with outer race and second in contact with inner race of bearing. These tapped connections are then linked to a standard electrical circuit for measurement of electrical resistance between inner and outer race of bearing which will be due to presence of lubricant in between balls and inner and outer race. The hook fitted to the second bolt at lower end of split copper sleeve is used to hang different radial load on the bearing. The

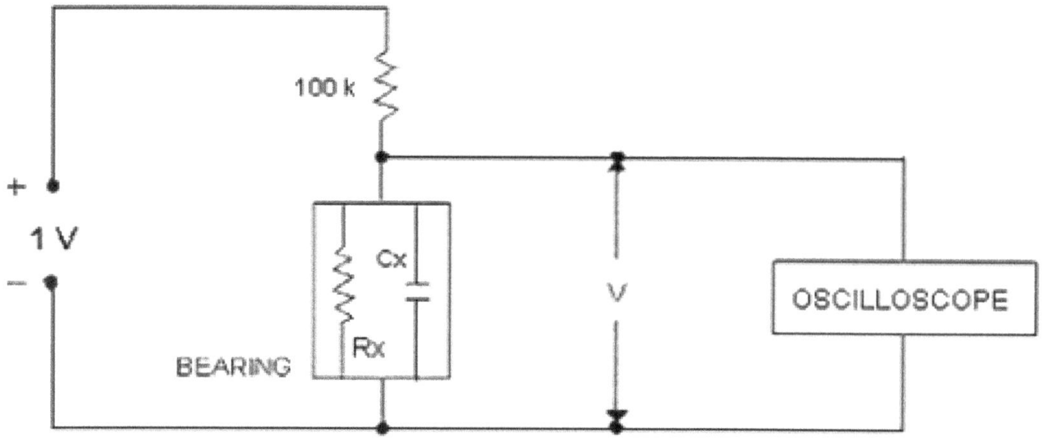

Figure 3. Circuit diagram for measurement of test bearing resistance.

technical specification of various components has been given in Appendix 1.

METHODOLOGY

The tapped points for electrical connections are connected to the electrical circuit shown in Figure 3.

The rolling element bearing is supposed to be a combination of electrical resistance and capacitance in presence of lubricant as shown in electrical circuit of Figure 3. A resistance of known value is connected in series with the bearing. This known resistance is used to determine the current flowing in the circuit. The bearing and a resistance R of known value (here 100 kΩ) is connected in series along with a constant DC voltage supply of 1 V (Vin). A DC voltage is used since the capacitance of bearing being non-polar will be inactive to DC current and hence the voltage drop will be only due to the bearing resistance. The voltage drop (say V) in the bearing can be read from the oscilloscope connected across the bearing terminals. The current flowing in the circuit is given by $I = (V_{in} - V)/R$ and the bearing resistance is given by $R_x = V/I$. This bearing resistance measured is the representative bearing resistance since it is obtained by overall voltage drop between inner and outer race of the bearing.

The bearing resistance thus obtained leads to determination of the lubricant film thickness by using the formula:

$$R_T = \rho h_0\{(1/A_1) + (1/A_2)\}$$

Where:

ρ = resistivity of lubricant in Ω mm,

h_0 = lubricant film thickness in mm,

A_1 = Elastohyrodynamic area of contact of ball with inner race in $mm^2 = \pi a_i^2$,

A_2 = Elastohyrodynamic area of contact of ball with outer race in $mm^2 = \pi a_o^2$,

R_T = Total resistance between inner and outer race of bearing in Ω.

a_i = Elastohyrodynamic radius of contact of ball with inner race in mm,

a_o = Elastohyrodynamic radius of contact of ball with outer race in mm,

Elastohyrodynamic radius can be calculated using the standard formulae

$$a_i = ((3\pi Q(k_1+k_2)r_i r)/(4(r_i+r)))^{1/3}$$

$$a_o = ((3\pi Q(k_1+k_2)r_o r)/(4(r-r_o)))^{1/3}$$

where:

$k1 = (1 - \nu_1^2)/\pi E_1$,

$k_2 = (1 - \nu_2^2)/\pi E_2$,

Q = radial force on ball,

ν_1, ν_2 = Poisson's ratio for the materials of ball and races respectively,

E_1, E_2 = Modulus of Elasticities of the materials of ball and races respectively,

r_i = outer contact radius of inner race,

r_o = inner contact radius of outer race,

r = radius of ball.

DISCUSSION

The formulae are based on the formulae used by Prashad (1966) and Harris (2001). The patterns of results are taken from Dowson et al. (1966). The technique, R-C oscillation was also explored (Peng-shun et al., 1991). The concept of Elastohydrodynamic lubrication was used. The results reported compare fairly well with the earlier available results of Dowson et al. (1966) and Cameron (1981).

The results reported for starved lubrication regime by researchers at laboratory were found to be satisfactory (Damiens et al., 2004). Even the experimental setup was run with no lubricant in the rolling element bearings and the results intimated were very encouraging (Cameron, 1981). The performance on thin lubricant films can also be done (Choo et al., 2004). The set up also tested some worn out bearings, the surface of balls being rough, the results communicated showed that the lubricant film thickness in the rough contacts come out to be less as compared to the smooth contacts (Guangteng et al., 2000).

Also the methodology used in the set up can test variation of lubricant film thickness with asperities making contact in rough bearings as done by Choo et al., 2004; Guangteng et al., 2000.

Experiments on the developed test rig, as informed, were conducted with ball bearings 6207 and 6307. The characteristics of representative film thickness reported agree with the findings of Dowson et al. (1966) and Prashad (2006). Thus the circuit created is adopted after its use is found to be satisfactory as communicated by the researchers.

Conclusion

The developed setup was used at Dynamics of Machines Laboratory of National Institute of Technology, Raipur, to perform experiments with rolling element bearing and it was reported that the results obtained for new (smooth) bearings follow the earlier published results. Similar pragmatic observations with rough (worn out) bearings were communicated which matches the earlier reported results. By changing the diameter of steel sleeve the setup can be used for testing of various rolling element bearings.

On the basis of reported results, it is also concluded that the reduction in quantity of lubricant inside the bearing will be well indicated by drop in voltage across the bearing races. At this stage more lubricant can be fed to the bearing thus maintaining a minimum level of lubricant which leads to better performance and long life of the bearing. Further, it is worth to mention that the setup can be effortlessly modified to incorporate axial loading as well. Journal Bearings can also be tested by suitable changes.

Overall facet of the setup developed was analyzed and thus it can be conveniently concluded that the test rig developed is simple, cheap, versatile, compact and robust. Performance evaluation and continuous condition monitoring of bearing can be achieved successfully by the use of the created test rig. This is imperative for achieving prolonged life and augmentation of the bearing-performance.

ACKNOWLEDGEMENTS

The authors wish to acknowledge their sincere thanks to All India Council of Technical Education, New Delhi and Directorate of Technical Education, Government of Chhattisgarh, Raipur and Government of India for providing sufficient financial support, under Research Promotion Scheme (RPS), without which the development of such test rig could not have been possible.

REFERENCES

Cameron A (1981). Basic Lubrication Theory, Third Edition, Wiley Eastern Ltd.

Choo JW, Olver AV, Spikes HA, Marie-Laure D, Eustathios I (2008). Interaction of Asperities on Opposing Surfaces in Thin Film, Mixed Elastohydrodynamic Lubrication, J. Tribol. April 130(2): 021505 (10 pages).

Damiens B, Venner CH, Cann PME, Lubrecht AA (2004). Starved Lubrication of Elliptical EHD Contacts, J. Tribol. 126(1): 105-111.

Dowson D, Higginson GR (1966). Elasto-hydrodynamic lubrication, The fundamentals of Roller and Gear Lubrication, Pergamon press.

Guangteng G, Cann PM, Olver AV, Spikes HA (2000). Lubricant Film Thickness in Rough Surface, Mixed Elastohydrodynamic Contact, J. Tribol. 122(1): 65-76.

Harris TA (2001). Rolling bearing analysis, John Wiley & sons, inc.

Prashad H (2006). Tribology in Electrical Environments, First Edition, Elsevier, Great Britain.

Prashad H (1988). Theoretical evaluation of impedance, capacitance and charge accumulation on roller bearings operated under electrical fields, Wear, 125: 223-239.

Prashad H (1987). Effect of operating parameters on the threshold voltages and impedance response of non-insulated rolling element bearings under the action of electrical currents, Wear, 117: 223-240.

Peng-shun Z, Shu-guang L, Wen-jie Z (1991). A new measurement method of oil film thickness in the EHL condition: R-C oscillation technique, Wear, 148: 39-46.

Appendix 1

S.No.	Name of equipment / instrument	Specifications	Current market price (Rs)
1	DC Motor	1 kW	8,000/-
2	Plain oscilloscope	60 MHz	15,000/-
3	Supporting shaft assembly	Robust enough to support radial loads of 1000 kg	5,000/-
4	Mild steel sleeves	As per the inner diameter of the bearing	3,000/-
5	Test bearings and lubricating oils	As per requirements	2,000/-
6	Digital Multimeter	Reputed Make, can measure resistances upto Megha-ohm	4,000/-
7	Weights along with mechanical loading arrangement	In steps of 10 kg	4,000/-
8	Microlab II with -12V to 12V DC supply	Having bread board, function generator, variable voltage supply, etc. of reputed make	18,000/-
9	Wiring and other small requirements	Wires should have low resistance	3,000/-
10	Tachometer	Non contacting type of reputed make	2,000/-
11	Voltage Stabilizer	5 kW, Reputed make with servo motor	12,000/-
12	Regulated DC supply	0 to 30 V DC, 2A, Digital, with separate adjustments of voltage and current, of reputed make	6,000/-
Total in Indian Rupees			82,000/-

5

Stability and accuracy of the finite difference time domain (FDTD) method to determine transmission line traveling wave voltages and currents: The lightning pulse

P. R. P. Hoole[1]* and S. R. H. Hoole[2]

[1]Department of Electrical Engineering, University of Malaya, 50603 Kuala Lumpur, Malaysia.
[2]Department of Engineering and Science, Rensselaer Polytechnic Institute, USA.

The numerical solution of the transmission line wave equation is examined using the transmission line model of a lightning return stroke (LRS) taken as a case study. The LRS is represented by a transverse electromagnetic wave travelling at a velocity close to the velocity of light along a highly ionized lightning leader channel. The leader channel is modeled as a lossy transmission line. Attempts to model the lightning return stroke by an electric circuit in order to determine the currents and radiated electromagnetic fields is categorized as the distributed transmission line model (DTLM). The DTLM method in turn may be subdivided into two categories: Distributed inductance (L), capacitance (C) and resistance (R) model (DLCRM) and the lumped LCR model (LCRM). In this brief communication, we indicate some pitfalls to watch out for in solving for the transmission line wave equationusing numerical methods, and how these may be avoided. Furthermore, we propose tests that may be applied, including the analytical solution of diffusion waves, to ensure the stability and accuracy of the numerical solution DLCRM simulations.

Key words: Finite difference time domain method, numerical solution of transmission line wave equation, lightning return stroke model, lightning radiated electromagnetic fields.

INTRODUCTION

The electric transmission line is used to model many electrical engineering phenomena in both man-made (e.g. overhead power line, underground cable, microelectronic circuit line) and natural (e.g. the nerve system of the human body and the thunder storm lightning channel) circuits or systems. Accurate, stable and fast solution of the transmission line wave equation is important to use the calculated currents and voltages of the wave in protection, shielding and filtering of the engineering systems threatened by high speed electromagnetic waves characterized by large transient currents that change within submicrosecond time intervals. Thus the accurate solution of both electric current and voltages using numerical methods such as the finite difference time domain methodis of great interest. In this brief communication, we explore the solution of the transmission line wave equation representing the LRS wave along the cloud-to-ground (earth flash) lightning channel. Earth flash LRS currents and radiated lightning electromagnetic pulses (LEMP) are electrically the most severe aspects of a thunder storm. The magnitude of its severity, quantified by LRS currents and radiated electromagnetic fields, and its mitigation are critical aspects of design, installation, operation and protection of power transmission and distribution systems, wind turbine generators with long turbine blades, aircraft and rockets. But direct measurements of natural LRS currents or LEMP at ground level and well above ground are not easily measured. Hence there is a need to correctly model the LRS channel in order to determine the currents and the magnitudes of the LEMP.

We do not in this paper discuss the physics of LRS, or

*Corresponding author. E-mail: prphoole@gmail.com.

compare the DLCRM (Hoole and Hoole, 1988, 1993; Hoole, 1993) and LCRM (Pierce and Price, 1977; Little, 1978) models. Our concern here is to alert and propose methods to avoid errors in computing the currents or the LEMP using the DLCRM. The search for a self consistent lightning return stroke (LRS) model has reached a fascinating and challenging point where many different approaches are explored to correctly model it: Price and Pierce (1999), Eyathi and Cooray (2005), and Baba and Rakov (2008) to mention a few. Lightning measurement and analysis groups - such as those in Europe (Berger, 1977) and in the USA (Baba and Rakov, 2009), have provided a massive data base of lightning measurements. Those who are working on the characterization of lightning and developing powerful software codes to simulate and study lightning need to exercise great care in the mathematical modeling and computational work done. Some aspects of the care needed to be exercised in modeling and simulating the LRS currents are presented herein. The distributed transmission line model (DLCRM) where the time-invariant line elements are calculated using basic electromagnetic theory, provides stable results for the voltage and time impulses which agree with current measurements at ground level (Berger, 1977), and yield computed LEMP that agree with measured LEMP (Baba and Rakov, 2009: 7; Little, 1978; Hoole, 1993; Moosave et al., 2009; Deindorfer and Uman, 1990).

DISTRIBUTED INDUCTANCE, CAPACITANCE AND RESISTANCE TRANSMISSION LINE MODEL (DLCRM)

Although the seed thought for the DLCRMwas planted by the lumped LCR model of Bruce-Golde followed by the work of many others on lightning engineering (Golde, 1977), it was most rigorously worked out and reported by Pierce and Price (1977), and Little (1978). In this earlier modeling work, however, lumped circuit parameters for each line segment (lumped-LCR) were used instead of distributed LCR parameters. Amongst others, Hoole and Hoole (1988, 1993), and Hoole (1993) reported a self consistent DLCRM, correcting some of the short comings of the earlier works. In an attempt to establish the scientific basis for the DLCRM (Hoole and Hoole, 1988), the electric plasma nature of the lightning channel was examined and it was shown that the LRS wave satisfies the essential characteristics of a quasi-transverse electromagnetic (quasi-TEM, not simply a TEM) wave along an un-magnetized plasma channel. The quasi-TEM wave is what is exactly approximated by a DLCRM transmission line travelling wave.

The FDTD approximation of the quasi-TEM wave equation is given by (Hoole and Hoole, 1993; Hoole, 1993)

$$\left(\frac{R\Delta t}{2}+L\right)V_n(z,t-\Delta t)+2\left(L-\frac{\Delta t^2}{C\Delta z^2}\right)V_n(z,t)$$

$$+\frac{\Delta t^2}{C\Delta z^2}\left(V_{n+1}(z+\Delta z,t)+V_{n-1}(z-\Delta z,t)\right.$$

... .(1)

for the voltages V_n along the DLCRM line (n= 1,2,.... N where N is the number of segments into which the channel has been divided into), with per unit length inductance, capacitance and resistance of L, C and R respectively. The time step is Δt and the distance step is Δz. With an initial current (e.g. lightning leader current) I_o assumed to be flowing along the entire length of the line, the current equation is approximated by (Hoole, 1993)

$$I_n(z,t)=I_0+\sum_{i=1}^{n}\left(\frac{V_i(z,t+\Delta t)-V_i(z,t-\Delta t)}{2\Delta t}\right)C_i$$

.........(2)

where, as noted previously, I_0 is the leader current that flows along the lightning channel before the initiation of the return stroke.

The line is subdivided into *n* segments of equal length. Each segment *i* has a capacitance of C_i per unit length, and the summation in Equation (2) is over the entire length of the lightning channel, representing integration over the whole length of the channel. Using L, C and R values that are representative of the lightning channel, it was shown that the DLCRM yields both currents and radiated electromagnetic fields (LEMP) that agree well with measured lightning currents and radiated LEMP (Hoole and Hoole, 1993).

TESTING THE ACCURACY OR RELIABILITY OF THE COMPUTER SIMULATED LIGHTNING

To start the time-stepping computations, initial values are required. The initial potential V_0 and current I_0 along the line are known (Golde, 1977). In solving the FDTD approximation, DLCRM transmission line wave, the so-called magic time step ($\Delta t = \Delta z/c$, where c is the velocity of light) and the Courant stability criterion (Taflove and hagness, 2005) alone are not adequate. In the DLCRM travelling wave, the L, C and R per unit length values play an explicit role in determining the stability and accuracy of the FDTD solution, as seen in (1) and (2). In the LCRM, there is this inherent weakness where the distributed line elements are lumped together for each segment: thus the L, C and R elements that appear in the simulation program as the per unit length elements multiplied by the length of each segment. The length of each segment does not play a direct role in the LCRM circuit solution, where as in DLCRM, it plays an explicit role, as for instance in (1) where it appears as (Δz^2).

It is our experience that the Equations (1) and (2) could be rapidly solved using finite difference time domain (FDTD) method in one spatial dimension (*z*), where the time step is kept small compared to $\sqrt{LC}$ and $L/2R$in order to obtain stable solutions with sufficient number of calculated points in the wave front. The distance step is chosen so to keep it longer than $\sqrt{2\Delta t/RC}$. Whence, to ensure stable solutions, the following two steps are adopted: (1) Choose Δt such that it is small compared to $L/2R$, and (2) choose Δz so that it is greater than both $\Delta t\sqrt{LC}$ and $\sqrt{2\Delta t/RC}$. These conditions ensure the stability of the time stepping and spatial stepping used in FDTD method to solve (1) and (2), whatever the magnitude of *R*, *L*, and *C* may be in (1). This is roughly verified by considering the ratios $\Delta z^2:2\Delta t/RC:\Delta t^2/LC$.

In the original (Price and Pierce, 1977) and more recent work on lightning modeling (Moosavi et al., 2009; Deindoffer and Uman, 1990; Theeyathi and Cooray, 2005), self consistent tests to ensure correct and stable solutions of Equations (1) and (2) are not discussed. It is rather uncertain that some of the complicated models used for the line elements such as using atime-varying LRS channel resistance, inductance or capacitance (Moosavi et al., 2009; Theeyathi and Cooray, 2005; Baba and Rakov, 2009), give better results. Recent work has reconfirmed that the simpler form of the DLCRM reported in Hoole and Hoole (1993) and Hoole (1993) is a reliable and an adequate model of cloud-to-ground lightning flash (Hoole et al., 2010).

RESULTS

In Figure 1a is given the lightning return stroke currents calculated using the DLCRM reported in Hoole and Hoole (1993) and Hoole (1993). The peak lightning return stroke current and the maximum rate of rise of current over the convex shaped part of the current wavefront, clearly and correctly captured in the DLCRM computation as seen in Figure 1a, are important parameters which determine the adverse effects such asheating, high pressure, electromagnetic interference and induced voltages due tothe lightning return stroke on aircraft structure, communication and control electronics, and power systems. In Figure 1b is shown theelectric field calculated at a distance of 10 km from thelightning flash, using the integral method (Hoole and Hoole, 1987). A knowledge of themagnitude and submicrosencond changes of lightning radiated electromagnetic pulse (LEMP) at various points in space is crucial when designing filters and shields to protect mission critical systems such as the military aircraft control system. The LEMP can cause changes in digital data and control signals that could result in destabilizing the control system sending the fast flying military, fighter aircraft spinning out of control.

The electric field is calculated from the LRS currents calculated using the DLCRM (Hoole and Hoole, 1993). The initial, sharp peak of the electric field is due to the second, convex section of the return stroke current wavefront. If the current versus time characteristics of the return stroke was correctly obtained from the DLCRM computer simulation, then the electromagnetic fields calculated from those current yields very helpful, lightning parameters above the ground and far from the LRS flash which are very difficult or impossible to measure.

The hump following the initial sharp peak in electric fields, as shown in Figure 1b, also matches with measured electric fields (Baba and Rakov, 2009; Hoole and Hoole, 1993). The computer simulations were limited to the first 20 μs, the LEMP period of interest in engineering design, the period over which very good agreement was found for the following parameters of the lightning: LRS current waveform at ground level, electric and magnetic field waveforms, LRS velocity and rate of rise of currents. In the future, we intend to examine the long term characteristics and computation of the LRS.

When solving Equations (1) and (2), the following three factors must be noted:

1. The L, C and R segment at the earth end must be rigorously worked out, and the values should be reasonable when compared to values obtained by using electromagnetic and plasma theory (Price and Pierce, 1977; Little, 1978; Hoole and Hoole, 1993). For instance, if the upward leader is assumed to be a resistive element, or if it is represented by C and R only, the numerical solution of the DLCR model will result in a singularity point at the earth end. When sketching the current waveform, this may get inadvertently suppressed by the computer plot of the current waveform. Thus it is important to ensure that points on the wave-front of the earth end segment were calculated by the FDTD computer code.
2. The length of eachLCR segment must explicitly appear in the numerical computation of the traveling wave. It ought not to be multiplied with the per unit length values of the R, L and C elements to form a lumped LCR network.
3. Care should be exercised in the time steps used. The time steps should be kept to a value well below the ratio L/R. Moreover, the time step should also be less than or equal to the length of each segment divided by the velocity of the wave (Taflove and Hagness, 2005). This ensures that at each point in time at which the computations of Equations (1) or (2) are done, the wave has not shot over into an adjacent segment of the DLCRM line. A figure for the time step may be worked out assuming a reasonable value for the velocity of the return stroke, say c/3 where c is the velocity of light.

Once the solution is done, take a look at the current wavefront to ensure that the solution obtained is not for aR >>2πfLdiffusion problem, where f is the significant and highest frequency of the LEMP spectrum (e.g. f = 1 to 6 MHz (Berger, 1977; Hoole and Hoole, 1996). In the diffusion case, the final current waveforms look rather smooth and restrained, and the velocity is about two orders less than the velocity of light (Hoole, 1993).

In order to check the proper working of the computer-coded DLCRM and to ensure reliable solutions of V and I, two helpful, analytical tests may be employed. First, set the values of the L, C and R elements to reflect a diffusion problem, for which there is an analytical solution (Hoole and Hoole, 1988). Now compare the current pulses obtained from the numerical solution of the DLCRM and the analytical solution of the same problem (Hoole and Hoole, 1993). Second, set the values of L, C and R elements to that of the LRS parameters that are to form the DLCRM to be computer simulated (Hoole and Hoole, 1993). Do two sets of computer simulation of LRS using the DLCRM. First, let the length of the lightning channel (e.g. 1 km long) be divided, say, into 10 segments (thus the length of each segment is 100 m long for a 1 km long channel). Obtain the voltage changes and the current pulses in each segment. Now repeat the solution with the channel divided into, say, 30 segments. Repeat the DLCRM simulation to compute the current and voltage pulses. Do the V and I values agree with time

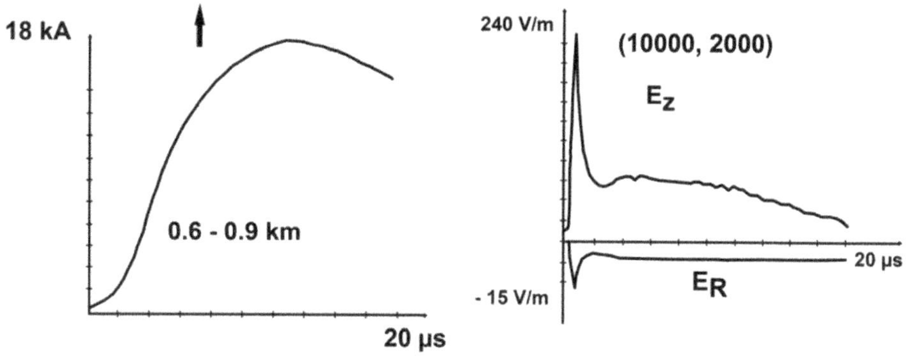

(a) (b)

Figure 1. (a) Lightning return stroke current calculated using the DLCRM, at above-the-ground height of 0.75 km along the lightning return stroke channel. Observe the three parts of the return stroke current wavefront: First a ramp like increase, second a sharply increasing convex wavefront. Thirdly, a slow rate of rising towards the peak of the current waveform; (b) Lightning radiated electric field calculated 10 km horizontally away from the lightning flash, and at a height of 2 km from ground level. E_z (peak at about 240 V/m) is the vertical, dominant component of the radiated electric field – a component that is important in engineering systems. E_R (peak at about -15 V/m) is the horizontal component (parallel to the earth) of the radiated electric field at (10 and 2 km).

domain characteristics of the LRS at the same height of the lightning channel?

SUMMARY

The transmission line wave equations that were numerically solved using Equations (1) and (2) is an approximation of the Maxwell's equations for time-varying electromagnetic fields. This brief communication has provided rules for discretizing the transmission line wave equation in time and spatial domains to obtain accurate and stable solutions of both voltages and currents along the distributed transmission line. Moreover, two specific methods to test the accuracy of the solutions have been proposed. On the solution of Maxwell's equations using a numerical method such as the finite difference method, we would like to sound the same note of caution as in (3) previously mentioned. It is pertinent that the time step should be so constrained that at each step, the wave remains within, say, a finite difference grid. This would result in a grid made up of a vast number of elements. Moreover, since the current (electric) wave travels along the LRS channel at a velocity (e.g. c/3) less than the velocity at which the radiated electric and magnetic waves travel out from the channel (at c, the velocity of light), care must be taken to account the existence of two waves at two different velocities: One wave moving at a velocity less than the velocity of light along the lightning leader channel modeled by a transmission line, and the second wave radiated out from the transmission line at the velocity of light.

REFERENCES

Baba Y, Rakov VA (2008). Applications of Electromagnetic Models of the Lightning Stroke, IEEE Trans. Power Delivery, 23(2): 800-811.

Baba Y, Rakov VA (2009). Electric and Magnetic fields produced by different electromagnetic models of the lightning g return stroke versus measured fields, IEEE Trans. Electromagn. Compat., 51(3): 479-487.

Berger K (1977). Earth Flash in Lightning: Physics of Lightning, R. H. Golde Ed., Academic, pp. 119-190.

.Deindorfer D, Uman MA (1990). An improved return stroke model with specified channel base current, J. Geophys. Res., 95: 13621-13664.

Golde RH (1997). Lightning, Vol. 1 and 2, Academic.

Hoole PRP, Hoole SRH (1987). "Computing Transient Electromagnetic Fields Radiated from Lightning", J. of Appl. Phys., 61(8): 3473-3475.

Hoole PRP, Hoole SRH (1988). Guided waves along an un-magnetized lightning plasma channel, IEEE Trans. Magn., 24: 3165.

Hoole PRP, Hoole SRH (1993). Simulation of Lightning Attachment to open Ground Tall Towers and Aircraft, IEEE Trans. Power Delivery, 8(2): 732-738.

Hoole PRP (1993). Modeling the Lightning Earth Flash Stroke for Studying its Effects on Engineering Systems, IEEE Trans. Magn., 29(2): 1839-1844.

Hoole SRH, Hoole PRP (1996). A Modern Short Course in Engineering Electromagnetics, Oxford University Press, USA, Chapter 7.

Hoole PRP, Pirapaharan K, Hoole SRH (2010). Electromagnetic Modeling of Lightning Return Stroke Currents: Waveguide and Circuit Models. Digests Asia Pac. Symp. Appl. Electromagn. Mech., pp. 152-153.

Price GH, Pierce ET (1997). The Modeling of Channel Current in the Lightning Return Stroke. Radio Sci., 12: 381-388.

Little P (1978). Transmission Line representation of a Lightning Return Stroke. J. Phys. D: Appl. Phys., 11: 1893-1910.

Moosavi SHS, Moini R (2009). Representation of a Lightning Return Stroke Channel as a Nonlinearly Loaded Thin-Wire Antenna. IEEE Trans. Electromagn, Compat., 51(3): 488- 498.

Taflove A, Hagness SC (2005). Computational Electromagnetics- The Finite Difference Time-domain Method, 3rd Edn, Artech House.

Theeyathi N, Cooray V (2005). On representation of the Lightning Return Stroke Process as a Current Pulse Propagating along a Transmission Line. IEEE Trans. Power Deliv., 20: 2.

6

Low-cost treatment for attenuation of nitrate from groundwater

B. S. Shankar

Department of Civil Engineering, East Point College of Engineering and Technology, Bidarahalli, Virgonagar Post, Bangalore–560 049, Karnataka, India. E-mail: shankar_bs1@yahoo.co.uk.

Elevated nitrate levels in drinking water by nitrate are an evolving public health concern, since nitrate can undergo endogenous reduction to nitrite, and nitrosation of nitrites can form N-nitroso compounds, which are potent carcinogens. Nitrates are also responsible for a number of health disorders, both in human beings and animals. Thus, nitrate reduction from water has become critical and the need of the hour. Water contaminated with nitrate needs to be treated so that it meets drinking standards. Boiling, softening, or disinfection will not reduce the nitrate content of water, and conventional measures to treat nitrate rich waters such as reverse osmosis, deionization and distillation are extremely cost-prohibitive. In this connection, studies were conducted to explore the avenues for low cost treatment for nitrate reduction in groundwater using selective filter media which are locally available and inexpensive. Media combinations of fly ash with pebbles and charcoal with brickbats were used. To assess their potential as a filter medium, a laboratory scale physical model study was constructed using the above two combinations and from the experimental studies, it was observed that 66% removal efficiency was achieved using flyash and pebbles, while 88% efficiency was achieved using charcoal and brickbats, for a simulated NO_3 concentration of 100 mg/L. Also it is seen that this optimum efficiency is achieved with a contact time of 3.5 h. Thus, the use of charcoal and brickbats for a contact period of 3.5 h is seen to produce an optimum degree of nitrate attenuation from groundwater.

Key words: Attenuation, brickbats, charcoal, flyash, groundwater, nitrate, pebbles, removal efficiency.

INTRODUCTION

The presence of high nitrate concentration (>45 mg/l) in groundwater would normally indicate pollution of groundwater. The Bureau of Indian Standards (BIS) has placed this limit of 45 mg/L for drinking water. Since the presence of excess nitrate ions in groundwater is harmful to health, their occurrence in high concentrations in ground-water is a matter of great concern. The leaching of nitrates from agriculture land has been a major research focus in the past two decades. Inorganic nitrogen fertilizers, septic tanks, poor dug wells and defective sewerage systems are the suspected major sources of nitrate in groundwater (Piskin, 1973; Canter, 1997; Lindsay, 1979). Nitrate is often considered an agricultural pollutant because, high concentrations above 10 mg/L have been found in aquifers located beneath agricultural fields where fertilizer and manure use is common. Soil bacteria may also increase nitrate levels in groundwater by oxidizing ammonia into nitrite and then to nitrate ion in oxygen-rich water (Dorsch et al., 1984). In addition to agricultural practices, there are other potential sources that can contribute to the increase of nitrate concentrations in groundwater. Most of these sources are related to urbanization and thus the types of sources are directly related to urban development. Urban sources of nitrate-N may have a high impact on groundwater quality because of the high concentration of potential sources in a smaller area than agricultural land (Wakida and Lerner, 2005). Nitrate sources can be grouped in diffuse sources (parks and gardens, atmospheric deposition), intense point sources (industrial chemical spills or landfill leachate) and multi-point sources such as leaking sewers and septic tanks (Lerner, 2003). The main sources of nitrate and other pollutants of urban groundwater is sewage, nitrate can reach the aquifer by sewer leakage and on-site disposal systems such as septic tanks. Leakage from sewage and water supply networks provides the highest percentage of water recharge to aquifers high volume of water loss. The major sources of nitrate in underlying many cities throughout the world. Mains leakage can be also a nitrate contributor due to the aquifers throughout the world are mostly related to wastewater disposal (on-site systems and leaky sewers) and solid waste disposal (landfills and waste tips).

HEALTH EFFECTS OF NITRATES ON HUMAN BEING

Nitrates and nitrites in water may cause methemoglobinemia in babies (Prakasa and Puttanna, 2000), where due to the oxidation of ferrous iron in haemoglobin to ferric state, the oxygen-carrying capacity of the red blood corpuscles is lost. Infants under six months are more susceptible to nitrate poisoning because, their undeveloped digestive tracts possess bacteria that convert nitrate into nitrite, which is toxic. When nitrite enters the bloodstream, it reacts with oxygen-carrying haemoglobin and forms a compound called methemoglobin. This compound reduces the blood's ability to carry oxygen. As oxygen levels decrease, infants may show signs of suffocation, a condition called methemoglobinemia. The most conspicuous symptom of methemoglobinemia is bluish skin, most noticeable around the eyes and mouth. If detected rapidly, methemoglobinemia can be successfully treated with an injection of methylene blue, which changes methemoglobin back to hemoglobin. Untreated, the condition is extremely serious and death may occur when 70% of the body's hemoglobin has been converted to methemoglobin.

Other health problems associated with nitrate toxicity include oral cancer (Badwai et al., 1998), cancer of the colon, rectum or other gastrointestinal cancers (Paul et al.,1990; Peter,1998; Leaf and Steven, 1996; Derache and Derache, 1997), Alzheimer's disease, vascular dementia of Biswanger type or multiple small infarct type (Tohgi et al., 1998), absorptive and secretive functional disorders of the intestinal mucosa and changes in maturation, differentiation and apoptosis in intestinal crypts (Ireneusz ,1998), neural tube defects, cytogenetic effect in children (Aspasia et al., 1996), Non-Hodkins's lymphoma (Helen,1994; Micheal, 1998) and hypertrophy of thyroid (Van et al.,1994).

EXPERIMENTAL STUDIES AND DISCUSSION OF MODEL OUTCOME

Treatment processes for nitrate reduction such as reverse osmosis, deionization and distillation, are extremely expensive, and boiling, softening, or disinfection will not reduce the nitrate content of water. Nitrate removal from groundwater may be accomplished by bacterially mediated denitrification, or chemically and physically based technologies. These treatment processes, however, require input of external energy sources (for example, electricity or organic carbon) and/or chemical additives and generate concentrated waste streams that must be carefully disposed. Therefore, they are often problematic and expensive. Micro algal treatment of wastewater has been investigated for over 4 decades as an environmentally sound alternative to remove nutrients and heavy metals from wastewater sources. However, very few investigators have considered applying this technology to the treatment of drinking water. The chemical composition (for example, the pH, dissolved inorganic carbon, nutrient levels and metals) of groundwater differs from that of wastewater, and the feasibility of growing micro algae (a term used for cyanobacteria and unicellular algae) in groundwater has yet to be evaluated.

In this connection, studies were conducted to explore avenues low cost treatment for nitrate reduction in groundwater using selective filter media. Media combinations of fly ash with pebbles and charcoal with brickbats were used, as they are known to have a very high potential for absorption of natural and manmade pollutants. A laboratory scale physical model study was constructed using a metal sheet column of size 0.40 m × 0.25 m × 1.58 m with coarse and fine layers of sand. The laboratory set up that was used is presented in Figures 1 and 2.

The front portion of the column used was of perplex glass 10 mm thick, so that the filter media could be clearly visible. The bottom portion of the sheet was fitted with nozzle for collection of samples. A wire mesh was placed at around three inches from the bottom, which serves as soil retainer and prevents small particles from leaching out of the column. A perforated plate was placed at the top to effect the uniform spreading of the solution over the entire area.

After the column was filled with sand, distilled water was applied for saturation of soil over a period of one day. After the complete draining of distilled water using different concentrations of nitrate, the experiment was started; the chemical compound used was potassium nitrate. Simulated solution of nitrate of varying concentration was induced to the column containing flyash and pebbles from the overhead placed tank. The constant head was maintained by using overflow outlet connected with the pipe. The samples were collected at the bottom at every 30 min interval and analyzed for nitrate concentration. Graphs were plotted for "percentage reduction versus time" and "percentage reduction versus initial concentration" as shown in Figures 3 and 4.

Similar procedure was repeated for another filter media in which charcoal and brick bats were used instead of fly ash and pebbles. The column setup for the two combinations is shown in Tables 1 and 2, while the results of the column studies are presented in Tables 3 and 4. From the experimental studies, it was seen that 66% nitrate removal efficiency was achieved using flyash and pebbles, while 88% efficiency was achieved using charcoal and brickbats, for a simulated NO_3 concentration of 100 mg/L. Also it was seen that this optimum efficiency

Figure 1. Photographic view of experimental set up.

Figure 2. Photographic view of experimental set up with fly ash and pebbles.

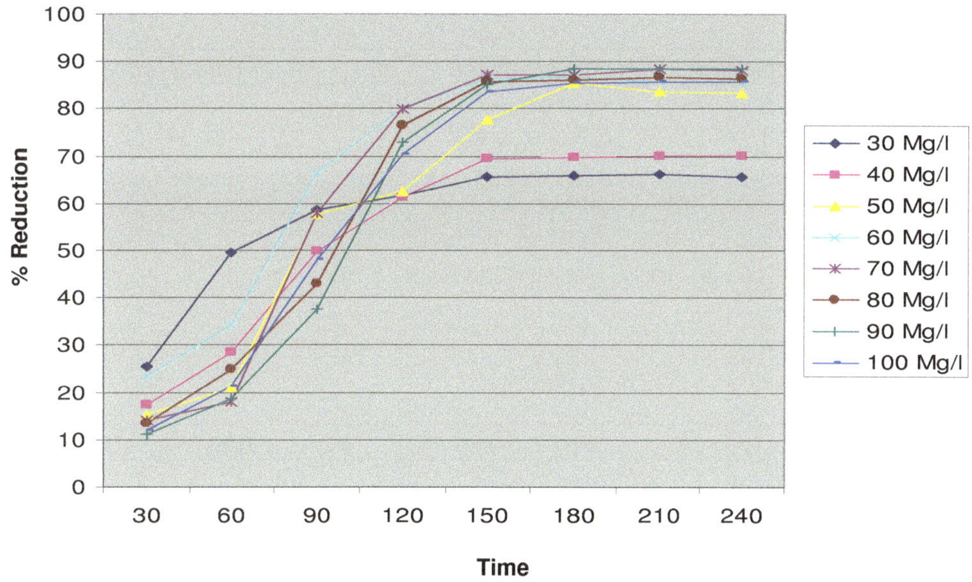

Figure 3. Percentage reduction vs. time using charcoal and brickbats.

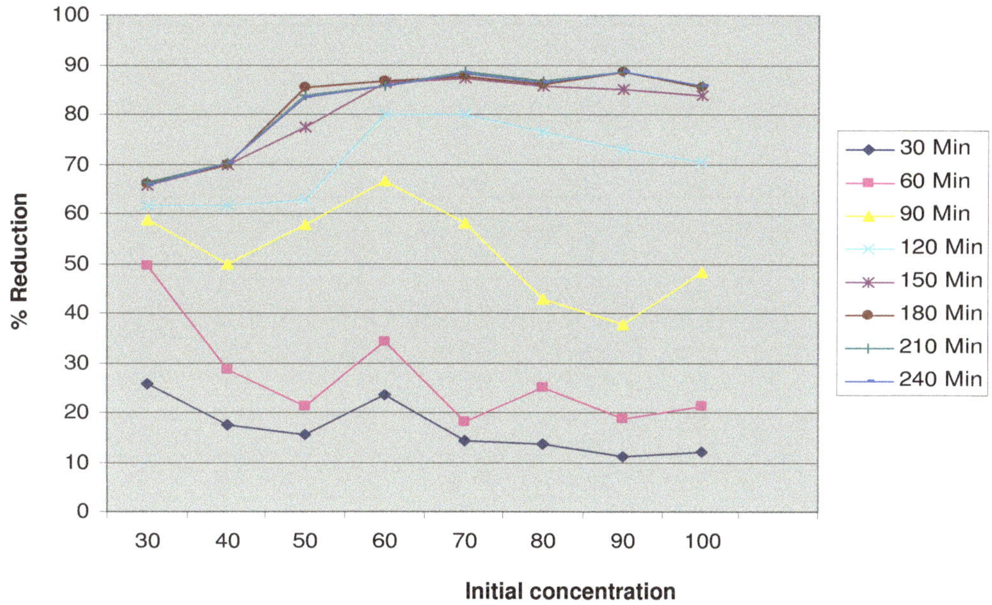

Figure 4. Percentage reduction vs. initial concentration using charcoal and brickbats.

Table 1. Filter media used in the Column Study–I (Fly ash + pebbles).

Serial number	Filter Media	Height (cm)
1	Fine sand	30
2	Coarse sand + Fly ash	30
3	Pebbles	10
4	Coarse sand	30
5	Red soil	5

Serial number 1, bottom most; serial number 5, top most.

Table 2. Filter media used in the column study–II (Charcoal + brickbats).

Serial number	Filter media	Height (cm)
1	Fine sand	30
2	Coarse sand	30
3	Charcoal	20
4	Brick bats	20
5	Red soil	5

Serial number 1, bottom most; serial number 5, top most.

Table 3. Variation in nitrate concentration using fly ash and pebble.

S/N	Initial concentration (mg/L)	30	60	90	120	150	180	210	240
		(min)							
1	30	26	24	22	21	18.5	16.8	16.6	17.1
2	40	34.2	25	23	22	20.6	20.1	19.8	19.9
3	50	38	26	22.3	22	20.5	20.3	20.2	20.2
4	60	46	32	27	26	22	21.7	21.5	21.8
5	70	54	38.6	30	26	25.2	24.8	24.7	24.7
6	80	63	46	34	30.8	27	25.9	25.8	26
7	90	71.4	52	39	30	28	27.6	27.5	27.9
8	100	91.5	71	53	41.1	35	33.8	33.6	33.9

Table 4. Variation in nitrate concentration using charcoal and brick bats.

S/N	Initial concentration (mg/L)	30	60	90	120	150	180	210	240
		(min)							
1	30	22.3	15.1	12.4	11.5	10.3	10.2	10.1	10.3
2	40	33	28.6	20	15.4	12.1	12	11.9	11.9
3	50	42.3	39.4	21.1	18.6	11.2	7.3	8.1	8.2
4	60	45.8	39.4	20	12	8.1	8	8.5	8.5
5	70	60	57.2	29.3	14.1	8.7	8.1	8	8.2
6	80	69	60.4	45.6	18.7	11.4	11.2	10.6	10.9
7	90	81.4	73	56	24.3	13.4	10.4	10.2	10.2
8	100	88.2	78.6	51.9	29.4	16.3	14.5	12.2	12.4

RESULTS AND CONCLUSION

In conclusion, the use of charcoal and brickbats for a contact period of 3.5 h is seen to produce an optimum degree of nitrate attenuation from groundwater. Thus the use of locally and cheaply available materials has been found to be mighty effective, simple and inexpensive for reducing nitrate from groundwater.

REFERENCES

Aspasia T, Kitsou-Izeli S, Galla A, Gourgiotis D, Papagiorgiou J, Mitron S , Molybdas PA, Sinaniotis C (1996). Arch. Environ. Health, 51:458–461.

Badawi AF, Gehen H, Mohamed EH, Mostafa HM (1998). Dis. Markers. 14: 91-97.

Canter LW (1997). Nitrate in groundwater, Lewis publishers. pp.263.

Derache R, Derache P (1997). Cah. Nutri. Diet. 32: 283–293.

Dorsch MM, Scragg RKR, McMichael AJ, Baghurst PA, Dyer F (1984). Congenital malformations and maternal drinking water supply in rural South Australia. J Epidemol., 119: 473–485.

Helen MW (1994). Non-Hodkin's lymphoma and drinking water nitrate: A case study, Johns Hopkins Univ, USA. pp. 289.

Ireneusz GP(1998). Roez Panstw. Zakl. Hig. 49: 321–330, CA130: 263274c.

Leaf CD, Steven RT (1996).The Role of Nitrate and Nitrite in Human Cancer In. Watson and Mufti. (Eds). CRC Press, Florida. pp. 317–324.

Lerner DN (2003). Estimating urban loads of nitrogen to groundwater. J. Chart. Institut. Water Environ. Manage., 17: 239-244.

Lindsay W L (1979). Chemical equilibrium in soils. John Wiley, New York. pp.149.

Micheal FD (1998). A population based case control study on the association between nitrate in drinking water and non-Hodgkin's lymphoma. Johns Hopkins University, USA. pp. 285.
Paul K, Ritva J, Jan D, Timo H (1999). Int. J. Cancer. 80: 852–856.
Peter WJ (1998). The Iowa Women's Health. Univ. Iowa, USA. pp. 255
Piskin R (1973). Evaluation of nitrate content of groundwater Hall County, Nebraska. Groundwater, 11: 4-13.
Prakasa EVS, Puttanna K (2000). Nitrates, agriculture and environment. Curr. Sci., 79: 1163-1168.
Tohgi H, Abe T, Yamazaki K, Murata T, Isobe C (1998). J. Neural Trans., 105: 1283-1291.
Van M, Johannes MS, Van D, Mulder K A, De Baets MH, Paul CAM, Vander HDM, Paul LJM, Jos CK (1994). Toxicol. Lett., 72: 365–374.
Wakida FT, Lerner DN (2005). Non-agricultural sources of groundwater nitrate: a review and case study. Water Res., 39: 3-16.

7

Determination of weathered and fractured zones in part of the basement complex of North-Eastern Nigeria

S. A. Arabi[1]*, B. B. M. Dewu[1], A. M. Muhammad[1], M. B. Ibrahim[1] and J. D. Abafoni[2]

[1]Centre for Energy Research and Training, Ahmadu Bello University, Zaria, Kaduna State, Nigeria.
[2]Department of Physics, University of Maiduguri, Maiduguri, Borno State, Nigeria.

Two electromagnetic (EM) traverses and seven Schlumberger Vertical Electrical Soundings (VES) were carried out using "*APEX MAXMIN" EM* machine equipped with a transmitter and a receiver and a "GEOPULSE" resistivity meter to study formation strata that could host groundwater within the overburden and fracture/fault zones of the basement complex along Gwoza-Damboa road in Borno State, North-eastern Nigeria. Seven anomalies were identified from the two EM traverses which were followed by vertical electrical sounding to confirm the conductivities of each of the anomalies. The VES results produced one KH-type curve at VES02 while the remaining were H- types. Most of the points sounded are three layered except VES05 which is four layered. The thickness of first layers ranges from 2 – 11 m with resistivity ranges of 15 – 38 Ωm, second layer thickness ranges from 11 – 22 m with resistivity range of 10 – 20 Ωm, the third layers was interpreted as the basement at VES01, 02, 03, 04 and 06 with resistivity ranging from 600 – 896 Ωm. The third layer at VES05 had resistivity of 18 Ωm and thickness of 19 m; it is also the thickest of all the layers interpreted for all the sounding points. While all the points sounded are good groundwater targets, VES05 appeared to produce the highest groundwater yield needed for the purpose of this study because it is the thickest of all the layers obtained. Base on the interpreted results, all the seven VES points are groundwater prospects. However, in order of suitability, VES05, VES01, VES06 and VES04 were recommended and four productive boreholes were drilled to an average depth of 55 m with VES05 producing the highest yield.

Key words: Vertical electrical sounding, basement, weathered zone, fractured zone, electromagnetic, traversing.

INTRODUCTION

Portable water is a scarce commodity especially around the study area where it occurs within fractured zones of the country rocks and in sub-surface porous layers where thick overburden exists (Nur and Kujir, 2006). Groundwater exploration in semi-arid zones in particular is pausing great challenges to water resources managers because both its quality and quantity is diminishing. Therefore, any technique that would furnish detailed information on targets where productive boreholes can be drilled is desirous.

Seven anomalies were detected by electromagnetic (EM) traversing using "APEX MAXMIN" electromagnetic machine and were confirmed to be a good target for groundwater exploration after vertical electrical sounding was conducted.

The study area is located along the Nigeria - Cameroon boarder on the South-eastern flank of the Mandara hills which trends northeast southwest from Gwoza in Borno State through Madagali, Michika and Mubi local Government of Adamawa State; It falls within longitudes 12° 45'E and 13° 15'E and latitudes 11° 19'N and 11° 27'N covering an area of about 9 km^2 (Figure 1).

The study area is underlain by the undifferentiated rocks of the Basement complex (Figure 2) which have been subjected to weathering that produced a fairly deep regolith (Kogbe, 1989). The rocks are of Precambrian origin and consist of metamorphic and igneous rock types

*Corresponding author. E-mail: arabisuleiman@gmail.com.

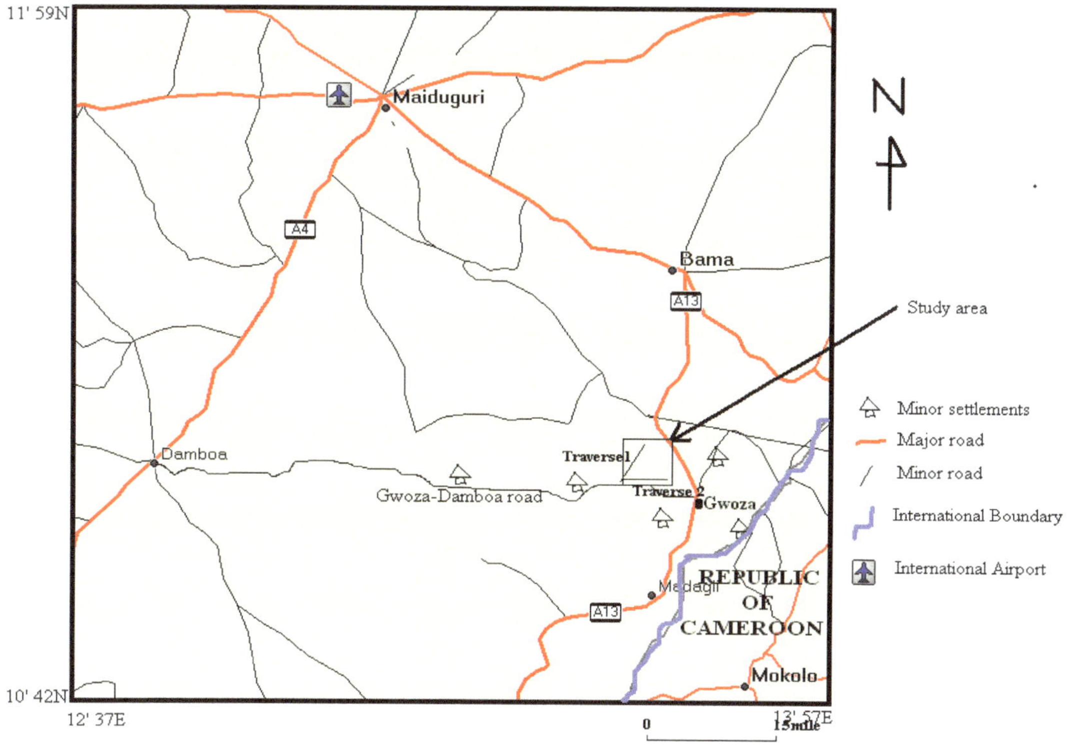

Figure 1. Location map of the study area (Map source, 2006).

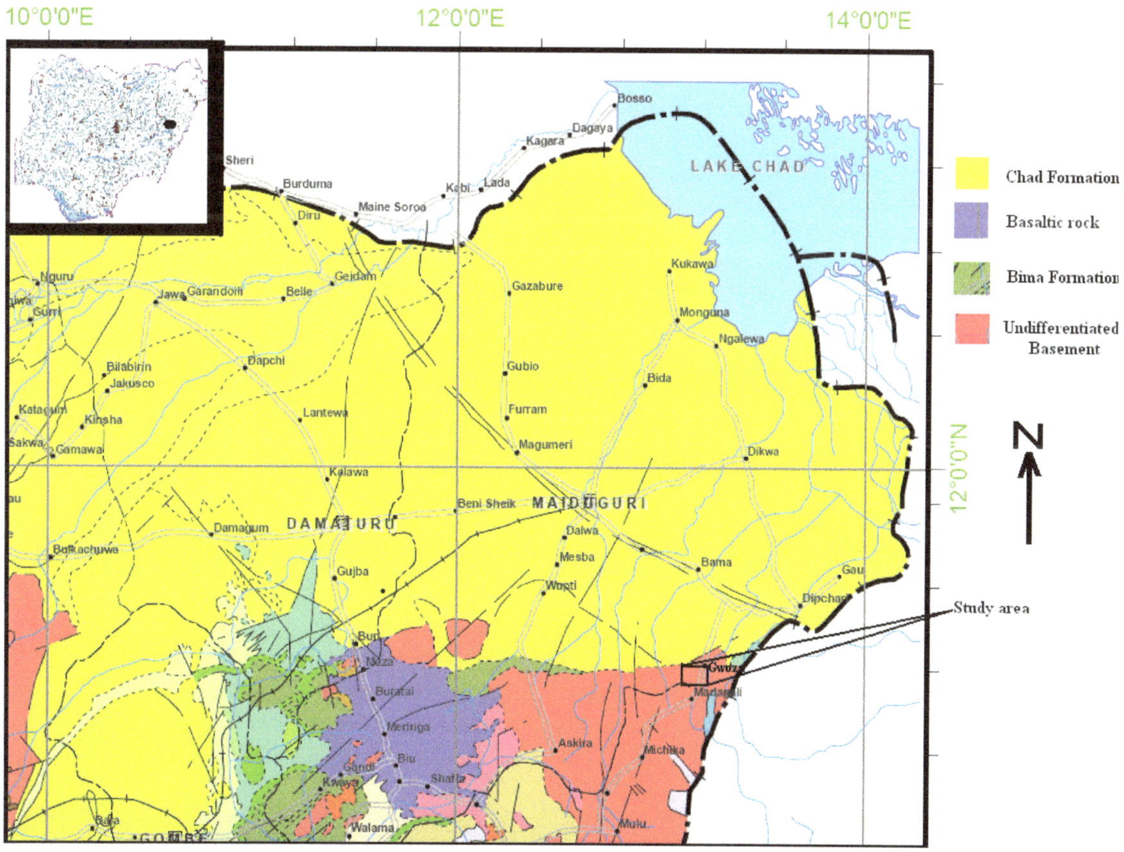

Figure 2. Geologic map of part of North-eastern Nigeria.

Table 1. Summary of the interpreted EM data.

Traverse no.	Number of anomalies detected	Anomaly no.	Traverse point (m)	Traverse length (m)
1	4	1	2	290
		2	14	
		3	25	
		4	28	
2	3	1	8	
		2	18	
		3	28	

common among which are the granite descriptions; migmatite, gneisses and phillites. The granite dominates the structures, hence, is refer to as the Older Granites (Carter, 1963). In most areas underlain by the Basement complex is a thin, discontinuous mantle of weathered rocks. The average thickness of the mantle is 15 m, but in some areas may extent to depths of up to 60 m. Groundwater is normally found if the original state of the rock has been deformed either by deep fracturing, weathering, faulting among others (Nur and Kujir, 2006). The presence of groundwater in any given area depends on the thickness and lateral extent of decomposition and where both occur, the groundwater condition is usually very good.

The Basement complex is generally a poor source of groundwater because the decomposed mantle is often too thin to harbour large quantities of water and sometimes too clayey to be highly permeable or the joints and fractured zones are poorly developed (Olorunfemi, 1993). Static water level in observed hand dug well in Gwoza town at the time of the survey ranges between 12 to 15 m and this can vary with seasonal fluctuation.

Though not published, different survey methods have been employed in the search for groundwater around the study area, but all were not successful. The present survey utilized both EM and VES and all assumed short comings in the previous surveys were taken care of. The Federal Government of Nigeria through the Federal Ministry of Water Resources (FMWR) under the Chad Basin Development Authority's Water Supply Intervention Scheme in 2001, carried out survey in six local government in Borno south senatorial districts (including Gwoza local government) but most of the boreholes drilled were either of low yield or abortive.

From the vertical electrical sounding, three layers were interpreted for sounding points VES01, VES02, VES03, VES04, VES06, VES07 and four layers for VES05 (Table 2). The first layer was interpreted as clay, the second layer as weathered basement and the third layer at VES05 as clayey sand which is a thick transition between the top clay and the weathered basement. Base on the depth and opening of the fractures detected (Figures 4 and 5), resistivities and thickness of the layers (Table 2), four points (VES05, VES01, VES06 and VES04) were recommended as targets for borehole construction and four productive boreholes with average yield of 2.7 l per second were drilled to an average depth of 55 m.

MATERIALS AND METHODS

An *"APEX MAXMIN"* electromagnetic (EM) machine equipped with a transmitter and a receiver was utilized during traversing to determine fractured zones within the basement, while a "GEOPULSE" resistivity meter with four current electrodes and two potential electrodes was used to confirm conductivities within the determined fractures. During the EM traversing, the transmitter transmits a current into the ground as in-phase (IP) and this is received from the receiver as out-phase (OP). Two traverses of 290 m length were carried out within the area using a 50 m cable separation with a station interval of 10 m. A graph of IP/OP versus distance (station interval) are plotted, one minima accompanied by two maxima gives the best points for further investigation. A Geopulse resistivity meter used to carry out Vertical Electrical Sounding (VES) transmits electrical signal into the ground via two current electrodes nailed into the ground some equal distance from the centre point and a potential difference between the two current electrodes is measured by the machine which displayed the result as ΔV/I. This value was multiplied by a geometric factor to obtain resistivity for each sounding. The VES data obtained was interpreted using a resistivity interpretation program "OFFIX" which produced resistivities and thicknesses of each layer encountered within the probing depth at each sounding point.

RESULTS

Details on the two EM traverses, number of anomalies, traverse distance and anomaly points are presented in Table1 while the resistivities and thicknesses of subsurface layers identified from interpretation of VES data is given in Table 2. Figure 3 shows one of the Vertical Electrical Sounding (VES) models obtained using "Offix" resistivity interpretation program. All these parameter formed the basis for final result that was utilized in selecting the points where productive boreholes were drilled.

The graph of IP/OP versus station interval were plotted (Figures 4 and 5), these clearly produced the anomalies and points at which they appeared.

Table 2. Summary of interpreted VES data.

VES	Layer no.	Apparent resistivity (ohm-m)	Layer thickness (m)
1	1	17	0 - 10
	2	10	10 - 32
	3	600	32 to -
2	1	34	0 - 4
	2	15	4 to 22
	3	750	22 to -
3	1	38	0 - 11
	2	14	11 - 22
	3	790	22 to -
4	1	21	0 - 8
	2	13	8 - 22
	3	872	22 to -
5	1	21	0 - 3
	2	20	3 - 15
	3	18	15 - 34
	4	786	34 to -
6	1	15	0 - 9
	2	11	9 - 28
	3	896	28 to -

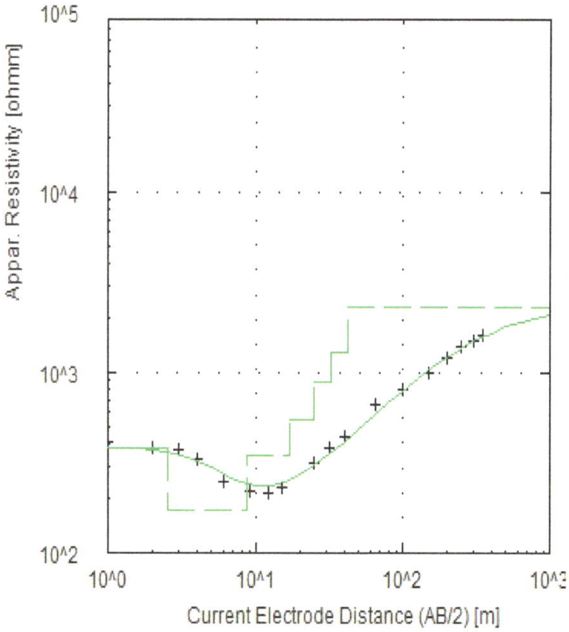

VES 01

Layer	Apparent Resistivity (ohm-m)	Thickness (m)
1	17	0 - 10
2	10	10 - 32
3	464	32 - inf

Figure 3. VES 01 model with interpreted resistivities and thicknesses.

DISCUSSION

From the results of Vertical Electrical Sounding (VES) only one sounding point (VES02) had KH curve types while the remaining are H-curve types. The thickness first layers ranges from 2 – 11 m with resistivity ranges of 15 – 38 Ωm. VES03 recorded the thickest first layer and highest resistivity of 11m and 38Ωm respectively. Thickness of second layer ranges from 11 – 22 m with resistivity range of 10 – 20 Ωm, the thickest second layer (22 m) was recorded at VES01 with resistivity of 10 Ωm. The third layers were interpreted as the basement at

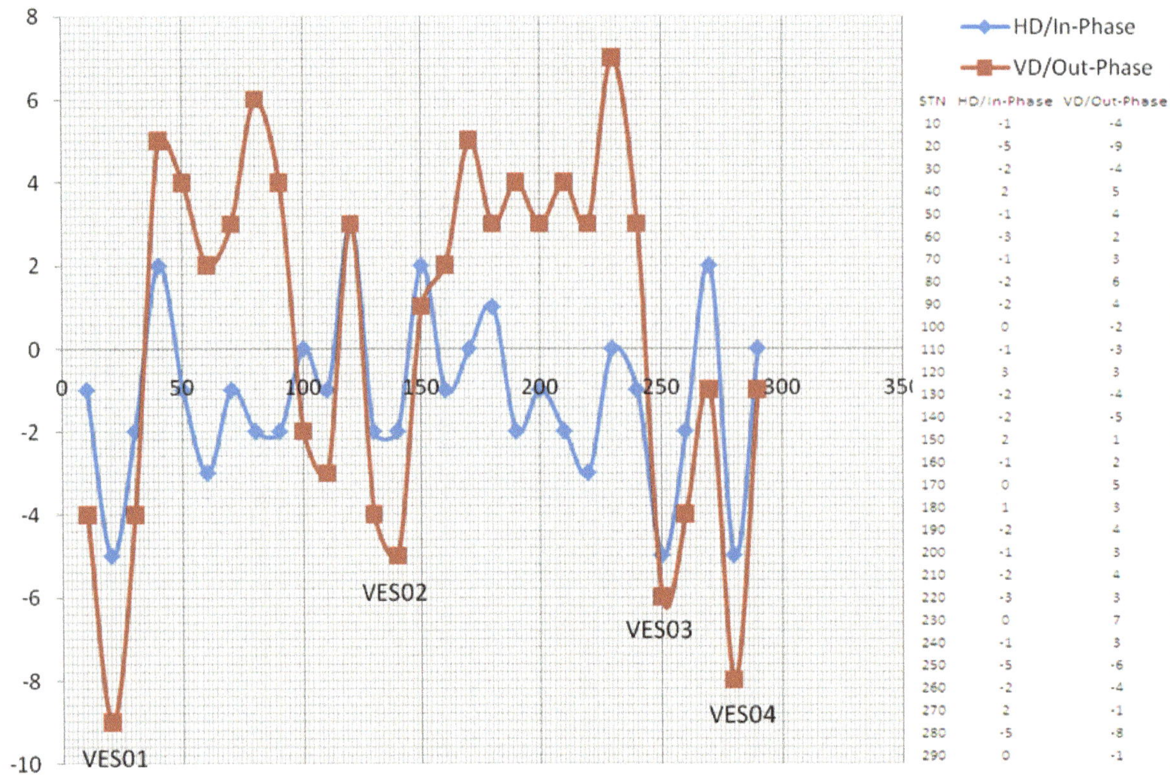

STN	HD/In-Phase	VD/Out-Phase
10	-1	-4
20	-5	-9
30	-2	-4
40	2	5
50	-1	4
60	-3	2
70	-1	3
80	-2	6
90	-2	4
100	0	-2
110	-1	-3
120	3	3
130	-2	-4
140	-2	-5
150	2	1
160	-1	2
170	0	5
180	1	3
190	-2	4
200	-1	3
210	-2	4
220	-3	3
230	0	7
240	-1	3
250	-5	-6
260	-2	-4
270	2	-1
280	-5	-8
290	0	-1

Figure 4. Plot of IP/OP against traverse distance 0f 290 m along traverse 01.

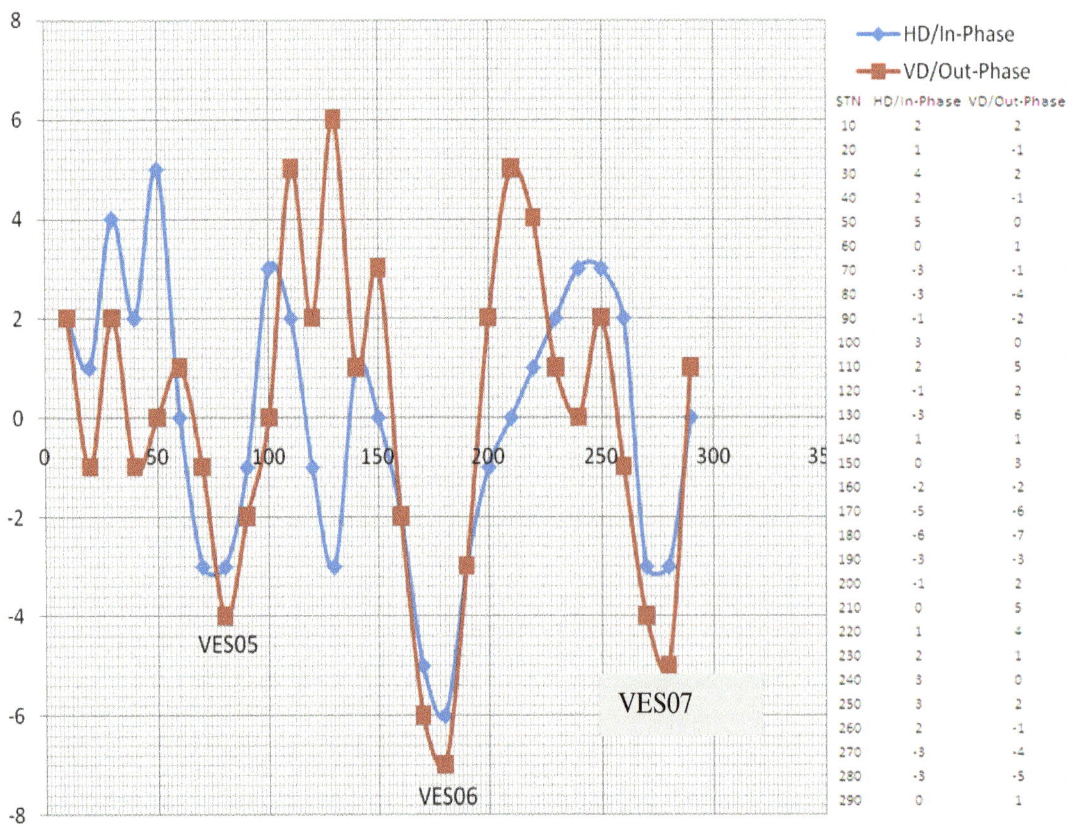

STN	HD/In-Phase	VD/Out-Phase
10	2	2
20	1	-1
30	4	2
40	2	-1
50	5	0
60	0	1
70	-3	-1
80	-3	-4
90	-1	-2
100	3	0
110	2	5
120	-1	2
130	-3	6
140	1	1
150	0	3
160	-2	-2
170	-5	-6
180	-6	-7
190	-3	-3
200	-1	2
210	0	5
220	1	4
230	2	1
240	3	0
250	3	2
260	2	-1
270	-3	-4
280	-3	-5
290	0	1

Figure 5. Plot of IP/OP against traverse distance 0f 290 m along traverse 02.

VES01, 02, 03, 04 and 06. The only four layered point (VES05) had resistivity and thickness of third layer as 18 Ωm and 19 m, respectively. The third layer at VES05 is also the thickest of all the layers interpreted for all the sounding points. From the results of VES and EM interpretation, all the points sounded are good targets for borehole construction and VES05 could produce the highest yield or volume of water needed for the purpose of this study because it has the thickest of all the layers interpreted.

Seven anomalies were detected (Figures 4 and 5) along the two Electromagnetic (EM) traverses conducted. Seven VES were carried out but only six were interpreted because the two anomalies on traverse 01 at station no. 250 and 280 were too close, therefore VES 04 was only reflected at station 28 (Figure 4). Base on the IP/OP versus distance (station interval) plot, stations 80, 180 and 280 has the widest anomalies among the others. Though all the anomalies had good groundwater potentials, a wider anomaly can produce larger quantity of water, but when added to the thickness and resistivity of the overlying layer, VES05 appears to be the best of all the targets.

The results of the seven Vertical electrical Sounding (VES) and two electromagnetic (EM) traversing has shown that sub-surface geologic structure along Gwoza – Damboa road are good sources of groundwater which when properly harnessed can provide portable water for Gwoza town and its neighbouring villages. It has also proved that the area is made of predominantly three layer earth model as reported in Dandy (2000) and Nur and Kujir (2006).

Conclusion

The results of electromagnetic (EM) traversing and vertical electrical sounding (VES) conducted along Gwoza-Damboa road have allowed the identification of subsurface layers, fractures and faults that were targeted for borehole construction for Gwoza town and its neighbouring village. The layers identified include the top clay, the sandy clay and the weathered and fractured basement. The sandy clay, weathered and fractured/ faulted basement produced the required groundwater yield as obtained from the drilled wells. This result has proved the effectiveness of pre-drilling EM and VES as an effective tool for avoidance of drilling of abortive boreholes in the basement terrain. The result has also discovered that these groundwater hosting structures (the sandy clay and the fractures and faulted basement) do exist within the subsurface along Gwoza-Damboa road.

ACKNOWLEDGEMENT

The authors wish to acknowledge the efforts of Mr. Salisu D. Tahir of Kano State Rural and Water Sanitation Agency (RUWASA) during both EM and VES exercise on the field.

REFERENCES

Carter JD, Barber W, Tait EA, Jones GP (1963). The geology of parts of Adamawa, Bauchi and Bornu Provinces in North-Eastern Nigeria. Geol. Surv. Niger. Bull., 30: 1-99.

Dandy AE (2000). Hydro Geo-electrical study of parts of Adamawa State (Gombi, Hong and Mubi LGAs) North-eastern Nigeria. MSc Theisis Federal University of Technology Yola. Adamawa State.

Kogbe CA (1989). Geology of Nigeria. Second Edn., Abiprint and Pak limited, Ibadan, Nigeria.

MapSource (2006). Rip and waypoint manager. Version 4.0 by Garmin Limited.

Nur AS, Kujir As (2006). Hydro geo-electrical study in the Northeastern part of Adamawa State, Nigeria. J. Environ. Hydrol., p. 14.

Offix (1990). Resistivity Interpretation Program. Interpex limited, Golden Colorado, USA.

Olorunfemi MO, Fasuyi SA (1993). Aquifer types and geo–electrical/hydrogeological characteristics of the central basement terrain of Nigeria. J. Afr. Earth Sci., 1: 209-317.

Reliability assessment of 3-phase three winding converter transformer of HVDC transmission system

L. Devi[1]*, C. H. Sai Babu[2] and S. Siva Nagaraju[2]

[1]Department of Electrical Engineering, Intellectual Engineering College, Anantapur, AP, India.
[2]Department of Electrical Engineering, Jawaharlal Nehru Technological University, Kakinada, AP, India.

High voltage direct current (HVDC) is today recognized as an effective and efficient means of transmitting bulk power over long distances through over head lines or cables. The reliability of HVDC links have always been of primary concern in planning and operating of power systems. The existence of spares and maintenance practices do affect the availability of HVDC stations. The HVDC converter transformer plays a relevant roll in HVDC converter station, not only due to the direct cost associated to them but also causes its availability affects the HVDC transmission capability (or) capacity. In this paper the effect of converter transformer failure rate, repair time on expected capacity level of three phase three winding converter transformer with and without spares is analyzed and compared. It also effect of number of spares on expected capacity level is presented.

Key words: HVDC transmission system, converter transformer, unavailability, Morkov modeling.

INTRODUCTION

High voltage DC transmission is growing much importance in evaluating reliability analysis of composite power systems due to its advantages and applications (Bhagawan, 2001). There is need to study the reliability analysis of equipment behaviors of each component in the HVDC system (CIGRE Joint Task Force, 2004). The HVDC system has been treated as a whole unit consisting of AC equipment, converter transformer, pole equipment, filters etc. It is well known that system reliability can be modeled by a continuous time homogeneous Morkov process whenever the components of the system have constant failure and repair rates (Singh and Billinton, 1977; Descrochrs, 1985).

The availability of HVDC systems has always been of primary concern in the planning and operation of power systems. The reliability requirements differ from project to project and are specified at a very early stage of project. The system is always designed and built with a certain degree of reliability i.e it is guarantee a particular line of availability of HVDC system (Hamoud et al., 2004; Direct Current Transmission, Volume 1). The total availability of the components in the line and at the stations. Among the various components and subsystems existing at an HVDC station, the converter transformer system (CTS) has a significant impact on the total availability of the HVDC system (Setres and Bertling, 2008: Hwan-Cha, 2006). To improve the availability of the HVDC station, spare transformers are usually provided so that in case of major failure, a switching to spare action can be taken to restore the link as soon as possible.

This leads to the fact that the availability of an HVDC station also depends on the existence of spare converter transformer (Billinton and Ronald). Moreover, spare transformers can be located at one or both stations, at supplier facilities located in the same country or overseas. Nevertheless, spare units are costly and this has to be taken into the account when the balance between the additional cost and the improved performance is considered (Leelaruji, 2007; Roy and Prasad, 1970).

Therefore, the evaluation of the expected availability of the CTS of HVDC stations is of most importance in the decision making.

In this paper Markov models for 3 phase three winding converter transformer system with and without spares is analyzed and compared.

*Corresponding author. E-mail: devika07_junnu@yahoo.com.

METHODOLOGY

In this paper Markov modeling approach is used to take account the distribution of component failure over time. It is possible to compute the reliability parameters as the long term probability of the system being in a given state by solving a linear differential equation system. The graphical model constructed to display all possible transition paths between states is called state space diagram.

A Markov model can be applied to systems whose random behavior varies either discretely or continuously with time. In this paper Markov modeling has been adapted. Figure 1 shows a basic example of two state systems.

In Figure 1 there are two states up state (State 1) and down state (State 2). There are two transitions associated with each component, the first L representing the failure process and the second M representing the repair process. This model implies that the component becomes operational and in service immediately following the repair action. The duration of the failure is limited to the time it takes to install the spare unit rather than to repair of the equipment itself.

In this paper, only the equations for calculating the steady state solution of the system are presented. This solution gives the probability of being in a certain state after an infinite time in the system. It can be shown that this steady state solution can be found by solving the equation 1.

$$PQ = 0$$
$$\sum P_i = 1 \qquad \text{............... (1)}$$

Here P_i is a row vector and
Q = transition rate matrix for the Markov process.

$$\begin{pmatrix} P_{11} & P_{12}\ldots\ldots\ldots\ldots\ldots P_{1n} \\ P_{21} & P_{22}\ldots\ldots\ldots\ldots\ldots P_{2n} \\ \ldots\ldots\ldots & \ldots\ldots\ldots\ldots\ldots\ldots \\ P_{n1} & P_{n2}\ldots\ldots\ldots\ldots\ldots P_{nn} \end{pmatrix} \qquad \text{.........(2)}$$

In this matrix P_{ij} is the transition rate from state i to state j. The diagonal element P_{ii} defined as:

$$P_{ii} = 1 - \sum P_{ij} \qquad \text{............(3)}$$

HVDC technology

High voltage direct current transmission is used to interconnection between two AC systems. These HVDC technologies are two types HVDC classic and HVDC light. HVDC Classic converters are driven by thyristor valves and HVDC light converters are driven by IGBTS. In this paper HVDC classic are designed with converter valves connected in 12-pulse configuration, in order to reduce the harmonics. In HVDC system converter transformer is connected in series with the 12-pulse converter at both ends.

The types of converter transformers used in HVDC transmission systems are:

1. Three Single Phase Three Winding (SPTW)
2. Two Three Phase Two Winding (TPTW)
3. Six Single Phase Two Winding (SSPTW)
4. One Three Phase Three Winding (TPTW)

In this paper the effect of failure rate and number of spares on expected capacity level of three phase three winding converter transformer at different capacity values is analyzed and compared.

State space model of converter transformer without spare

State space model of three phase three winding transformer is as shown in Figure 2.

State 1 represents operating or up state, when the transformer is up. When the transformer fails, a transition to state 2 occurs, with a rate of L, and the operation of CTS will be shut down. When any unit fails immediately a spare unit is ordered to be transported to the site. The transportation rate for the transformer to arrive is M, when the transformer arrives it is installed with the rate of G.

Using STPM approach:

$$\alpha P = \alpha \qquad \text{...........(4)}$$

Where α is a row vector.

$$[P_1\ P_2\ P_3] \begin{pmatrix} 1-L & L & 0 \\ 0 & 1-M & M \\ G & 0 & 1-G \end{pmatrix} = [P_1\ \ P_2\ \ P_3\ ] \qquad \text{...... (5)}$$

Writing the state equations for Equation 5,

$$P_1\,[1 - L] + P_3G = P_1 \qquad \text{..........(6)}$$
$$P_1L + P_2\,[1 - M] = P_2 \qquad \text{.........(7)}$$
$$P_2M + P_3\,[1 - G] = P_3 \qquad \text{......(8)}$$

We know that $P_1 + P_2 + P_3 = 1$ ......(9)

Putting Equations 6, 7 and 9 in matrix form:

$$\begin{pmatrix} -L & 0 & G \\ L & -M & 0 \\ 1 & 1 & 1 \end{pmatrix} \begin{pmatrix} P_1 \\ P_2 \\ P_3 \end{pmatrix} = \begin{pmatrix} 0 \\ 0 \\ 1 \end{pmatrix} \qquad \text{... (10)}$$

Applying Cramer's rule to the above Equation 10; The obtained probabilities are:

$$P_1 = \frac{GM}{GM + LM + LG}$$

$$P_2 = \frac{LG}{GM + LM + LG}$$

$$P_3 = \frac{LM}{GM + LM + LG}$$

For selected values of failure rates the expected capacity level of converter transformer without spare is tabulated as shown in Table 1.

State space model of converter transformer with one spare

Consider now the situation that could exist it one spare was normally available. In this case, when a transformer fails, it can be replaced by the spare. Provided that either the spare has not already been used or the component it has replaced has been repaired and is available as a spare. This provides the state space model of three phase three winding converter transformer with one spare as shown in Figure 3.

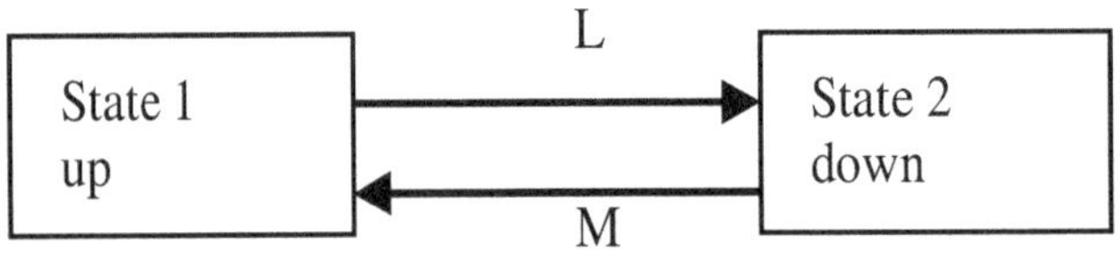

Figure 1. State space diagram of two state Markov process.

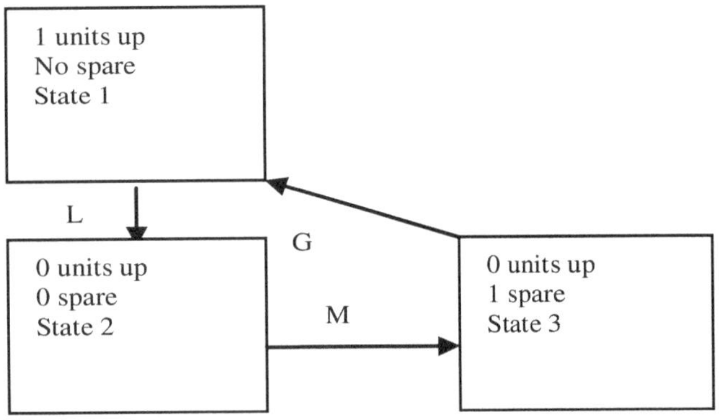

Figure 2. State space model of three phase three winding converter transformer without spare. Where L = Transformer failure rate, M = Transformer repair time, G = Transformer installation time.

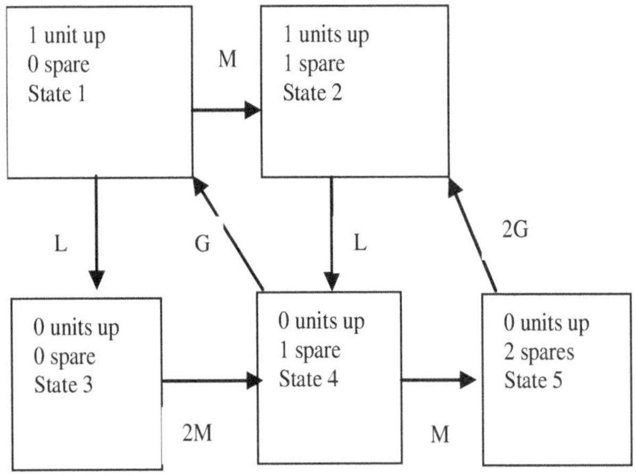

Figure 3. State space model of three phase three winding converter transformer with one spare.

In Figure 3, States 1 and 2 are up states, states 3 to 7 are down states. For selected values of failure rates (ref from CIGRE) expected capacity level of converter transformer at different capacity values is evaluated and tabulated in Table 2.

When number of spares increase number states also increases so it is difficult to find out the limiting state probabilities. Therefore, software has been developed in CPP to find the limiting state probability values of three phase three winding converter transformer with two spares. Figure 4 shows the state space diagram of three phase three winding converter transformer with two spares.

Similarly for selected failure rates expected capacity level of three phase three winding converter transformer with two spares at different capacity values are evaluated and tabulated in Table 3.

Description of tables

The failure rate data for the three phase converter transformer is based on data from CIGRE joint task force.

In Table 1 first column represents failure rates of converter transformer column 2 reliability values and columns 3, 4, 5 and 6 represents expected capacity level of three phase three winding transformer without spares at different capacity values. Similarly Tables 2 and 3 represents effect of failure rates on expected capacity level of three phase three winding transformer with one and two spares at different capacity values.

RESULT ANALYSIS

In this paper effect of failure rate and number of spares on three phase three winding converter transformer is analyzed. From the report of CIGRE maximum major failure rate is taken as 2.0 f/yr (percentage). The repair time in worst case is taken as 90 h. Installation time is taken as 48 h. The results from three tables are represented graphically in the following Figures 5 to 8.

Figures 5, 6 and 7 shows the variation in expected capacity level due to variation in failure rate at 120, 150, 180 and 300 MVA capacity values. In Figures 5, 6 and 7 curve 1 represents expected capacity level at 120 MVA. Curves 2, 3 and 4 represents expected capacity levels at 150, 180 and 300 MVA, respectively.

In Figures 6 and 7 variations in expected capacity level is less as compared to Figure 5. Figure 8 shows the effect of number of spares on expected capacity level of three phase three winding converter transformer at failure rate 0.1f/yr, repair time 90 days and installation time 48 h with and without spares.

In Figure 8 curve 1 shows variation of expected capacity level at 120 MVA at spares one, two, three and without spares. Similarly curves 2, 3 and 4 at 150,180 and 300 MVA, respectively.

Conclusion

This paper has shown the benefit of using Markov models in the evaluation of expected capacity level of three phase three winding transformer with and without spares.

Table 1. Expected capacity level of three phase three winding converter transformer without spare.

Failure rate	Reliability	Expected capacity level of three phase three winding transformer			
		120 MVA	150 MVA	180 MVA	300 MVA
0.01	0.997452	119.69	149.61	179.53	299.23
0.02	0.994916	119.38	149.23	179.0	298.47
0.05	0.987387	118.48	148.10	177.73	296.21
0.09	0.977524	117.302	146.63	175.95	293.26
0.1	0.975088	117.91	146.26	175.51	292.53
0.5	0.886729	106.4	133.0	159.61	266.01
0.8	0.8303	99.636	124.54	149.45	249.09
1.0	0.796508	95.58	119.48	143.37	238.95
1.5	0.722951	88.754	108.44	130.13	216.88
2.0	0.661831	79.42	99.24	119.13	198.549

Table 2. Expected capacity level of three phase three winding converter transformer with one spare.

Failure rate	Reliability	Expected capacity level of three phase three winding transformer			
		120 MVA	150 MVA	180 MVA	300 MVA
0.01	0.997748	119.99	149.99	179.99	299.99
0.02	0.999878	119.8	149.98	179.98	299.98
0.05	0.959777	119.96	149.96	179.96	299.98
0.09	0.99832	119.91	149.89	178.89	299.78
0.1	0.998157	119.64	149.55	179.45	299.0
0.5	0.98956	118.86	148.58	178.3	297.17
0.8	0.96453	117.57	146.96	176.36	293.9
1.0	0.70031	116.52	145.65	174.77	291.29
1.5	0.942207	113.4	141.75	170.1	283.5
2.0	0.90682	109.85	137.31	164.77	274.62

Table 3. Expected capacity level of three phase three winding converter transformer with two spares.

Failure rate	Reliability	Expected capacity level of three phase three winding transformer			
		120 MVA	150 MVA	180 MVA	300 MVA
0.01	0.9997295	119.99	149.99	179.99	299.99
0.02	0.999944784	119.99	149.99	179.99	299.99
0.05	0.99862093	119.98	149.98	179.98	299.98
0.09	0.999747	119.96	149.96	179.96	299.96
0.1	0.999718	119.96	149.96	179.96	299.96
0.5	0.998214	119.78	149.78	179.78	299.78
0.8	0.99642	119.57	149.46	179.35	298.72
1.0	0.994804	119.37	149.22	179.06	298.44
1.5	0.989075	118.68	148.36	178.3	296.72
2.0	0.980797	117.69	147.11	176.58	294.24

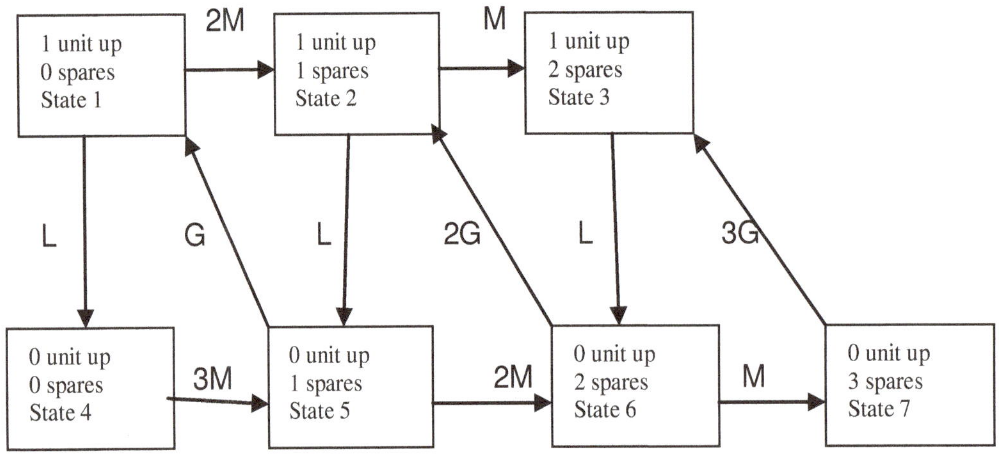

Figure 4. State space model of three phase three winding converter transformer with two spares.

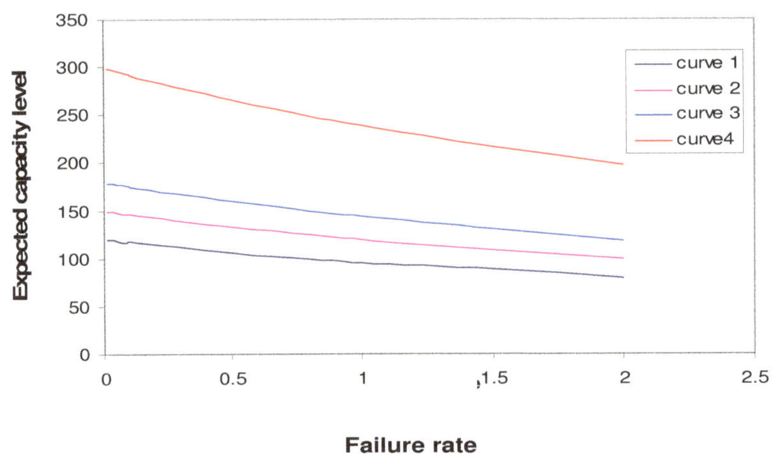

Figure 5. Expected capacity level of converter transformer without spare.

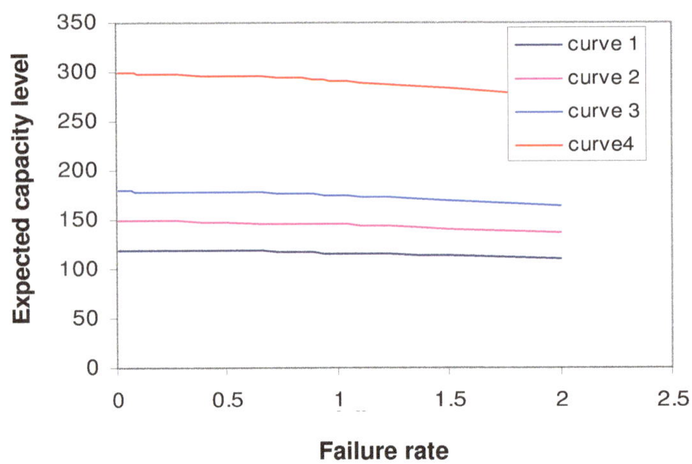

Figure 6. Expected capacity level of converter transformer with one spare.

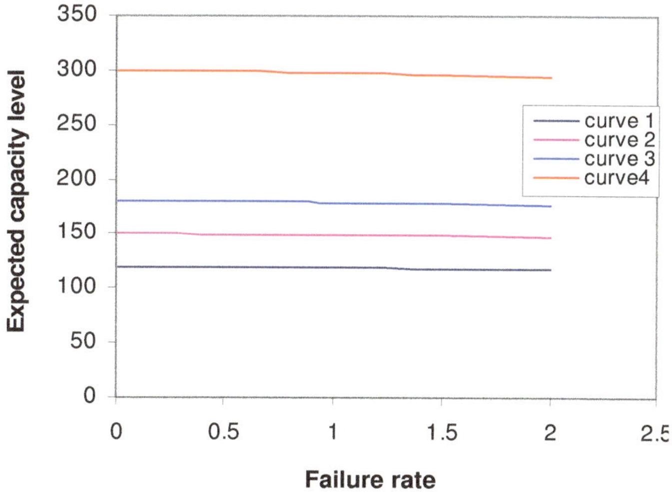

Figure 7. Expected capacity level of converter transformer with two spares.

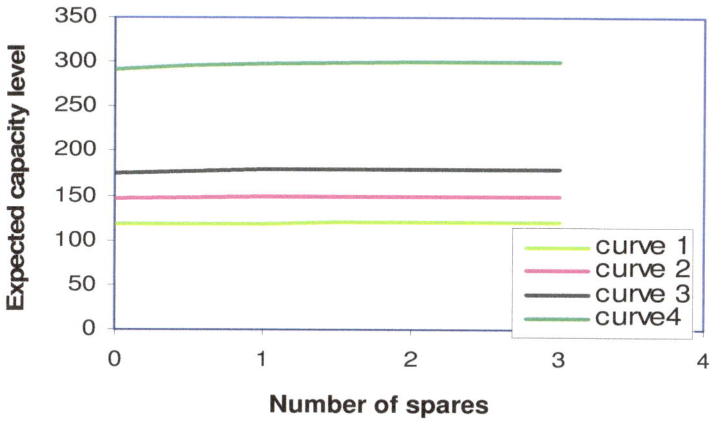

Figure 8. The effect of number of spares on expected capacity level of converter transformer.

According to the results it has been observed that increase in failure rate decreases the expected capacity level of the transformer. An attempt was also taken to reduce the effect of failure rate by increasing the number of spares which also concluded an increase in number of spares which increases the expected capacity level of the transformer.

REFERENCES

Bhagawan R (2001). Application of Markov Modeling concepts for multi bridge and composite HVDC system reliability analysis" international conference on " Quality reliability and control " organized by IETC, Mumbai center India and IIT Bombay, Mumbai, India.

CIGRE Joint Task Force B4.04/A2-1 (2004). Analysis of HVDC Thyristor converter transformer performance," Feb.

Singh C, Billinton R (1977). System reliability Modeling and Evaluation, Hutchinson Educational Publishers, London, UK, ISBN:0-091-26500-2.

Descrochrs C (1985). V.A Method for the reliability evaluation of DC Links: CIGRE REPORT" .14:01-105.

Hamoud F, Qureshy A, Elen LL (2004). Assessment of high voltage auto-transformer spare requirement in bulk transmission systems" 8th international conference on probabilistic methods applied to power system (PMAPS), Lowa state university, Ames lowa.

Direct Current Transmission volume 1: by EDWARD WILSON KIMBARK, Sc. D., Fellow I.E.E.E.

Hwan-Cha J (2006). "An extended model for optimal burn-in procedures," IEEE Trans. Reliab., 44:2.

Setres J, Bertling L (2008). "Introduction to the HVDC technique for reliable electrical power systems" to be presented at the 10th international conference on probabilistic methods applied to power system (PMAPS), Rincin,Puerto Rico.

Leelaruji R (2007). "Availability assessment of converter transformers using Makov modeling" Maste of science theses, scholl of electrical engineering, Division of electromagnetic engineering KTH.

Roy B, Prasad V (1970). " Quantitative Reliability Analysis of HVDC transmission System" Part 1 : Spare valve and assessment in mercury arc bridge configurations. IEEE Trans. Power Apparatus Syst., 90: 1034-1053.

9

The effect of magnetic water on strength parameters of concrete

M Gholizadeh[1] and H Arabshahi[2*]

[1]Chemistry Department, Ferdowsi University of Mashhad, Mashhad, Iran.
[2]Physics Department, Ferdowsi University of Mashhad, Mashhad, Iran.

A machine for magnetic water treatment is designed and its application on the ions contained in water has been investigated to examine the effect of magnetic water on concrete parameters. Strength parameters of concrete have been studied for more than 104 concrete samples, including the non-magnetic samples (made by ordinary water) and magnetic samples (made by magnetic water), with slump and compressive strength experiments. Based on slump experiments, magnetic samples were 7 centimeters more than non-magnetic group and the average compressive strength of samples made by magnetic water was 23% more than that of samples made by ordinary water. The experimental results show the advantages of magnetic samples in concrete industry because of increase in plasticity, the efficiency and quality of concrete boosts in comparison with nonmagnetic samples.

Key words: Magnetic water, compressive strength, slump, plasticity, surface tension decrease.

INTRODUCTION

Magnetic water treatment machine was first invented in 1945 by a distinguished Belgian engineer (Vermieren, 1957). This machine which is currently utilized in more than fifty countries around the world such as America and Russia was practically introduced to the market since 1950 and as a result of a series of persistent and continuous studies; it was developed into its present improved and perfect form, Aqua Correct (Baker et al., 1996). This machine contains a powerful magnet which is the source of production of magnetic energy. This powerful magnet is fixed in a small tube with such high technology that can create a magnetic field reaching a high rate of about 6500 Gauss (Eshaghi et al., 2004).

This machine simply works based on the electromagnetic induction principle which was introduced by Mickeal Faraday in 1832 (Castro et al., 2004). Water passing through this machine undergoes the magnetic energy present there, which influences the colliding particles, calcium carbonate molecules($CaCO_3$), salts and other saline in water and changes their physical and electrical properties and creates a particular pattern (Strum, 1940). This change in the electron pattern in ions leads to a lack of activity in forming compositions and no precipitation occurs (Gholizadeh et al., 2005). Also, magnetic energy reduces the surface tension of water by 10%. However, because of such changes, no suitable environment is provided for the formation of viscous deposits (adherence of calcium crystals and other substances) and in this way magnetic water with its unique and special properties is produced (Figure 1).

According to the Lorentz force $QV \times B$ in this machine, the output of magnetic anti-precipitation depends on the power of magnetic field imposed, velocity of water flow, magnetic force, number of particles with positive or negative charge in water and angle between particle velocity and applied magnetic field which is always upright and hence, maximum effect of magnetic field is applied (Davenport et al., 2006). Water temperature has no significant effect on the magnetic anti-precipitation process and the machine itself is quite resistant toward high temperatures (Welder, 1954).

CONCRETE STRUCTURE

Concrete is a proportionate mixture of four basic ingredients, sand, gravel, cement and water (under some conditions, other ingredients are added to achieve a certain objective but these are not among the basic ones). In fact, the main bulk of concrete is made of small and large pebbles (sand and grit) which are the stone construction

*Corresponding author. E-mail: arabshahi@um.ac.ir.

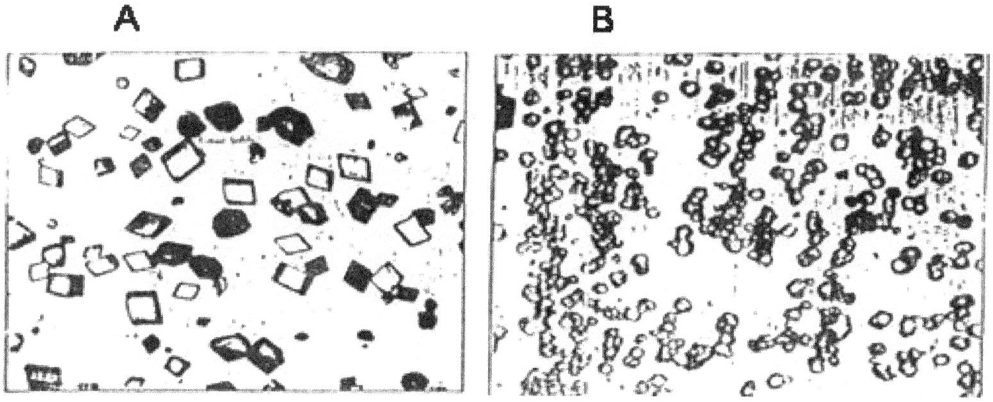

Figure 1. (a) Microscopic representation of calcium carbonate crystals in water which is not filtered (magnified 600 times); (b) microscopic representation of calcium carbonate crystals in electromagnetic water (magnified 600 time).

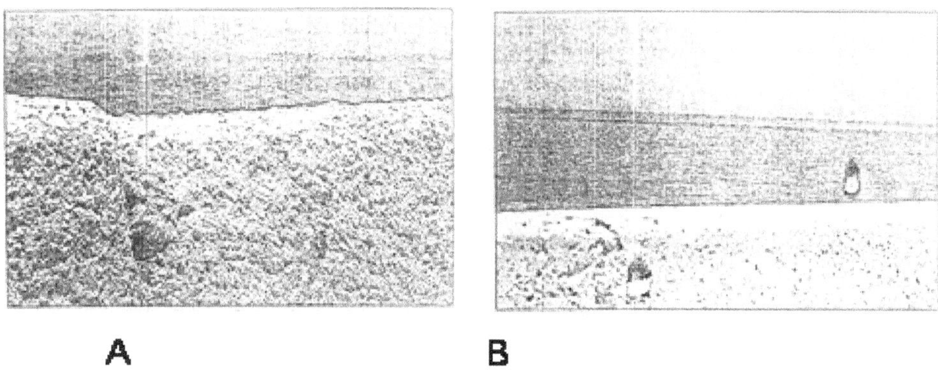

Figure 2. Black sea coasts, (a) before repair (ordinary water); (b) after repair (magnetic water).

materials and the chemical interactions between cement and water form a mixture which surrounds stone construction materials and adheres and consolidates them together.

To initiate chemical reactions in cement, a limited amount of water is needed, whereas, the water used in concrete composition is always much more than that (Furu et al., 2006). The water added is to increase the necessary utility and flexibility of concrete for filling all the angles of the mould and surrounding all the reinforcing bars. By choosing different proportions of the ingredients of concrete, a wide variety of concrete with different resistance is gained. Indeed, the impact of water is very remarkable. That is why most recommended methods for mixture design of concrete have regarded water with importance and great sensitivity for its hydration nature. In order to make concrete, the usual natural water is suitable such as rainwater, stream water, etc. River water can also be used if the chemical wastes of factories have not polluted it (Gil et al., 2002).

Water with extra chlorine is harmful to reinforced concrete as it corrodes the armature framework. Other kinds of water which are heavily polluted by oil, fat or other pollutants which negatively affect on the solidification or other properties of concrete must not be applied. The residue of plants, even a small amount is harmful to concrete and muddy water lessens the compressive strength of concrete. Water of swamps which are dark, stale and stinking and fizzy such as Co_2 and water containing gypsum, sulfate or chloride have all negative effects on concrete and must not be used. In dubious cases the water sample must be sent to laboratory for chemical analysis of water to determine its suitability for the construction of concrete (Strum et al., 1940).

Generally, water suitable for drinking is suitable enough for concrete construction (This is not always true such as water containing sugar which is drinkable but may not be suitable for concrete construction). However, taste, smell or source of water may not reject a kind of water by themselves. If the impurities in water go beyond a certain level, this may greatly affect the time span of consolidation resistance and volume stability of concrete or may also corrode the steel.

Water used for concrete construction must be pure of any oil, acids bases, salts, organic or any other materials harmful to concrete. Using the standard water, the machine for magnetic water treatment has been applied in many countries to make high quality concrete. Figures 2 and 3, show two examples of magnetic concrete applications.

RESULTS AND ANALYSES

Generally, the essential condition for a good concrete is a satisfactory compressive strength. A satisfactory compressive strength does not signify a certain degree of tolerance against compressive tensions, but this

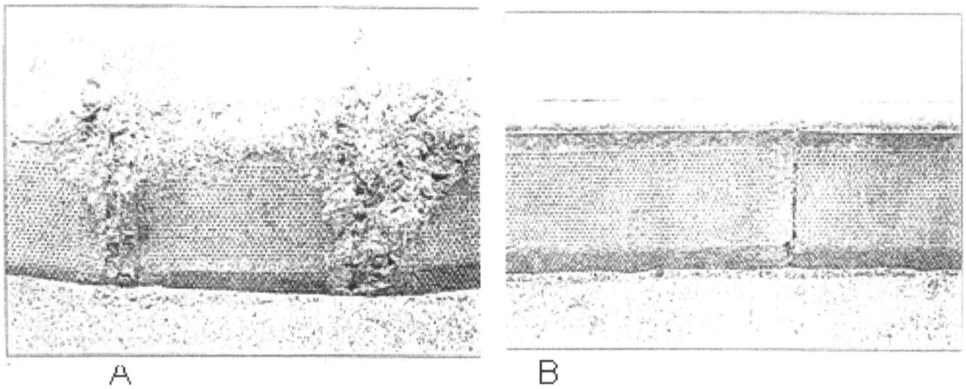

Figure 3. Superstructure of rostove nodonove way, (a) before renovation (ordinary water); (b) after renovation (magnetic water).

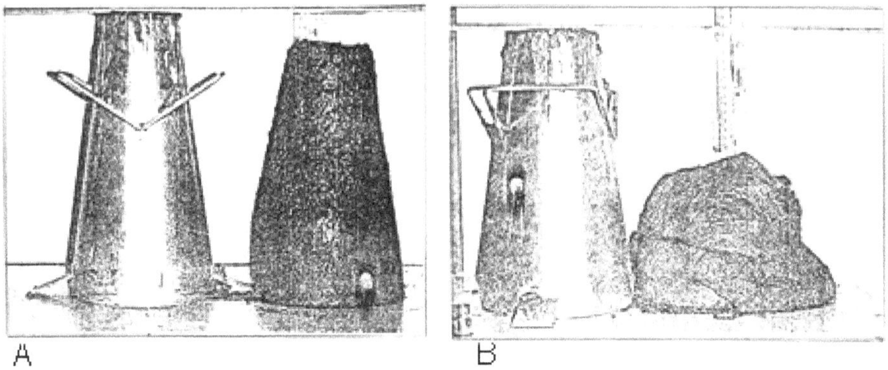

Figure 4. Subsidence rate of samples in slump experiment, (a) group A which are made by ordinary water; (b) group B which are made by magnetic water.

condition is because of the fact that other properties of concrete are relevant to compressive strength. Regarding this fact, slump experiments were applied to examine concrete efficiency and compressive strength experiment to determine maximum strength of the samples against pressure. Therefore, 104 concrete samples were categorized within two groups; group A, non-magnetic samples which were made by ordinary water and group B, magnetic samples which were made by magnetic water. Each group contained 52 samples both in cube and cylinder form to measure and analyze their strength separately (Coey, 2000). Based on the mixture design of concrete, some concrete samples with 35 kg/m^3 standard were provided for both groups. These samples had the same condition in construction materials, grading and kind of cement. The only difference was the water applied which was drinkable water with two different states, ordinary and magnetic. As soon as concrete was prepared, slump experiment was carried out (Figure 4).

Due to lack of enough moulds, concrete was made in the same mixture design in several phases and slump experiment was repeated. Table 1 shows the experimental results carried out for subsidence rate of group A and B samples in slump experiment.

After making both kinds of concrete with high efficiency and smoothness which is achieved in slump experiment, samples were placed in cube and cylinder moulds in three layers by special mallets. After 24 h, the samples were brought out of the moulds and were exposed by 25 temperature water. The water of this pond was ordinary water for both groups. In the new series of experiments different conditions of water of the pond are going to be applied. After seven days, some of the samples from both groups were placed under pressure. After 28 days, compressive strength experiment was carried out on the rest of the samples. The results from the average rates of these experiments are included in Table 2.

The graph resulted from these changes is illustrated in Figure 5. It can be seen from Figure 5 and Table 2 that, the magnetic water does increase plastic properties of concrete. Hence, it is the advantages of this method in comparison to other methods which increase plasticity of concrete. Also, our results of slump experiments show that, magnetic field has a considerable effect on clusters of water molecules, which means water molecules scattered more from each other. This causes participation

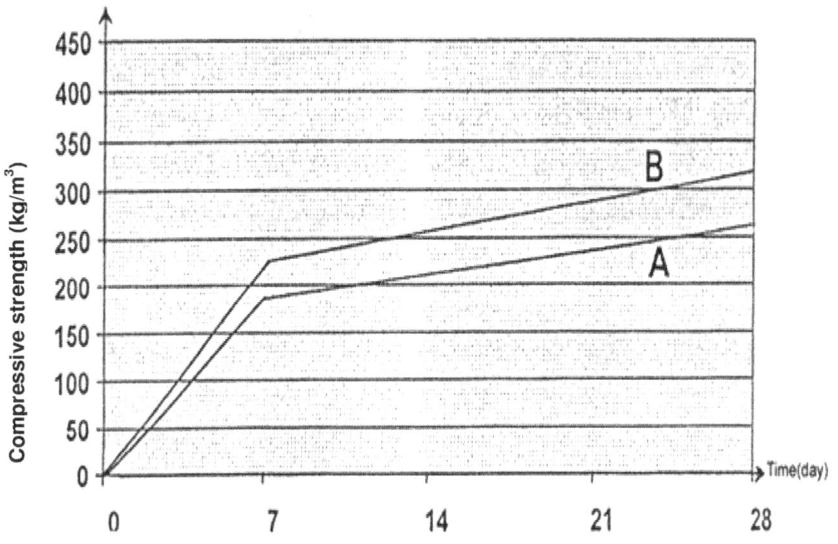

Figure 5. Compressive strength changes for cylindrical samples for group A and B.

Table 1. Subsidence rate of samples in slump experiment.

Samples	Slump (cm)
Nonmagnetic group A	4
Magnetic group B	11

Table 2. Compressive strength of cylindrical and cubic samples within seven and twenty eight days for non-magnetic and magnetic group.

Samples	Compressive strength (kg/m^3)			
	Cylindrical samples		Cubic samples	
	7 days	28 days	7 days	28 days
Group A, non-magnetic	180	265	125	185
Group B, magnetic	220	325	160	225

of water molecules in cement hydration reaction. So cement particles are scattered inside concrete mix in form of complex of molecules and hydration reaction initially take place in surface of cement particles. Therefore, a soft layer of hydration reaction products forms on cement particles, those prevent them from the penetration of water molecules to material bulk (Figure 4b)

Concrete made by magnetic water, in comparison with concrete made from plastic substances, has several advantages such as its low expense, limitation of added substances and high durability of the machine without decrease in its magnetic energy. Also, there is no need for repair, service or any other operations. Based on our experimental results, the magnetic concrete has the following advantages as well:

1. It increases the efficiency of concrete.
2. Decrease in the volume of water becomes quite possible.
3. It prevents the particles from accumulation in one place and this in its turn leads to a better distribution of them.
4. As the most important advantage, it increases the compressive strength of concrete by 25% and its traction strength by 40%.
5. Concrete made in this way is more resistant against freezing and more impermeable.
6. Decrease in water surface tension in magnetic water by 10% in comparison with ordinary water.

Conclusions

We have investigated the effect of magnetic water on strength parameters of concrete. Based on our research, magnetic water that caused the average compressive strength of samples has 23% more than that of samples made by ordinary water. The experimental results have also shown the advantages of magnetic samples in concrete industry because of increase in plasticity, efficiency and resistant in comparison with non-magnetic samples.

REFERENCES

Baker JS, Judd SJ (1996). Water Res., pp. 247-255.
Baker JS, Parsons SA (1996). Industrial process. Water Treatment, pp. 36-42.
Coey JM (2000). Magnetic Water Treatment. J. Magn. Magn. Mater., 209: 71-78.
Castro FR, Sellars CM (2004). Deformation behavior in metals at high temperature range. Phil. Mag., 60: 487-493.
Davenport SB, Sparks CN, Sellars CM (2006). Proc. Conf. Thermo-Mechanical Processing in Theory, Modeling and Practice (TMP), Stockholm, Sweden, pp. 249-255.
Eshaghi Z, Gholizadeh M (2004). Talenta, 64(2): 558-562.
Furu T, Schercliff HR, Ashby MF (2006). The interaction between the microstructural variables subgrain size and metals. Mater. Sci. Forum, 43: 217-222.
Gholizadeh M, Arabshahi H, Benam R (2005). Int. J. Appl. Chem., 1: 84-89.
Gil J, Van Houtte P, Kuhlmann-Wilsdorf D (2002). Nano particle simulation at high temperature. Prog. Mater. Sci., 25: 69-77.
Vermieren T (1957). Societe Belgedes Inginieurs et des Industriels, pp. 3-9.
Welder B, Partridge EP (1954). Industrial and Engineering Chemistry. 46: 954-961.

Appraising the structural geology of Kakuri Sheet 144: Implications for the tectonic evolution of the basement complex

Goki N. G.[1], Amadi A. N.[2*], Olasehinde P. I.[2], Dada S. S.[3], Ikpokonte E. A.[4] and Adekeye J. I. D.[1]

[1]Department of Geology and Mining, Nasarawa State University, PMB 1022 Keffi, Nigeria.
[2]Department of Geology, Federal University of Technology, PMB 65, Minna, Nigeria.
[3]Department of Earth Sciences, Salem University, Lokoja, Nigeria.
[4]Department of Geology, Ahmadu Bello University, Zaria, Nigeria.

The appraisal of the structural geology of Kakuri Sheet 144 using integrated analyses of remotely sensed lineaments, aeromagnetic anomaly, micro and macro structures show a tectonic framework expressed by a mainly NW – SE ductile vertically dipping foliation of transpressed tectono-metamorphic possibly pre-Pan African (S1) trend which has been overprinted by a set of NE – SW (L2) fracture system cut obliquely by an E-W (L3) sinistral shearing with a dextral sense. Magmatic activity might have exploited these north-easterly fracture systems which are consistent with the regionally deep seated north-easterly fault systems that define the limits of the schist belts, the younger granites and Benue Trough.

Key words: structural geology, structural fabric, geological map

INTRODUCTION

Olasehinde (1989) observed that elucidating structural fabrics, especially the fracture pattern in the Nigerian basement is a problem due to poor exposure. Beyond this, elucidating them is useful in mineral and water exploration programs. Aeromagnetic data are useful in overcoming such problems and aeromagnetic coverage of continental areas has been helpful in reconstructing the tectonic evolution on a global basis (Mita and Anand, 2003). The structural and tectonic framework of the Nigerian Basement Complex has been reported as comprising northeast – southwesterly and northwest – southeasterly lineaments superposed over a dominant north-southerly trend (Olasehinde et al., 1990), and NW – SE and NE – SW pair superposed on a N - S joint set (Annor et al., 1990; Annor and Freeth, 1985). Olasehinde et al. (1990) confirmed Ball's (1980) presence of NW – SE aeromagnetic signature. These conclusions have chiefly been based on aeromagnetic anomaly studies on a more or less regional scale and which coincides with the regional patterns recorded on the basement (Figure 1).

Tying these structural patterns to particular episodes of deformation has been difficult in nearly all the basement exposures in Nigeria. This paper attempts to study the structural and tectonic history of the basement by reappraising the magnetic anomaly and satellite data, and comparing these with field structural and micro-structural signatures of the basement rocks using the Kakuri Sheet 144 which represents a contact area between the meta-sediments to the west and a transition into the gneissic basement to the right (Figure 1). The Nigerian Basement Complex is located in the Pan African Mobile belt sandwiched between the West African Craton to the west and the Congo Craton to the southeast. The basement is predominantly amphibolite - grade quartz-feldspar-biotite + hornblende gneisses and migmatites (Wright et al., 1978). It is a polycyclic ensemble of heterogeneous migmatites and gneisses, meta-sediments and granites that have undergone a complex evolutionary history spanning through the Archaean (>2.50 Ga) to the Pan-African (550+100 Ma) (Rahaman, 1988; Ogezi, 1977; Ajibade et al., 1981; Dada, 2000, 2008). Several unanswered questions continue to trail the

*Corresponding author. E-mail: akoamadi@yahoo.com.

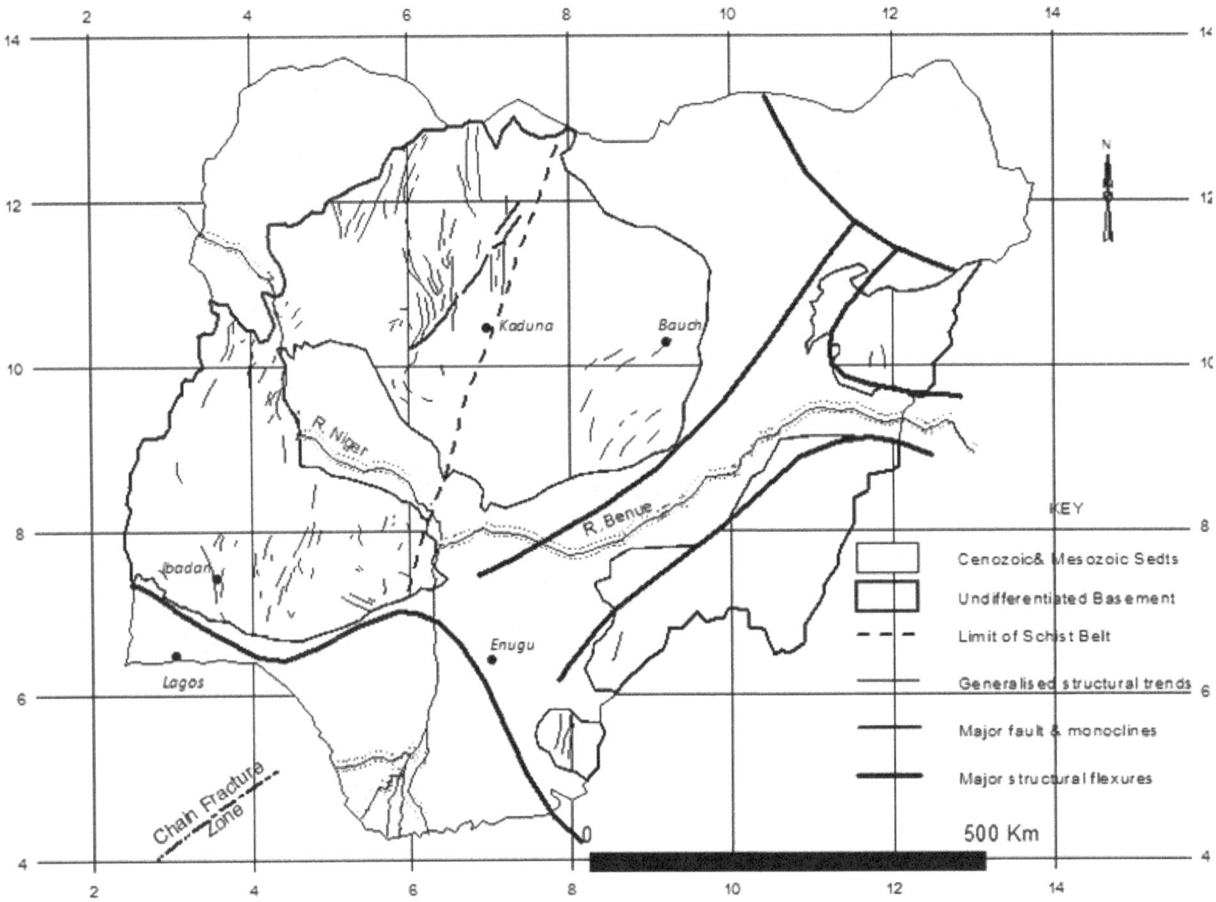

Figure 1. Generalised geological map of Nigeria showing the western Nigerian metasedimentary trends, after Oyawoye (1972) and location of Kaduna.

tectonic history of the basement such as whether they were formed as a result of ensialic or ensimatic processes related to plate tectonics (Wright et al., 1978; Rahaman et al., 1988; Ajibade, 1987) and the thermo-tectonic control on the distribution of the supra-crustal and granitic rocks within the basement. Using major and trace element geochemistry, Dada (1999) studied the petrogenesis and evolution of the basement around Kaduna. He assigned the amphibolite facies related to subduction-related magma generation of volcanic arc settings for the gneisses of the basement based on the high Rb, Sr, K/Rb, K/Sr, low Ca and chondrite normalized negative anomalies of Nb, P, Ti coupled with a high Al_2O_3/TiO_2 while the amphibolites are of the low Mg-tholleites and poorer in LILE but enriched in P. This pattern tends to support a multistage evolution that involved an early basaltic crust transformed to amphibolite.

Field relations as well as regional tectonics seem to favour the gneisses as the primary basement that was in part intruded by basic dykes that were subsequently sheared and deformed. This complex tectonic history has not only resulted in a complex rock composition but a complex structural pattern. Such structural patterns are a major control to mineral resources in Nigeria and elucidating the history of these fractures will help our understanding of the basement metallogeny. Figure 2 is a regional structural map produced from satellite imagery. The tectonic patterns such as the shear zones highlighted, correspond well with mapped structures such as Kwaga gold belt (west of Kaduna) which was mapped by Truswell and Cope (1963) and modified by Garba (2002). This is the southern prolongation of the Birnin Gwari metasedimentary belt. These patterns which is replicated in the Kakuri area of study therefore requires a closer reappraisal in confirmation of the theory of the late Pan-African brittle deformation proposed by Ball (1980) and Garba (2002).

METHOD OF STUDY

Datasets used in this study include satellite imagery, aeromagnetic data, geological map and field measured data all of which were processed in that order, to correspond to the conventional reconnaissance - detailed field - scale approach. The Kakuri Sheet studied lies within Latitudes 10°00"N and 10°30"N and Longitude 7°00"E and 7°30"E (Figures 1 and 2). Geometrically rectified dry

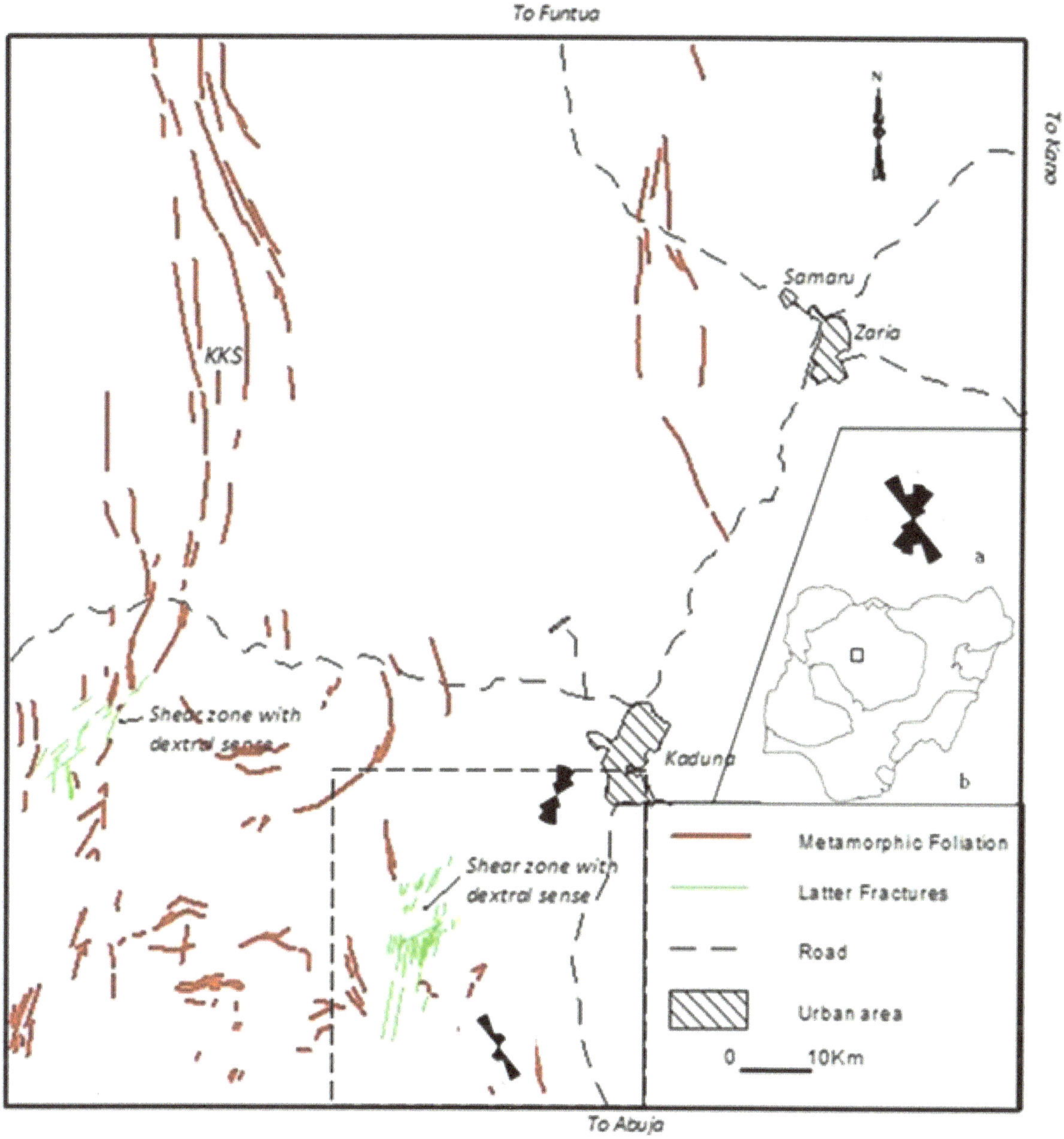

Figure 2. Regional fracture patterns of Kaduna - Zaria area generated from Landsat5 TM imagery, Kakuri Sheet 144 in set.

season Landsat5 TM imagery acquired from National Centre for Remote Sensing Jos, was subset from the Landsat5 main scene, to correspond to the Kakuri 144 Sheet. RGB 543 composite provided sufficient enhancement from which lineaments were screen digitized in Ilwis Academic 3.1 utilizing the UTM projection, Clark 1880 Spheroid and Minna datum all of which served as the georeference parameters for subsequent maps. Lineament generation automatically creates their orientations and saved in an attribute table from which the frequency azimuth rose was obtained.

Lineament density map utilized 2 x 2 km grid, which calculates the total length of lineaments within this box. The map returns a graduated scale of least to highest lineament density in meters.

Published aeromagnetic map for the area was screen digitized using 1.85 squared grids. The resultant point map with a value domain was interpolated, using the nearest neighbour algorithm in which each output pixel returns Euclidean distances towards all points. The value of the point with the shortest distance towards an output pixel is assigned to this output pixel. Subsequently, the raster map was sliced along bounding values of < 7,789 (Low), 7,780 to 7,860 (intermediate) and > 7,860 (high). This procedure is analogous to the modified upward continuation filtering of Ajakaiye et al. (1986). At least 100 fracture measurements were plotted and compared with the imagery and aeromagnetic trends for the sheet.

RESULTS AND DISCUSSION

The meta-sedimentary lithologies dominated by migmatized and schistose rocks (Figure 2) expectedly gave darker shades on the satellite imagery while the

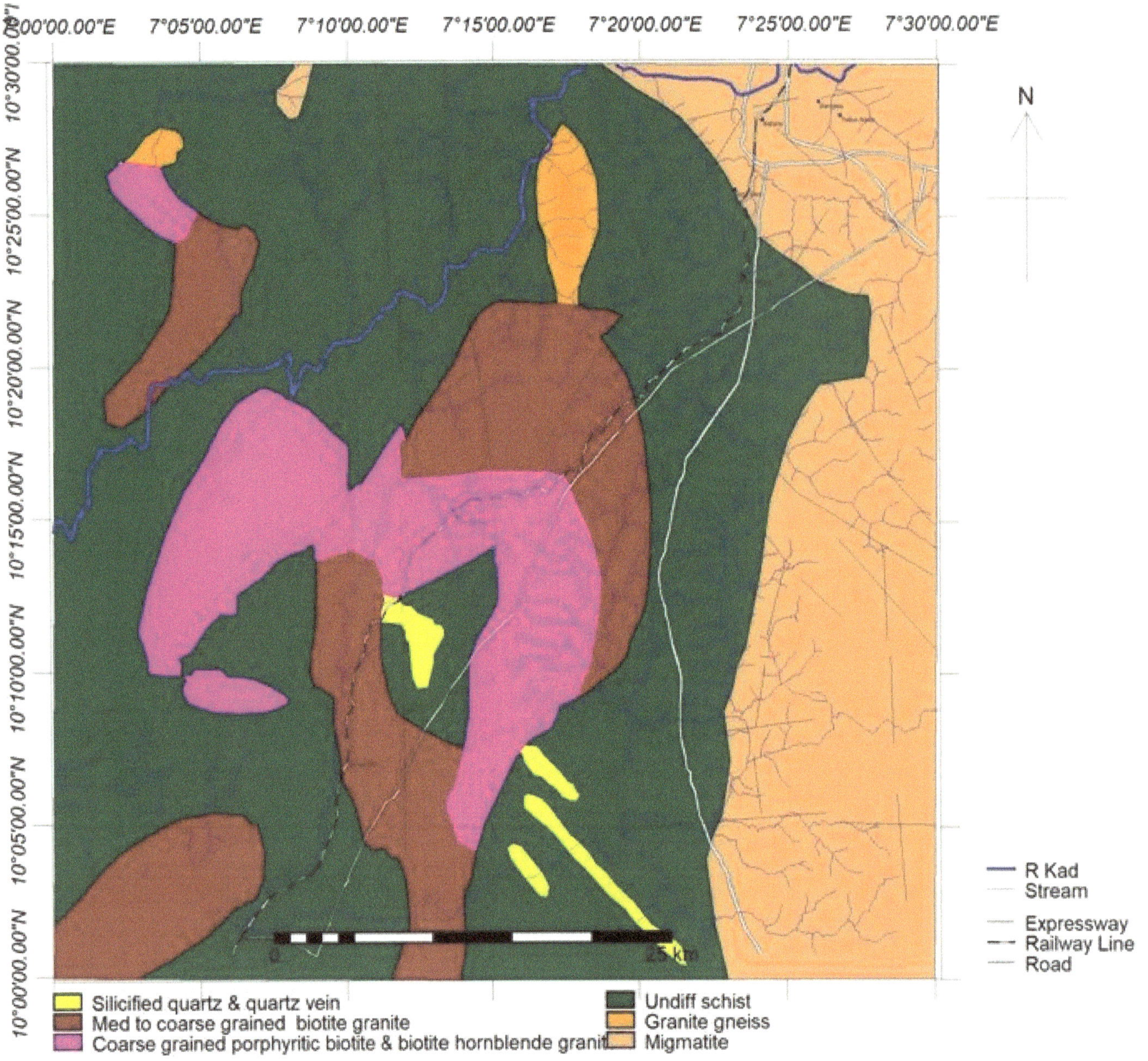

Figure 3. Geology of Kakuri Sheet 144 (modified after geological survey).

more granitic terrains gave lighter shades (Figure 3). This is because the meta-sediments have more of the mafic minerals which together with the lateritized clays contribute to the staining of band 3 and absorptions of the clayey components at or around 2.2 µm of the EM spectrum which corresponds to band 5 of Landsat5 TM. It is apparent from the imagery (Figure 4) that the lineaments display a wide scatter in orientation. NW – SE trends that stretch from the lower central area and extend to the north-western corner, represents the uplifted metamorphosed gneisses and quartzite ridges. These bodies and trends were mapped by the geological survey published map. This ridge appears to have been truncated by apparently NE – SW shearing with a dextral sense. This stepped NE – SW trends have been exploited by the River Kaduna system.

The Landsat5 TM generated lineament map shown in Figure 5, shows a preponderance of two major conjugate trends. Approximately 285 lineaments that summed up to 523,814.5 m, but some were considered negligible and thus discarded leaving a total of 111 orientation data which gave a mean resultant direction of 028-208. The NE – SW trend and the NW – SE trend and the sub-optimal N-S and E-W trends were apparent. As explained earlier, the NW – SE trend represent metamorphosed bodies that are sufficiently exposed at the surface to be captured by satellite imagery. The high density of lineament at the central portion represents a fracture

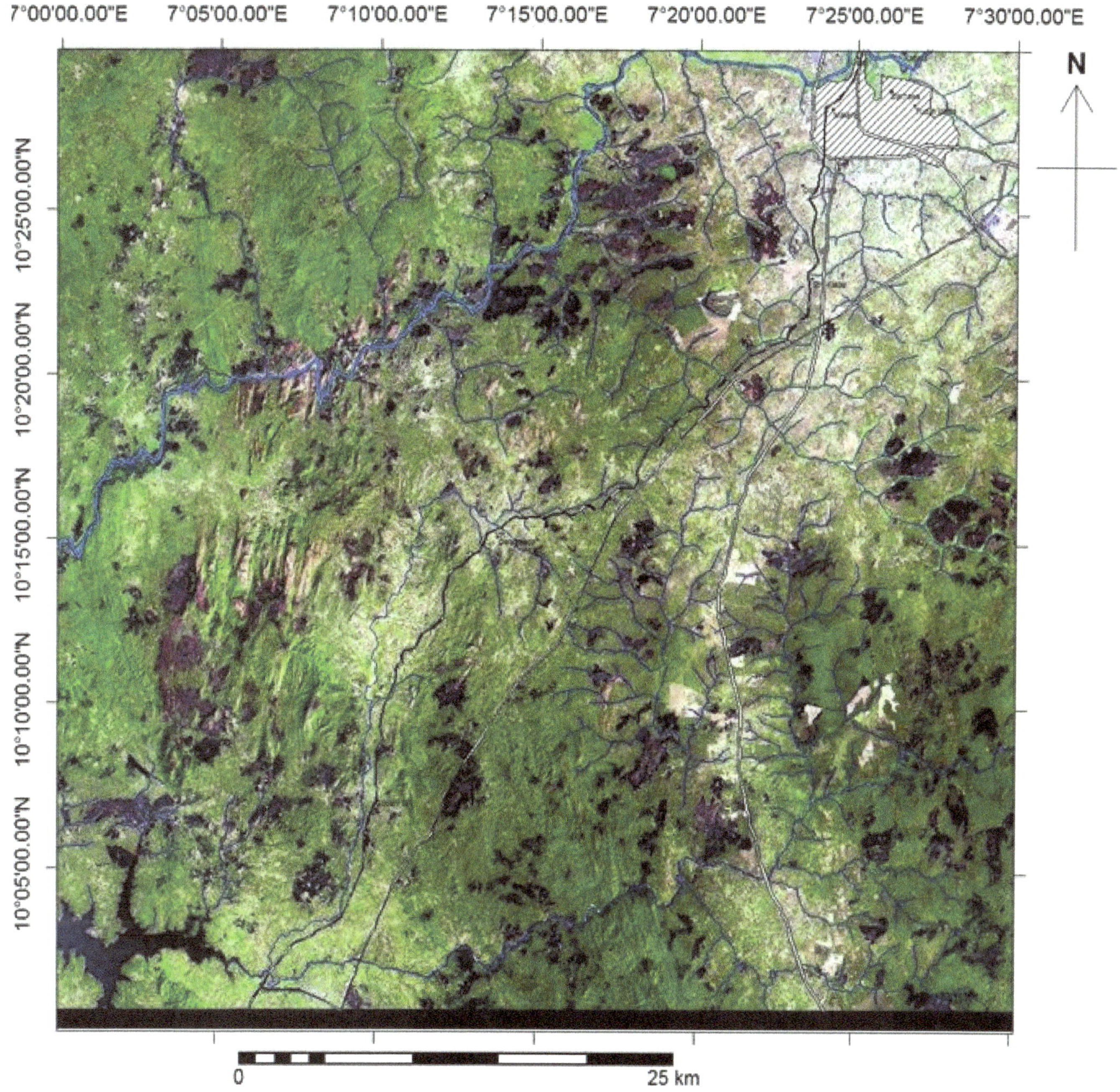

Figure 4. RGB543 colour composite of Landsat5 Tm imagery of Kakuri Sheet 144.

system that is predominantly NE – NW trend.

The 3-D analysis of the lineament shows that, three major successions of trends were preserved. The main older NW-SE metamorphic lineament which is expressed by banded and quartzitic ridges. The second is the suboptimal NNE to N-S heading of linear fractures which have been obliquely truncated by a NE-SW trends which have been here related to dextral faulting. The trend of these faults has been reported by Oyawoye (1972). These last set appear interestingly deeper and which have been exploited by the River Kaduna and Sarkin Pawa depressions. For exploration purposes, these last set may be the most interesting. This has been confirmed by the deep seated magnetic trends (Figures 6, 7, 8 and 9). Even though lineaments generated from the satellite imagery recorded a wide scatter, two bimodal peaks can be depicted as follows (Figure 8):

a) 0° - 10° (8.1% of data) and 010° – 019° (7.2% of data) = NNE – SSW trend.

b) 060° -070° (8.1% of data) and 050° – 060° (7.2% of data) = ENE – WSW trend.

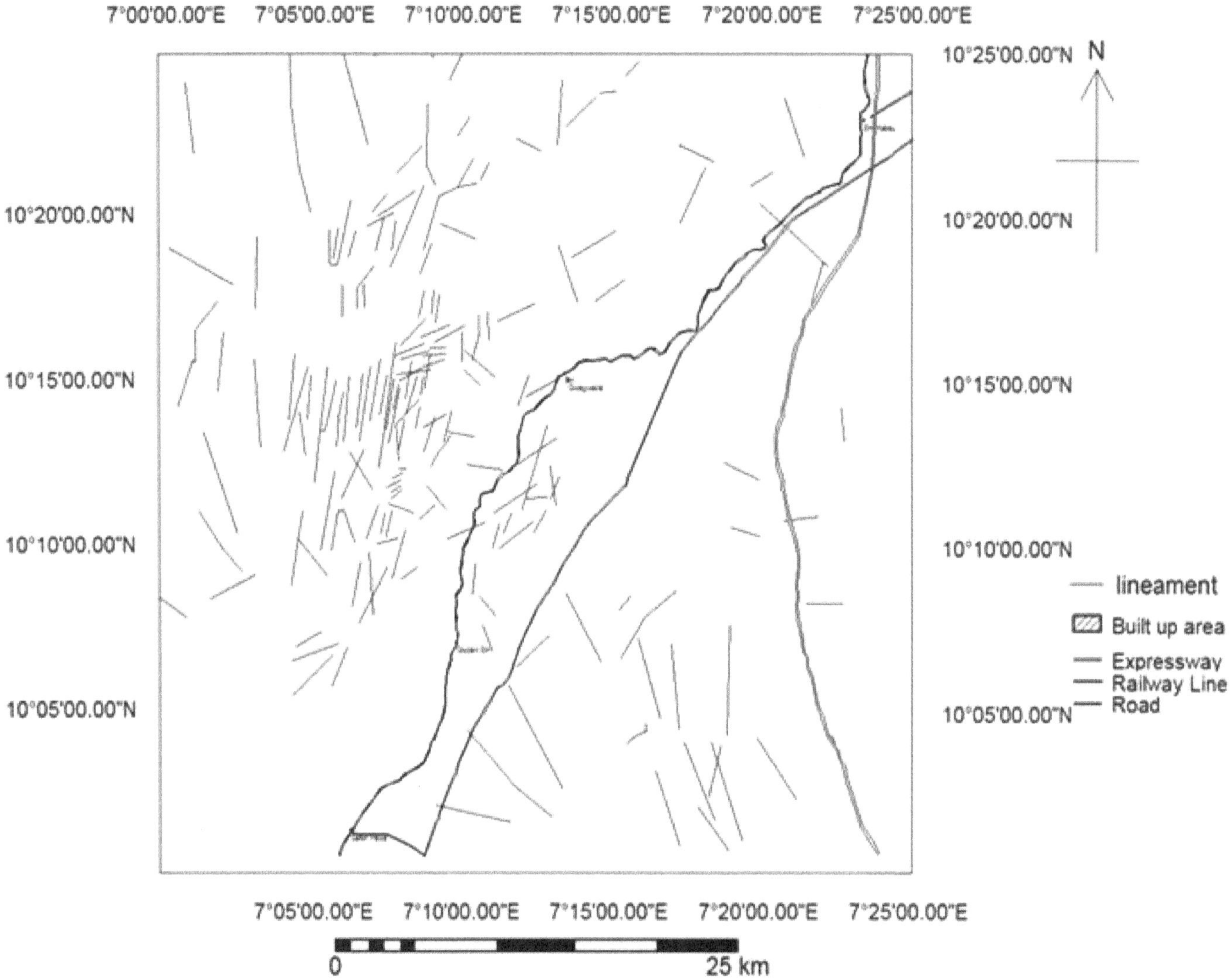

Figure 5. Landsat5 TM generated lineaments including both fractures, as well as foliation surfaces for Kakuri Sheet 144.

The rose plotted for the Landsat5 TM generated orientations is generally comparable to that obtained by Ananaba and Ajakaiye (1989) for the basement rocks of Nigeria. The field data plot gave a single peak uni-modal trend with a NNW – SSE (Figure 7b). A statistical summary of the frequency of orientations plotted against number of lineaments (Figures 8,9,10) gave about 43 lineaments in the ENE (60-90), constituting 38.73% of the total data plotted. NE-SW trends are apparent in Figure 7 which is consistent with the observations of Ajaikaye (1986), Olasehinde et al. (1990) in which they related such trends with deep seated basement lineaments. The near absence of NW – SE trends is very instructive especially when magnetic signatures are supposedly deep seated signatures. Deep seated fractures are zones of weaknesses where granititic materials such as the younger granite trends (Ajakaiye et al., 1986) take advantage of. The high magnetic materials are the migmatites, gneisses and meta-sediments. The Intermediate range is the granitoids while the low range represents the more basic geological materials (Olasehinde et al., 1989). In this terrain, geological mapping have not revealed such materials but the possibility of mapping such materials at deeper depths is here discovered.

An attempt was made at assessing the deformation recorded on these rocks by carrying out preliminary strain measurement on the rocks. Caution is here advocated because the number of data is not sufficient at the scale of the work considered, but it sure gave an idea of the recorded deformation at the micro-structural level. Axial ratio of the ellipsoids range between 1.71 and 2.34 with only one ellipsoid having an extreme 3.8 (Figures 11, 12 and 13). This indicates moderate to intermediate strains. The trends are generally north-westerly, since strain measures were performed on the steeply and near vertically dipping gneisses. These interpretations were however non-conclusive, since it was not conducted on

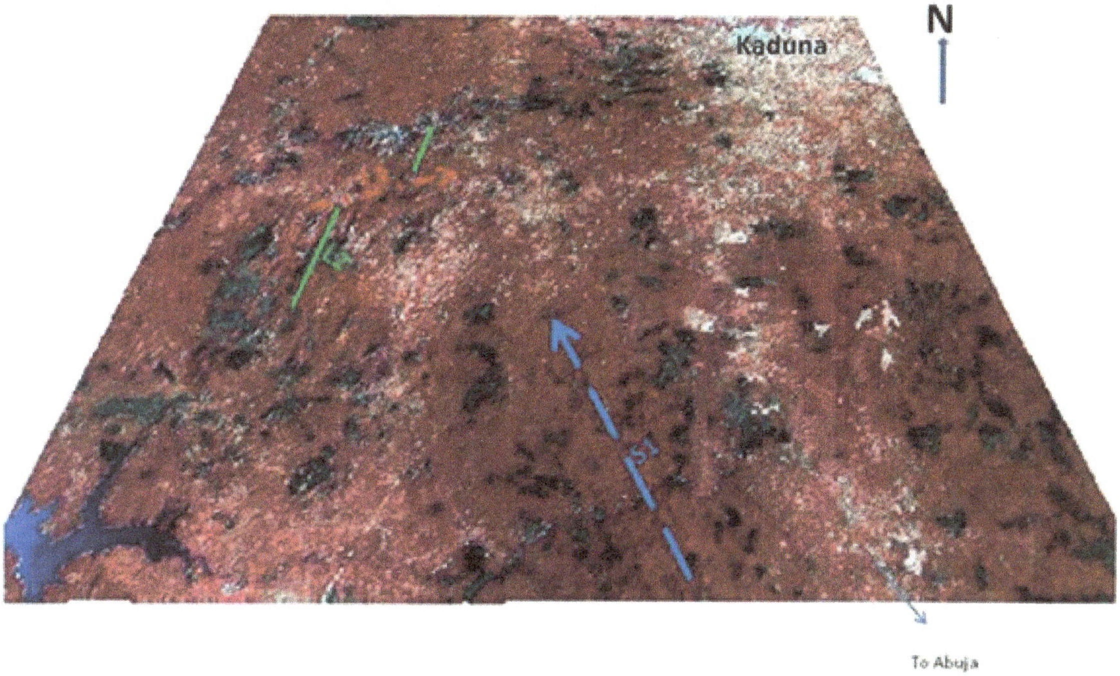

Figure 6. 3D view of the area showing the three major trends and episodes. S1 is gneissose foliation trends on ridges of gneisses and quartzites, L1 is fracture trends and L3 is fault systems.

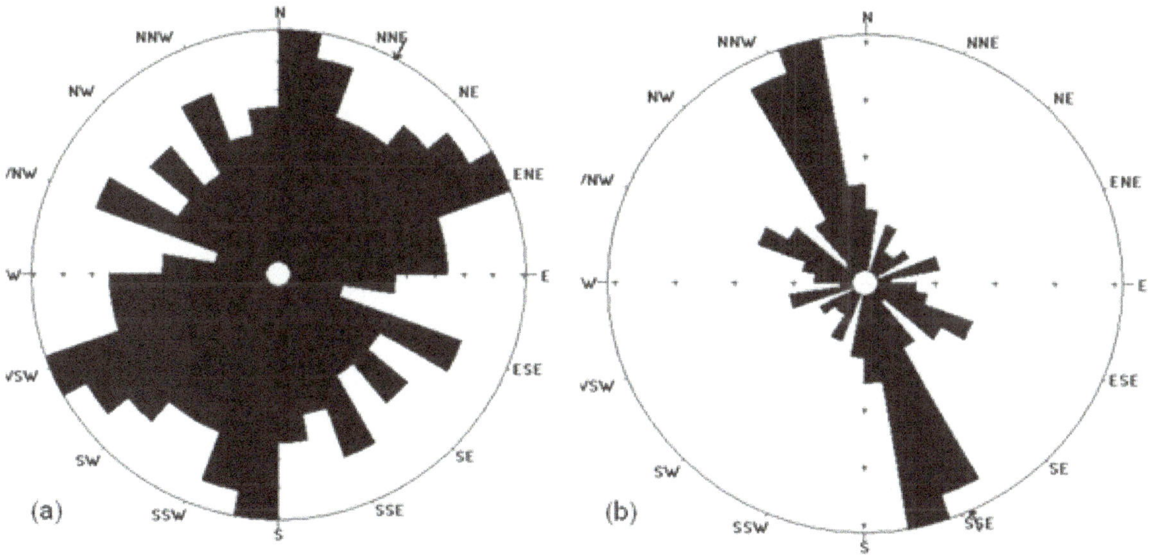

Figure 7. A comparison of the Frequency azimuth rosette diagrams for fracture trends of Kakuri sheet 144 (a) generated from Landsat5 TM (n=111) and (b) field measurements scattered over the same area (n = 92).

non-deformed granites, to ascertain the homogeneity of the original bodies. It however gave useful orientation data for comparison.

Summary of findings

A good understanding of the structural patterns in the Nigerian Basement Complex is critical to the understanding of the tectonic evolution of this part of the Pan-African terrain. The problem associated with mapping these fractures include poor exposure and masking by surface weathering processes. This problem is further accentuated by a rather unclear chronology of these patterns. Though regional fracture patterns have

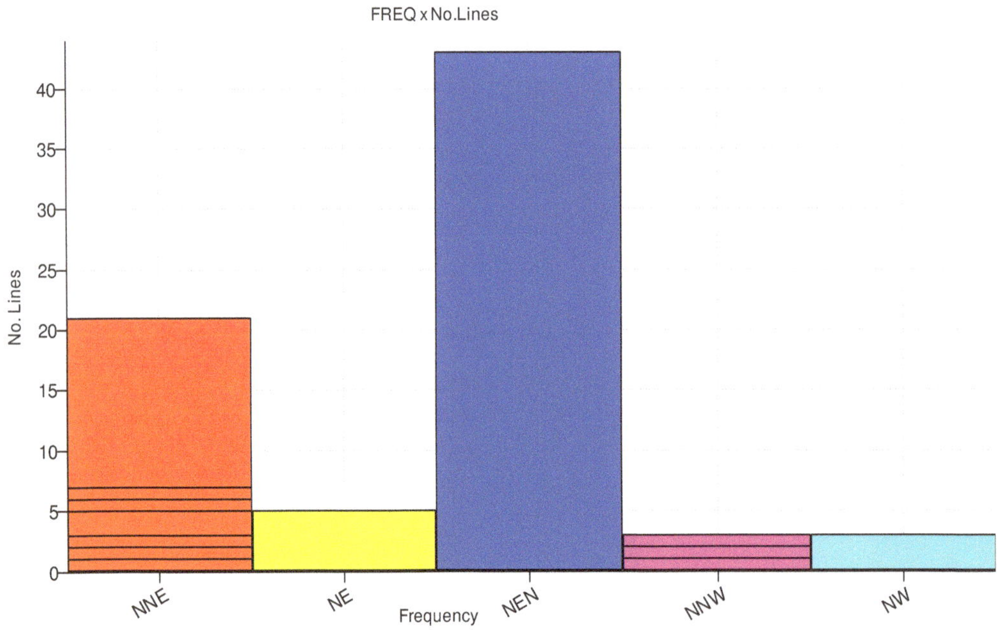

Figure 8. Statistical histogram showing the variation of the trends.

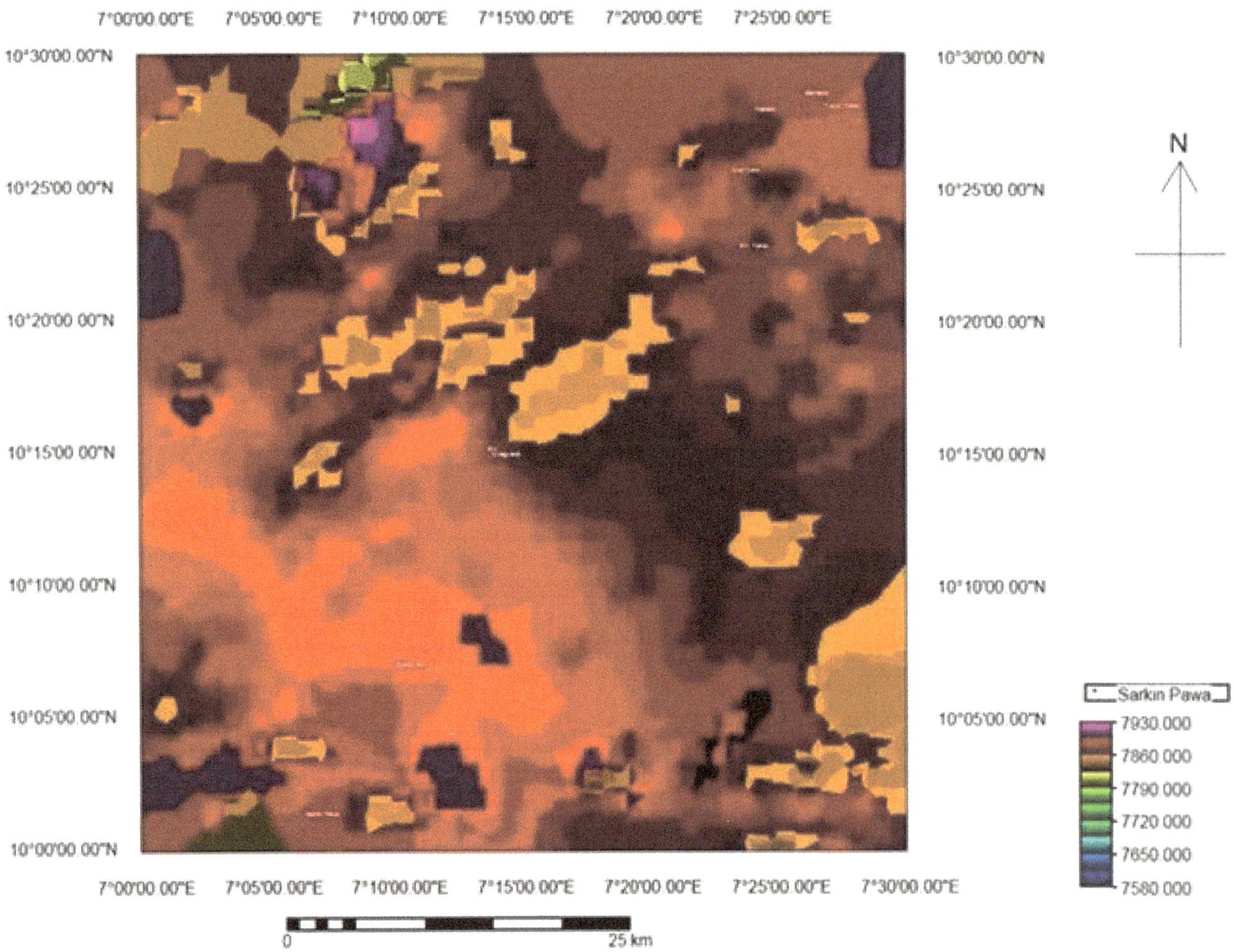

Figure 9. Digitized aeromagnetic map of Kakuri sheet 144.

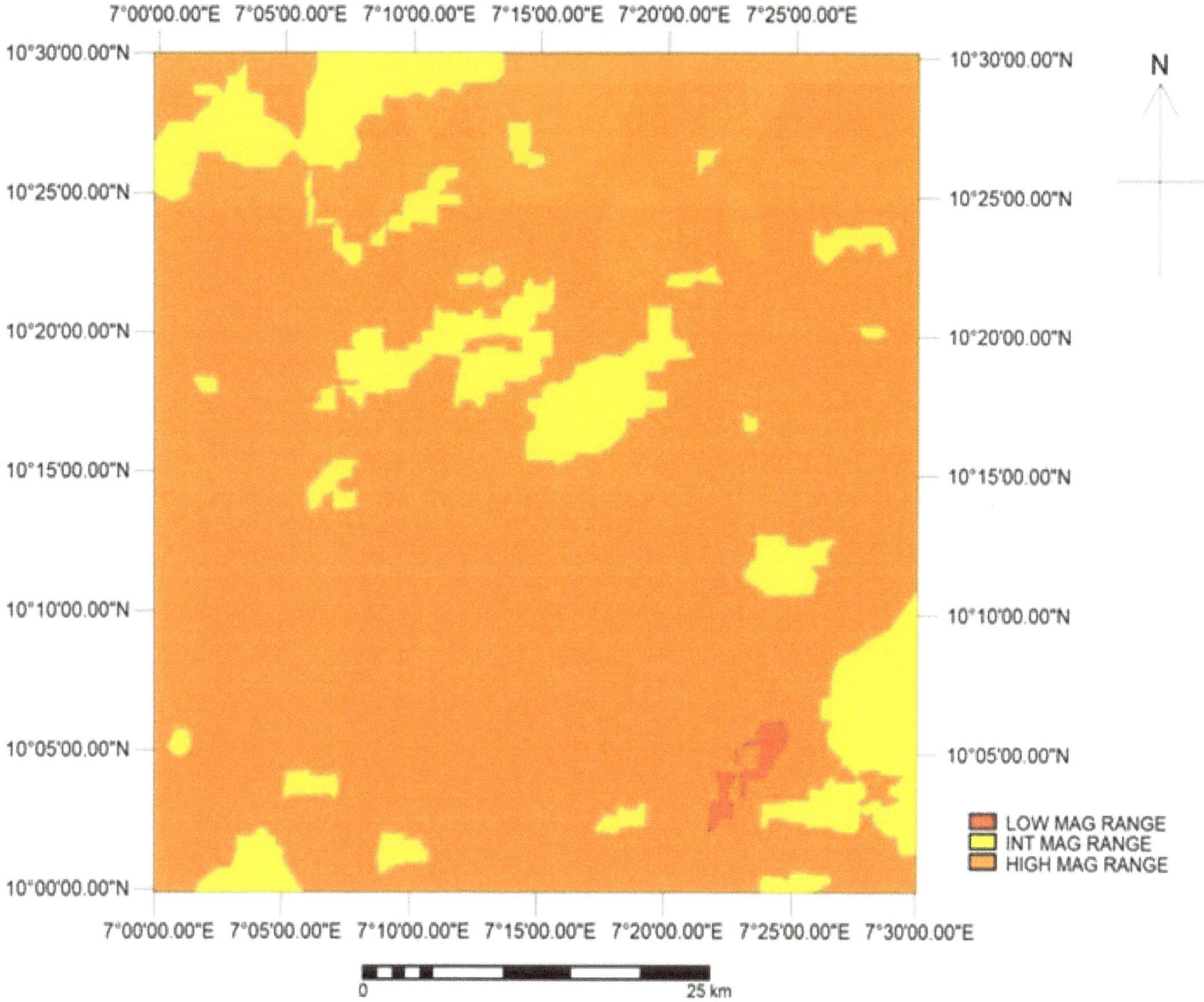

Figure 10. Reclassified colour map of magnetic range.

been amply documented (Ball, 1980; Ajakaiye, 1986; Annor et al., 1990, Annor and Freeth, 1985, Olasehinde et al., 1990, Olasehinde, 1989), these patterns have not been closely compared with micro tectonic patterns. Aeromagnetic lineaments are normally deep seated and often, non – fracture lineaments such as metamorphic trends are not taken into account. Additionally, field geological mapping of structures are extrapolated. These observations have informed the focus of this research.

Two main trends have been captured by the Landsat5 TM study and include NW-SE longer lineaments and a second pair of NE-SW and NNE-SSW trends which appear statistically abundant but spatially discontinuous and shorter. These trends coincide well with deep aeromagnetic trends, which have been attributable to an extension of Atlantic fracture zone that extends into the Nigerian Basement Complex (Olasehinde et al., 1990). The north-westerly trends are however not apparent in the aeromagnetic trends, even though they are very pronounced in the landsat5 TM study. These in the authors' opinion may be supra-crustal signatures and not related to deep seated aeromagnetic trends of Ball (1980). The micro-structural trends confirm the near absence of the north-easterly trends in the gneisses. A close study of regional map (Figure 2) which represents about nine 1:100,000 sheets, show that, the same pattern is replicated in the metasedimentary trends which obviously predate the subsequent shearing with a dextral sense which possibly were exploited by the Pan-African magmatism. If we subscribe to the suggestion that deformation may not have been synchronous with magmatism (Grant, 1978; Rahaman et al., 1991), then we equally subscribe to the assumption that, these NW – SE pre- Pan African did not probably assume significant

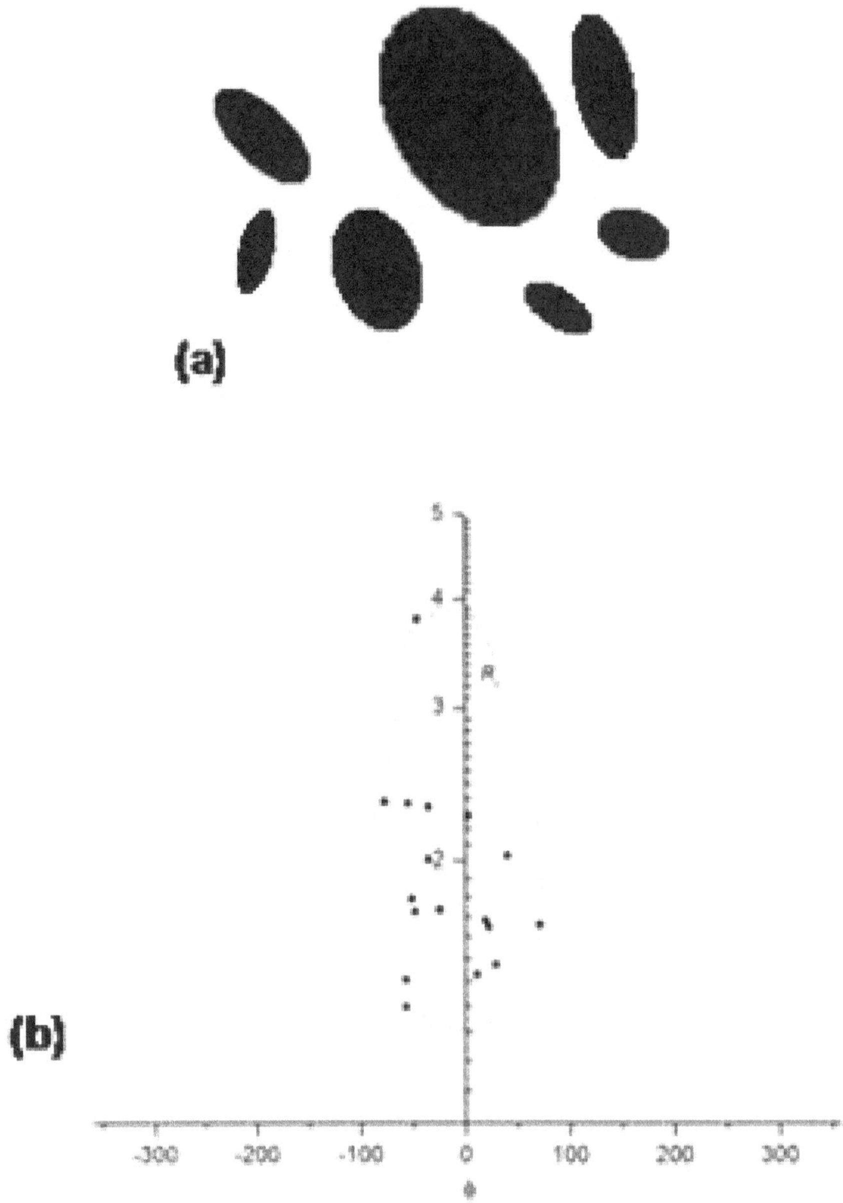

Figure 11. (a) Strain ellipsoids obtained from gneisses showing inconsistent orientations, (b) Rf / θ diagram of the axial ratios of ellipsoids in the same area.

proportions.

Regionally, it appears D1 deformation is typified by F1 surfaces oriented approximately 3500 to N-S both in the gneisses and shaft-like enclaves of the concordant schist with nearly vertical and gentle dips. Locally, intense shearing developed D2 penetrative surfaces which cut the N-S axial folds creating N-S, NW-SE crenulation surfaces.

Meso- and micro-structural data and significance

A similar scatter was also noticed in the strain ellipsoids which generally gave a NNW SSE trends with strain axial ratios approximately.

Conclusions

The following conclusions can be drawn about the tectonic framework:

1. An early, mainly NW – SE ductile vertically dipping foliation of transpressed tectono-metamorphic possibly pre-Pan African (S1).
2. A latter overprinting by a set of NE – SW (L2) fracture

Figure 12. XPL of the banded and strained gneiss cut normal to foliation planes.

Figure 13. Dark thick bands are thick shafts of biotite schist. Pen at the top central part of the picture is north oriented.

trend cut obliquely by an E-W (L3) sinistral shearing with a dextral sense.

3. Magmatic activity exploited these north-easterly fracture systems which are consistent with the regionally deep seated of north-easterly fault systems

REFERENCES

Ajakaiye DE, Hall DH, Millar TW, Verheijen PJT, Awad, MB, Ojo SB (1986). Aeromagnetic anomalies and tectonic trends in and around the Benue Trough, Nigeria. Nature, 319: 582-584.

Ajibade AC, Okoye NO, Wright JB (1981). Structural Relationship in the Schist Belts of North Western Nigeria. In P.O Oluyide et al (Eds).Precambrian geology of Nigeria. A publication of Geological Survey, pp. 103-109.

Ajibade AC (1987). Proterozoic Crustal Development in the Pan-African Regime of Nigeria. In: C.A. Kogbe (Editor) Geology of Nigeria Published by Rock View (Nigeria) Ltd., pp. 57-63.

Ananaba SE, Ajakaiye DE (1989). Evidence of tectonic control of mineralisation in Nigeria from lineament density analysis. Geology of Nigeria, Kogbe, C. A. (Ed), pp. 531-538.

Annor AE, Freeth SJ (1985). Thermotectonic evolution of the basement complex around Okene, Nigeria, with special reference to deformation mechanism. Precamb. Res., 28: 269-281.

Annor AE, Olasehinde PI, Pal PC (1990). Basement fracture patterns in the control of water channels – An example from central Nigeria. J. Min. Geol., 26(1): 5-12.

Ball E (1980). An example of very consistent brittle deformation over a wide intra-continental area, the late Pan African fracture system of the Taureg and Nigerian shield. Tectonophysics, 61: 363-379.

Dada SS (1989). Geochemistry and petrogenesis of the reworked Archaean gneiss complex of North Central Nigeria: Major and trace element studies on Kaduna amphibolites and migmatitic gneisses. Global J. Pure Appl. Sci., 5: 543.

Garba I (2002). Late Pan-African tectonics and origin of gold mineralisation and rare-metal pegmatites in the Kushaka schist belt, Northwestern Nigeria. J. Min. Geol., 38(1): 1-12.

Olasehinde PI (1989). Elucidating fracture patterns of the Nigerian Basement Complex using electrical resistivity methods. Z. Angew. Geowiss, 8: 109-120.

Olasehinde PI, Pal PC, Annor AE (1990). Aeromagnetic anomalies and structural lineaments in the Nigerian Basemen Complex. J. Afr. Earth Sci., 11(3 and 4): 351-355.

Oyawoye M. (1972). The basement complex of Nigeria. In: T. F. J. Dessauvagie and A. J. Whitement (eds). Ibadan. Afr. Geol., pp. 66-102.

Rahaman MA (1988). Recent advances in the study of the basement complex of Nigeria. Precamb. Geology of Nigeria, Geol. Surv. Nig. Publ., pp. 11-43.

Rahaman MA, Ajayi TR, Oshin IO, Asubiojo FOI (1988). Trace element geochemistry and geotectonic setting of Ile-Ife schist belt. In Precamb. Geol. Nig. Geol. Surv. Nig. Publ., pp. 241-256.

Wright JL (1978). The Geology of Parts of Niger and Zaria Province Northern Nigeria. Bulletin. No. 29. Published by the Geological Survey of Nigeria Proterozoic crustal development in the Pan-African Regime of Nigeria. In: C. A. Kogbe (Editor) Geology of Nigeria. Published by Rock View (Nigeria) Ltd,, pp. 57–63.

Wright JL, Ajibade AC, Mc-Curry P (1987). The Geology of the Precambrian to Lower Paleozoic Rocks of Northern Nigeria. A Review, in Geology of Nigeria, Edited by C.A. Kogbe. Published by Elizabethan Co. Lagos, pp. 15-39.

11

Strain energy distribution in an auxetic plate with a crack

S. K. Bhullar[1]*, J. L. Wegner[1] and A. Mioduchowski[2]

[1]Department of Mechanical Engineering, University of Victoria, Victoria, B.C. Canada V8W 3P6, Canada.
[2]Department of Mechanical Engineering, University of Alberta, Edmonton, AB, Canada, T6G 2G8, Canada.

Compared with non-auxetic materials, auxetic materials have special and desirable mechanical properties. For example, if the material has a crack, when it is being pulled apart, it expands and closes up the crack. In other words, this type of material has more crack resistance to fracture. Also, it has high material resistance to shear strain. Shear resistance is particularly important in structural components such as sheets or beams in buildings, cars and aircraft. In the present paper strain energy density, temperature variation, thermal stresses and total stresses in an infinite plate of auxetic material with a line crack, subjected to the mechanical and thermal loading in the context of thermo elastic theory are studied.

Key words: Auxetic behavior, thermal stresses, strain energy function, temperature variation.

INTRODUCTION

Materials with negative Poisson's ratio characterized as auxetic materials have been of significant scientific interest and also have considerable practical applications. A number of potential applications of the auxetic behavior for the design of innovative materials and structural elements or for new processing techniques have been outlined in the engineering literature. The importance of using cork, which is a material with nearly zero Poisson's ratio, for sealing wine bottles, is well known. Another class of applications is based on the sound-absorbing properties of auxetic materials, which make them interesting for both civil and military applications. Furthermore, several authors speculate auxetic behaviour in biomechanics (for example, for the spongy part of the bones), with obvious implications for the efficient design of prostheses. Recently, European patents have proposed annuplastry prostheses that provide plastic repair of a cardiac valve.

LITERATURE REVIEW

A number of potential applications of the auxetic behaviour for the design of innovative materials and structural elements or new processing techniques have been outlined in the engineering literature. Lam et al. (1992) studied the perturbation effects caused by the presence of a crack on thermal stresses, displacements and stress intensity factors in an isotropic linear elastic medium with varying crack surface heat conductivity under uniform heat flow. Minguez (1993) analyzed elastic behavior and some fracture mechanics concepts, such as the geometry factor and the fracture toughness of an infinite plate and a finite plate with a central crack. Shindo et al. (1998) analyzed the scattering of time harmonic flexural waves by a thorough crack in a symmetric piezo-electric laminated plate subjected to electric field loading in the context of dynamic theory of linear piezoelectricity.

The transient thermal conduction problem of a finite plate with multiple insulated cracks and a dynamic problem for two equal rectangular cracks in an infinite elastic plate are studied by Chang and Ma (2001) and Itou (2002) respectively. Coker et al. (2003) presented experimental and numerical results for a dynamic crack growth along the interface of a fiber-reinforced polymer composite-Homalite bimaterial subjected to impact shear loading. A periodic group crack problems in an infinite plate is studied by Chen and Lin, (2005) while Yetmet and Gecit (2005) investigated normal and shearing stress distributions and the stress intensity factors at the edges

*Corresponding author. E-mail: sbhullar@uvic.ca.

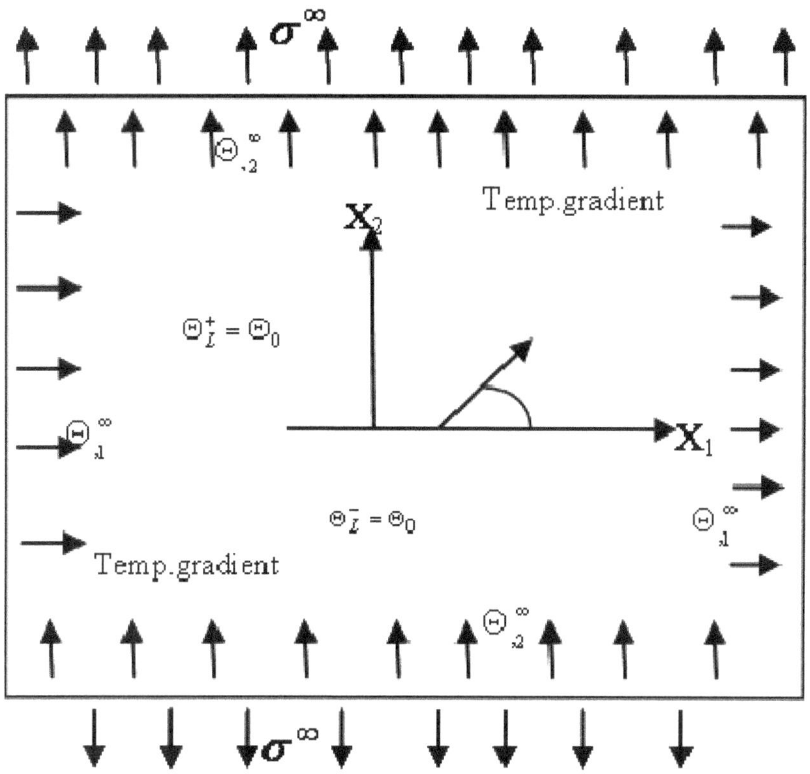

Figure 1. The geometry of the problem.

of the crack and at the corners of the finite strip. Ayatollahi and Aliniaziazi (2006) investigated the effect of the load applied parallel to the crack (or the lateral load) in prestressed square plate containing a center crack. Wu (2006) evaluated stress intensity factor distribution for a surface crack under Mode-I loading conditions. An exact closed-form stress intensity factor solutions have been developed through-thickness cracks in an infinite plate by Lee-Tan et al. (2007) and Feng and Su (2007) obtained solution for the dynamic anti-plane problem for a functionally graded magneto-electro-elastic plate containing an internal or an edge crack parallel to the graded direction. De Matos and Nowell (2008) presented three-dimensional finite element analyses of a plate with a central crack to investigate the effect of corner point singularities. Bogdan (2009) studied a problem of a thin plate made of a piezoelectric ceramic material containing a crack perpendicular to its surfaces. Also, Guo et al. (2009) obtained an analytical solution for an eccentric crack loaded by shear forces in a finite width plate using crack line analysis method.

There are two basic ways in which a volume element can store energy, by dialation which is associated with change in volume and distortion which is associated with the change in shape. When an element of the material exceeds a certain energy threshold fracture could occur. Auxetic materials possess more crack resistance to fracture. Thus, if this type of material has a crack, when it is being pulled apart, it expands and closes up the crack. In this paper strain energy density, temperature variation, thermal stresses and total stresses in an infinite plate of auxetic material with a line crack, subjected to the mechanical and thermal loading in the context of thermo elastic theory is studied.

MODEL AND METHODS

An analytical solution is obtained under the boundary condition that the constant temperature is retained on the crack surfaces while the remote uniform heat flow is applied. For this purpose an infinite plate with a central crack of length $2a$, is considered and a uniform heat flow and the uniform tensile stress σ^{∞} is applied at infinity. There is an angle between the directions of heat flow and crack line. The heat flow vector can be resolved into two parts, respectively, along two co-ordinate axis. Two temperature gradient components along two co-ordinate axis at infinity are denoted by $\Theta_{,1}^{\infty}$ and $\Theta_{,2}^{\infty}$ and the heat flow is proportional to the negative temperature gradient. The geometry of the problem is shown in Figure 1.

In this case thermal loading only induces the model stress intensity factor. The heat flow along the vertical direction has no influence on the thermal stress intensity factor. The problem is typical model crack problem. There is no heat source in the plate, thus governing differential equation in this case is:

$$\nabla^2\Theta = 0 \quad (1)$$

Assume that the temperature on the upper and lower surfaces of the crack remains constant

$$\Theta_L^+ = \Theta_L^- = \Theta_0, \text{ on L.} \quad (2)$$

The symbol "L" denotes the crack and "+" and "-" stand for, respectively, the upper and lower crack surfaces. Θ_0 is the constant temperature on the upper and lower crack surfaces. Further, the full field temperature solution is given by Tang (2009):

$$\Theta = \frac{1}{2}\Theta_{,1}^{\infty}\left(\sqrt{z^2-a^2}+\sqrt{\bar{z}^2-a^2}\right)-\frac{i}{2}\Theta_{,2}^{\infty}(z-\bar{z})+\Theta_0 \quad (3)$$

and the asymptotic solution of the temperature near the crack tip is expressed as

$$\Theta = \frac{1}{2}\Theta_{,1}^{\infty}\sqrt{2ar_1}\cos\frac{\theta}{2}+\Theta_{,2}^{\infty}r_1\sin\theta+\Theta_0 \quad (4)$$

Where, R is the distance from the crack tip, z and $\bar{z}$ are complex conjugate.

Also, the asymptotic solution of the thermal stresses and the expression of the stress intensity factor are:

$$\sigma_{11}^{\Theta} = \frac{K_1^{\Theta}}{\sqrt{2\pi r_1}}\cos\frac{\theta}{2}\left(1-\sin\frac{\theta}{2}Sin\frac{3\theta}{2}\right) \quad (5)$$

$$\sigma_{22}^{\Theta} = \frac{K_1^{\Theta}}{\sqrt{2\pi r_1}}\cos\frac{\theta}{2}\left(1+\sin\frac{\theta}{2}\sin\frac{3\theta}{2}\right) \quad (6)$$

$$\sigma_{12}^{\Theta} = \frac{K_1^{\Theta}}{\sqrt{2\pi r_1}}\sin\frac{\theta}{2}\cos\frac{\theta}{2}\cos\frac{3\theta}{2} \quad (7)$$

$$K_1^{\Theta} = \frac{\alpha E a\Theta_{,1}^{\infty}}{\kappa+1}\sqrt{\pi a}, \quad \kappa = 3-4\nu, \quad (8)$$

$$K_2^{\Theta} = 0 \quad (9)$$

Where, α, E, ν, K_1^{Θ} and K_2^{Θ} are linear expansion coefficient, Young's modulus, Poisson's ratio and stress intensity factor, respectively. The superscript Θ denotes the thermal effect. As this problem is a typical crack problem of mode I, the stress intensity factor is only related to the heat flow along x_1- direction. The heat flow along x_2- direction does not produce thermal stresses. Further under the combination of thermal and mechanical loading, the asymptotic solution of total stress field in the vicinity of crack tip is:

$$\sigma_{11} = \frac{K_1}{\sqrt{2\pi r_1}}\cos\frac{\theta}{2}\left(1-\sin\frac{\theta}{2}\sin\frac{3\theta}{2}\right) \quad (10)$$

$$\sigma_{22} = \frac{K_1}{\sqrt{2\pi r_1}}\cos\frac{\theta}{2}\left(1+\sin\frac{\theta}{2}Sin\frac{3\theta}{2}\right) \quad (11)$$

$$\sigma_{12} = \frac{K_1}{\sqrt{2\pi r_1}}\sin\frac{\theta}{2}\cos\frac{\theta}{2}Cos\frac{3\theta}{2} \quad (12)$$

$$\sigma_{33} = \nu(\sigma_{11}+\sigma_{22})-\alpha E\Theta, \quad \sigma_{13}=\sigma_{23}=0 \quad (13)$$

Where, K_1 total stress intensity factor is the sum of the mechanical and thermal parts, r_1 and θ are the polar coordinates measured from the crack tip as shown in Figure 1. The term $\alpha E\theta$ in equation (13) is nonsingular because the temperature field is nonsingular at the crack tip in view of equations (3)-(4). It can be omitted.

$$K_1 = K_1^{\sigma}+K_1^{\Theta}, \quad K_1^{\sigma} = \sigma^{\infty}+\sqrt{\pi a}, \quad K_1^{\Theta} = \frac{\alpha E a\Theta_{,1}^{\infty}}{\kappa+1}\sqrt{\pi a}$$

$$K = \left(\sigma^{\infty}+\frac{\alpha E a\Theta_{,1}^{\infty}}{\kappa+1}\right)\sqrt{\pi a}. \quad (14)$$

Equation (14) indicate that the temperature gradient $\Theta_{,1}^{\infty}$ can be regarded as an equivalently applied stress at infinity denoted by σ_{Θ}^{∞} which is written as

$$\sigma_{\Theta}^{\infty} = \frac{\alpha E a\Theta_{,1}^{\infty}}{\kappa+1} \quad (15)$$

Therefore, the total applied stress at infinity is equal to

$$\sigma = \sigma^{\infty}+\sigma_{\Theta}^{\infty} \quad (16)$$

Where, σ^{∞} is the mechanical part and σ_{Θ}^{∞} is the thermal part. Bodies having large volume to surface ratio dissipate more energy by dilatation in contrast to distortion. The dominance of dilatation tends to exhibit brittle behavior. As the volume to surface ratio is decreased, energy dissipation via distortion would overtake that of dilatation and a unit volume of the same material become more ductile. The ductile-brittle transition depends on the material microstructure, specimen shape and size, temperature and rate of deformation.

Regardless of nonlinearity, the proportion of energy

Table 1. Material properties.

Material parameter	Copper foam	Steel alloy
E, Young's modulus	30 (GPa)	200GPa
Linear thermal expansion α	0.55×10-6 (1/0C)	12.94 × 10-6 (1/0C)
Poisson's ratio	-0.7	0.33

expanded by dilatation and distortion can be assessed from the stationary values of the energy density function. The strain energy density criterion can be applied to predict the crack initiation and extension in isotropic materials. Making use of the stress and strain components σ_{ij} and ε_{ij} the strain energy density given by Carloni and Nobile, 2008:

$$\frac{dW}{dV}=\int_0^{\varepsilon_{ij}}\sigma_{ij}d\varepsilon_{ij}$$

Further, for linear elasticity the above equation is expressed as

$$\frac{dW}{dV}=\frac{1}{2}\sigma_{ij}\varepsilon_{ij} \tag{17}$$

Constitutive relation of linear thermoelasticity is

$$\varepsilon_{ij}=\frac{1+\nu}{E}\sigma_{ij}-\frac{\nu}{E}\sigma_{kk}\delta_{ij}+\alpha\Theta\delta_{ij}$$
(18)

Where, Θ stands for the temperature to the reference temperature. The expressions obtained (see Appendix I) for strain energy density, dialation part and distortional part of strain energy density are as follows:

$$\frac{dW}{dV}=\frac{(1+\nu)(1+\cos\theta)(3-4\nu-\cos\theta)}{8\pi Er_1}K_1^2 \tag{19}$$

$$\left(\frac{dW}{dV}\right)_v=\frac{(1-2\nu)(1+\nu)^2(1+\cos\theta)}{6\pi Er_1}K_1^2 \tag{20}$$

$$\left(\frac{dW}{dV}\right)_d=\frac{(1+\nu)(1+\cos\theta)\left[2(1-2\nu)^2+3(1-\cos\theta)\right]}{24\pi Er_1}K_1^2 \tag{21}$$

In the linear elastic fracture mechanics theory, the strain energy density factor S is defined as $\frac{dW}{dV}=\frac{S}{r_1}$

And the strain energy density factor S is in form of (Appendix I)

$$S=A_{11}(\sigma^\infty)^2+2A_{12}\sigma^\infty\Theta_{,1}^\infty+A_{22}(\Theta_{,1}^\infty)^2 \tag{22}$$

$$A_{11}=\frac{(1+\nu)(1+\cos\theta)(3-4\nu-\cos\theta)a}{8E} \tag{23}$$

$$A_{12}=\frac{(1+\nu)(1+\cos\theta)(3-4\nu-\cos\theta)\alpha a^2}{32(1-\nu)} \tag{24}$$

$$A_{22}=\frac{(1+\nu)(1+\cos\theta)(3-4\nu-\cos\theta)\alpha^2Ea^3}{32(1-\nu)} \tag{25}$$

RESULTS AND CONCLUDING REMARKS

The problem of an infinite plate with a central crack subjected to the remote uniform heat flow along any direction has been solved by the complex function method. To study the auxetic behavior of an infinite plate with a central crack of length 2a the results for temperature variation, stress field, strain energy distribution-the dilatational and distortional part of strain energy function and strain energy factor are obtained and are shown graphically. The copper foam and steel alloy are the material chosen to perform the numerical calculations and the material parameters are given in Table 1. The temperature gradient components along two co-ordinate axis at infinite are taken as $\Theta_{,2}^\infty$ =100K/m= $\Theta_{,2}^\infty$, constant temperature $\Theta_o=100K$, $a=10mm$ and the uniform tensile stress applied at infinite, σ^∞ =10MPa . The temperature variation versus angle θ (radians) in the crack region due to non-dimensional distance, $r=\frac{r_1}{a}$, is illustrated in Figure 2. By the combination of mechanical and thermal loading the total stress field in the neighborhood of crack tip is shown in Figures 3 to 6 and symmetrical behavior is observed. It is noticed that for the stress component σ_{33} in Figure 6, the pattern of curves is the same as in the case of temperature distribution but in opposite direction.

The contours of strain energy density, distortional part and dilatational part of strain energy density are shown in

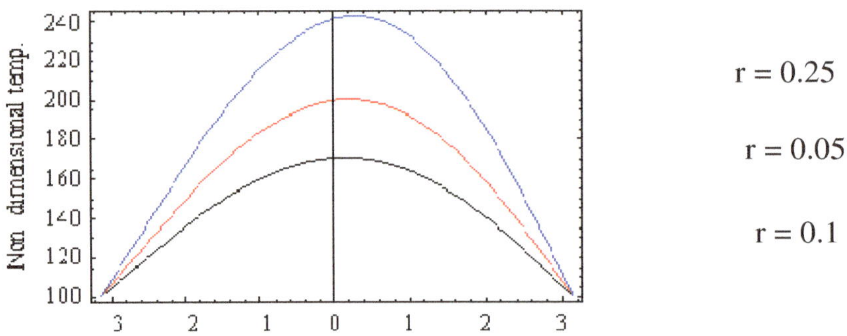

Figure 2. Temperature variation near the crack due to normalized distance.

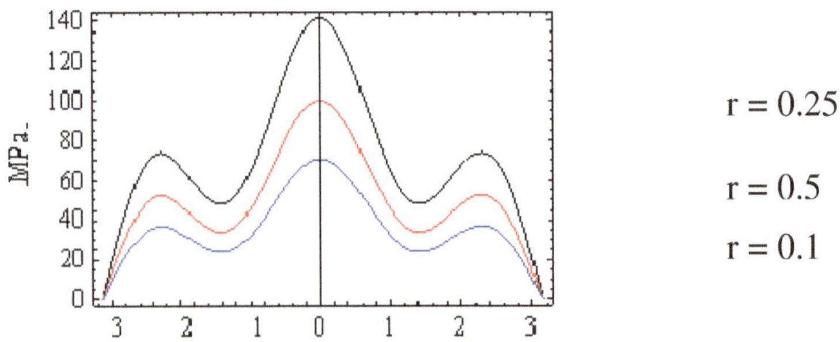

Figure 3. Stress component σ_{11}, near crack tip.

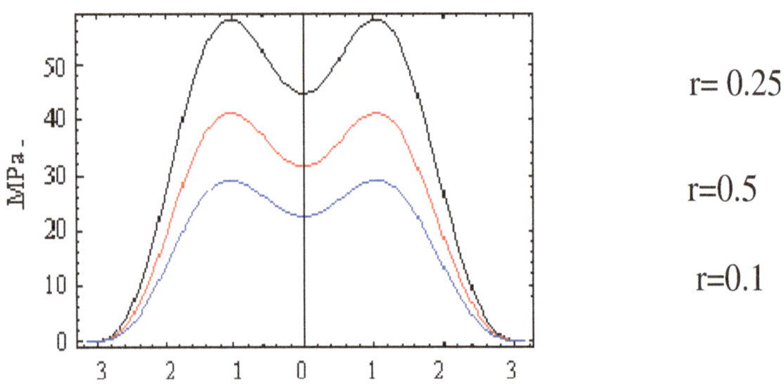

Figure 4. Stress component σ_{22}, near crack tip.

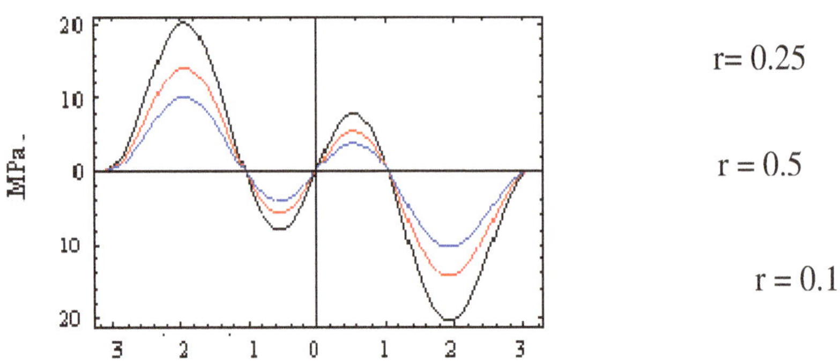

Figure 5. Stress component σ_{12}, near crack tip.

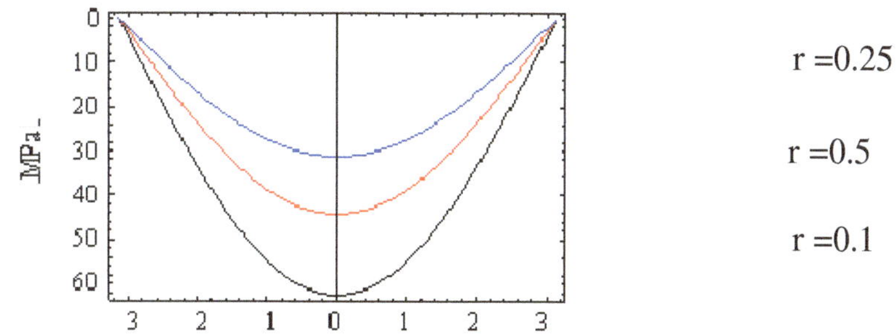

Figure 6. Stress component σ_{33}, near crack tip.

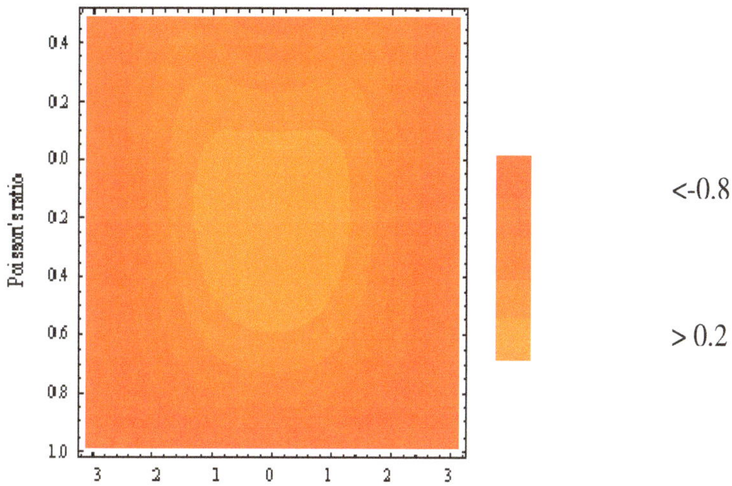

Figure 7a. Strain energy density crack tip.

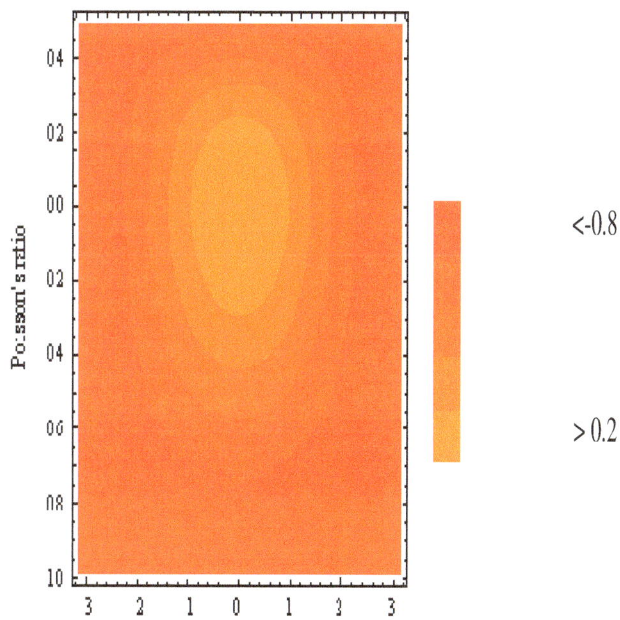

Figure 7b. Strain energy density (dilatational) near crack tip.

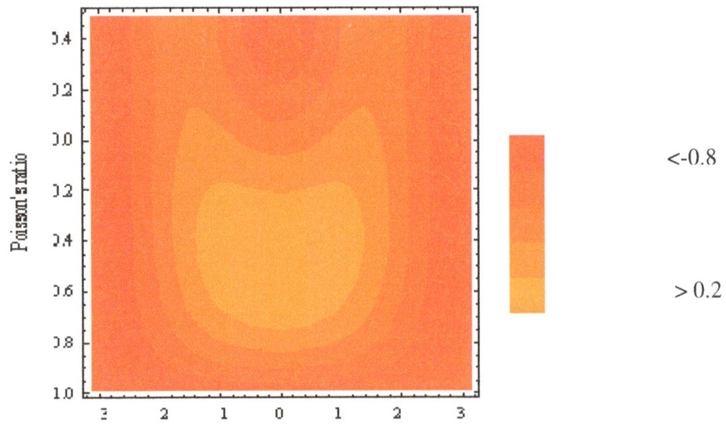

Figure 7c. Strain energy density (distortional) near crack tip.

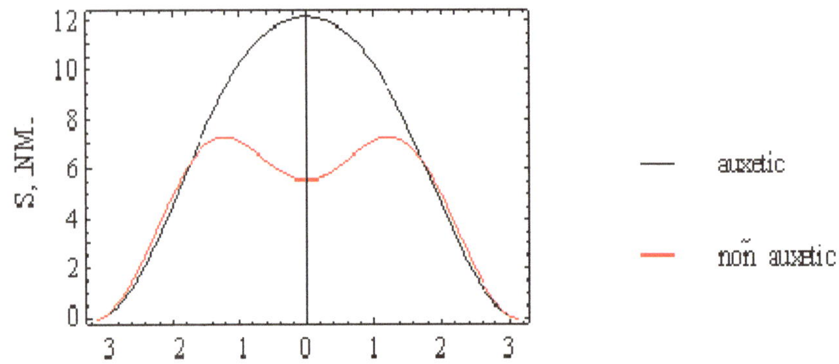

Figure 8. Comparison of strain energy density factor, S.

Figure 7a-c. It is clear, the contours of $\left(\frac{dW}{dV}\right)$, $\left(\frac{dW}{dV}\right)_d$ and $\left(\frac{dW}{dV}\right)_v$ are not affected by the thermal effect.

The comparison of strain energy factors *S* for non-auxetic and auxetic material is shown in Figure 8. It is observed that maximum value of *S* occurs at angle, $\theta = 0$ in the case of auxetic material and the local minimum value of *S* in the case of non-auxetic material. It is clear when compared with non-auxetic materials that auxetic materials have almost twice crack resistance to fracture.

REFERENCE

Ayatollahi MR, Aliniaziazi A (2006). Effects of lateral load on warm pre-stressing in a center crack plate, J. Mater. Sci. Eng., 441: 170-175.
Bogdan R (2009). The transient thermo-electro-elastic fields in a piezoelectric plate with a crack, Int. J. Pressure Vessels Piping 86: 384-394.
Carloni C, Nobile L (2008). Crack initiation behavior of orthotropic solids as predicted by the strain energy density theory, Theoret. Appl. Fract. Mech. 38: 109-119.
Chang C-Yi, Ma C-C (2001). Transient thermal conduction analysis of a rectangular plate with multiple insulated cracks by the alternating method, Int. J. Heat Mass Transfer, 44: 2423-2437.
Chen YZ, Lin XY (2005). Periodic group crack problems in an infinite plate, Int. J. Solids Struct., 42: 2837-2850.
Coker D, Rosakis AJ, Needleman A (2003). Dynamic crack growth along a polymer composite–Homalite interface, J. Mech. Phys. Solids, 51: 425-460.
De Matos PFP, Nowell D (2008). The influence of the Poisson's ratio and corner point singularities in three-dimensional plasticity-induced fatigue crack closure a numerical study, Int. J. Fatigue, 30: 1930-1943.
Feng W, Su R (2007). Dynamic fracture behaviors of cracks in a functionally graded magneto-electro-elastic plate, Eur. J. Mech. - A/Solids, 26: 363-379.
Guo C, Zhou XP, Zhang YX, Yang HQ, Li XH (2009). Elastic–plastic near field solution of an eccentric crack under shear in a finite width plate, Theoret. Appl. Fract. Mech., 51: 174-180.
Itou S (2002). Dynamic stress intensity factors around two rectangular cracks in an infinite elastic plate under impact load, Mech. Res. Commun., 29: 225-234.
Lam KY, Tay TE, Yuan WG (1992). Stress intensity factors of cracks in finite plates subjected to thermal loads, J. Eng. Fract. Mech., 43: 641-650.
Lee-Tan JM, Fitzpatrick ME, Edwards L (2007). Stress intensity factors

for through-thickness cracks in a wide plate: Derivation and application to arbitrary weld residual stress fields, J. Eng. Fract. Mech., 74: 2030-2054.
Minguez JM (1993). Study of elastic behaviour of plates containing cracks by finite element analysis, J. Comp. Struct., 47: 917-925.
Yetmet M, Gecit MR (2005). Finite strip with a central crack under tension, Int. J. Eng. Sci., 43: 472-493.
Shindo Y, Domon W, Narita F (1998). Dynamic bending of a symmetric piezoelectric laminated plate with a through crack, Theoret. Appl. Fract. Mech., 28: 175-182.
Tang XS (2009). Mechanical thermal stress intensification for mode I crack tip, Theoret. Appl. Fract. Mech., 50: 92-104.
Wu Z (2006). The shape of a surface crack in a plate based on a given stress intensity factor distribution, Int. J. Pressure Vessels Piping, 83: 168-180.

Appendix 1

By making use of equations (17) - (18) we have obtained following equation:

$$\frac{dW}{dV}=\frac{1+\nu}{2E}\left(\sigma_{ij}\sigma_{ij}-\frac{\nu}{1+\nu}\sigma_{kk}^2\right)$$
$$=\frac{1+\nu}{2E}\left[\sigma_{11}^2+\sigma_{22}^2+\sigma_{33}^2-\frac{\nu}{1+\nu}(\sigma_{11}+\sigma_{22}+\sigma_{33})^2+2\sigma_{12}^2\right] \quad (1)$$

Upon using equation (13), above equation takes the form for plane strain

$$\frac{dW}{dV}=\frac{1+\nu}{2E}\left[\sigma_{11}^2+\sigma_{22}^2-\nu(\sigma_{11}+\sigma_{22})^2+2\sigma_{12}^2\right] \quad (2)$$

Equation (1) can also written as

$$\frac{dW}{dV}=\frac{3}{2}\sigma_m\varepsilon_m-\frac{1}{2}\alpha\Theta\sigma_{kk}+\frac{1}{2}s_{ij}e_{ij}\left[\sigma_{11}^2+\sigma_{22}^2-\nu(\sigma_{11}+\sigma_{22})^2+2\sigma_{12}^2\right] \quad (3)$$

Where, $\sigma_m=\frac{\sigma_{kk}}{3}$ and $\varepsilon_m=\frac{\varepsilon_{kk}}{3}$ are hydrostatic stress and strain; s_{ij} and ε_{ij} are deviatoric stress and strain. The dilatational part and distortional part of strain energy density are:

$$\left(\frac{dW}{dV}\right)_v=\frac{3}{2}\sigma_m\varepsilon_m-\frac{1}{2}\alpha\Theta\sigma_{kk} \quad (4)$$

$$\left(\frac{dW}{dV}\right)_{dv}=\frac{1}{2}s_{ij}e_{ij} \quad (5)$$

By application of equation (18) in equations (4)-(5) we get:

$$\left(\frac{dW}{dV}\right)_v=\frac{1-2\nu}{6E}(\sigma_{11}+\sigma_{22}+\sigma_{33})^2=\frac{(1-2\nu)(1+\nu)^2}{6E}(\sigma_{11}+\sigma_{22})^2 \quad (6)$$

$$\left(\frac{dW}{dV}\right)_d=\frac{1+\nu}{6E}\left[(\sigma_{11}-\sigma_{22})^2+(\sigma_{22}-\sigma_{33})^2+(\sigma_{33}-\sigma_{11})^2+6\sigma_{12}^2\right]$$
$$=\frac{1+\nu}{3E}\left[\sigma_{11}^2+\sigma_{22}^2-\sigma_{11}\sigma_{22}-\nu(1-\nu)(\sigma_{11}-\sigma_{22})^2+3\sigma_{12}^2\right] \quad (7)$$

12

Road and urban storm water drainage network integration in Addis Ababa: Addis Ketema Sub-city

Dagnachew Adugna Belete

Addis Ababa, Ethiopia. E-mail: dagnachew2@gmail.com

Urbanization along with its impermeable structures is the major causes of flooding in urban areas. Urban storm water influences the service life of urban infrastructures. The rainfall intensity and characteristics of catchment area are the major factors for designing urban storm water drainage facilities. These facilities have a paramount advantage to safely dispose the generated floods to ultimate receiving system. This study has assessed the integration of road and urban storm water drainage infrastructure with the help of topographic map and also the condition, pavement type and hierarchy of every road and drain were assessed in Addis Ketema Sub-city. This study area, particularly, is bounded in between Addis Ababa Municipality (East), Addis 'Ketema' high school and General bus terminal (West), 'Yohannes' Church (North-East) and Bethel high school (North-west). The objectives of this study includes: to identify sites most prone to flooding problems, to assess the existing condition of road and urban storm water drainage infrastructure, to identify the extent of integration of urban storm water drainage infrastructure in road projects provision, to examine the impacts of Urban storm water drainage infrastructure integration on road performance and related environment issues and to make recommendations on road and Urban storm water drainage infrastructure integration and their provision and management. An exploratory and descriptive type of methods were used to describe and investigate the existing condition and coverage and level of integration between road and Urban storm water drainage infrastructure infrastructures respectively. Data collection methods were carried out using both primary and secondary data sources, but the secondary data source was only relevant to reinforce the primary data, which was accomplished with the help of topographic map and a check list. The collected data were analyzed and presented using Microsoft-excel, AutoCAD and ArcGIS and tables, graphs and percentages respectively. The findings of this study includes: the major causes of flooding which was found to be the blockage of urban storm water drainage lines along with inadequate/poor integration between road and urban storm water drainage infrastructures. This study strongly recommends improvement in the integration of road and urban storm water drainage infrastructure and integrated solid waste management to prevent over flowing of flood as a result of blockage of drains.

Key words: Drainage, irrigation, urban areas, flooding.

INTRODUCTION

Infrastructure is one of the indispensable elements in the process of urbanization and emergence and continuity of an urban growth. It is considered as motor/engine for economic development (World Bank, 2006). Infrastructure is important in eradicating poverty through various job creation opportunities and by so doing, it enables to speed up economic development and ultimately ensures improved quality of life.

According to World Development Report (2006), infrastructure development in Africa is abysmal, lagging behind the rest of the world in terms of quality, quantity, and access.

The situation is worse in poor countries like Ethiopia, where a myriad of problems have made the supply of physical infrastructure and services to continually lag behind the urban population growth rate.

The high urban population growth rate on the other hand produces features associated with urban expansion. For example, the increase in population generates greater infrastructure demand, including road and urban storm water drainage facilities. When road and urban storm water drainage facilities are not properly provided nor integrated in a given urban area, the performance of road infrastructure can be limited. Problems like flooding and erosion when persistent can shorten the life span of road infrastructures and other urban utilities.

Other urban utilities like water supply lines, telephone and electric lines can also be affected since road infrastructure often influence the locational pattern.

With urbanization, impermeability increases because of the increase in impervious surfaces. This in turn changes the drainage pattern, increases overland flow resulting in flooding and related environmental problems. The impact of this is severe on spatial structures like road. This is because, flooding and its related environmental problems like sheet and gulley erosion, surface inundation tends to affect road services and its life span. Given the significance not only in socio-economic development, but also a path way for the location of other infrastructure, issues that affect its performance and longevity are critical areas of research.

Addis Ketema Sub-city is a major economic nexus of Addis Ababa City. It is also the oldest and most economically significant part of the city. The largest merchandize market in Africa, Merkato (AACA, 2006), is also located in this sub-city.

Unfortunately, street flooding, over topping and other environmental related problems are common in this area. This is particularly severe in areas where road infra-structure appears to be without adequate storm water drainage infrastructure.

Problem statement

Inadequate urban storm water drainage problems represent one of the most common sources of compliant from the citizens in many towns of Ethiopia (GTZ-IS, 2006), and this problem is getting worse and worse with the ongoing high rate of urbanization.

The pattern of urbanization and modernization in Ethiopia has meant increase densification along with urban infrastructure development. This has led to de-forestation, use of corrugated roofs and paved surfaces. The combined effect of this results in higher rain drop intensity and consequently accelerated and concentrated runoff.

Due to inadequate integration between road and urban storm water drainage infrastructure provision and poor management significant proportion of the area is exposed to flooding hazards/risks. This has resulted in negative impacts on urban storm water drainage provision and management.

The location of Addis Ababa being at the base of mountain "Entoto and Yeka" is also a challenge. These mountains are bare and predominantly covered by eucalyptus trees, which further exposes the ground for degradation and thus contributes to the generation of more flood to the city of Addis Ababa, because eucalyptus trees do not let shrubs and bushes and grasses to grow underneath due to its soil nutrient depletion nature and/or high competition for water and nutrients.

Erosion and flooding problems have become pronounced instead of hindering the phenomenon.

Therefore, this research analysis the extent of integration between road and urban storm water drainage net-work and associated problems.

Research objectives

General objective

To assess road and urban storm water drainage net-work integration and the challenge of their provision in Addis Ketema Sub-city.

Specific objectives

1. To identify sites/areas most prone to flooding hazard/problems.
2. To assess the existing condition of road and urban storm water drainage net-work infrastructure.
3. To identify the extent of integration of urban storm water drainage infrastructure in road projects provision.
4. To examine the impacts of urban storm water drainage integration on road performance and related environment issues like flooding.
5. To make recommendations on road and urban storm water drainage integration and their provision and management.

Significance of the research study

This study, generally, contributes the following major significances:

1) The sub-city will use it as reference while they are preparing their annual plans in relation to spatial and financial plans for roads and urban storm water drainage infrastructure.
2) It has a special relevance to use it as a reference for infrastructure inventory. So that the new added and upgraded infrastructure will easily be upgraded to a recent data report through appropriate software.
3) Policy makers and any organization working in the area of roads and urban storm water drainage infrastructure can use it as a further reference to fill the existing gap between road and urban storm water drainage demand and supply.
4) It will be an alternative means of solution in ensuring sustainable development in Addis Ababa by strengthening the environmental and socioeconomic

activities in regarding to road and urban storm water drainage infrastructures.

Scope of the study

This study is geographically limited to Addis Ketema Sub-city of Addis Ababa. Generally, it will address issues related to urban storm water drainage and its integration with road provision. The specific focus of it includes: extent of integration of urban storm water drainage infrastructures in road provision, existing condition of road and urban storm water drainage net-work, impacts of integration of urban storm water drainage on road performance and related environmental issues and flood prone areas.

The engineering design aspects of roads and urban storm water drainage net-work, in general, are not part of this study.

General description of the study area

The study area, Addis Ketema Sub-city (Figures 2 and 3), is located in the capital city of Ethiopia, Addis Ababa. It is located between 38°50′ latitude and 8°45′ longitude. It is situated at the centre and is accessible to all parts of the city, Addis Ababa.

LITERATURE REVIEW

About thirty two valuable materials, which were published by various authors, have been employed in this research work as a reinforcing input. These materials along with their authors include:

1. Development of road and its origin (World Bank, 2006).
2. The historical development of road (Addis Ababa City Roads Authority, 2008).
3. The development of modern roads (Rebekah, 2006).
4. The history and origin of road development in Ethiopia/Addis Ababa (Addis Ababa City Roads authority, 2008; National High Way Institute (2001).
5. The history and origin of urban storm water drainage development (Addis Ababa City Roads Authority, 2006).
6. The history and development of urban storm water drainage infrastructure in Ethiopia (Federal Urban Planning Institute, 2006).
7. Current approaches in managing urban drainage system (GTZ-IS, 2006).
8. Policy issues in managing urban storm water drainage (Federal Urban Planning Institute, 2008).
9. Urbanization and urban storm water run-off (Leopold, 1968; Hollis, 1975; Neller, 1988; Booth, 1990, 1991; Glazner, 2001; GTZ-IS, 2006).
10. The importance of urban storm water drainage infrastructures/facilities (Thomas and Andrew, 2002).
11. The importance of road and urban storm water drainage integration (Fabian and Barry, 2003).

RESEARCH METHODOLOGY

Both descriptive and exploratory types of researches were employed. The descriptive type was used to describe the existing condition and coverage of roads and urban storm water drainage facilities. Whereas, the exploratory type was particularly used to explore the existing condition and coverage of urban storm water drainage facilities which were not found in the base-map was collected from the sub-city.

Data types and sources

Quantitative as well as qualitative data types were employed. Of the total data about 95% of the research data was collected from primary sources. Where as the rest 5% was collected from secondary data sources-this was employed to reinforce the primary data source.

Data collection methods

The following data collection methods were used.
About 10% of the primary research data was collected through questionnaires and interview. The rest 90% was collected through field survey.

1. Questionnaires were prepared for kebele and sub-city professionals to identify: areas prone to flooding, major causes of flooding, techniques of urban storm water drainage contract and budget allocation to road and urban storm water drainage infrastructure.
2. Interview was employed to collect data related to flooding hazards and causes of flooding through household survey.
3. Field survey was the dominant method of data collection in this study using base map and check list. The sizes of urban storm drains and roads have been measured and checked for their length, width, height and/or radius. It was also employed to identify the existing hierarchy, condition and pavement of road and urban storm water drainage infrastructure.

Data analysis

The collected data was analyzed with the help of Microsoft excel, AutoCAD and ArcGIS softwares.

Data presentation

The analyzed data was presented in tables, graphs, charts and percentages. Besides, GIS and CAD figures and field survey photos were also incorporated.

DATA PRESENTATION AND ANALYSIS

Here, which is the core of the study, focuses on data presentation and analysis of primary and secondary data composed from the households (Figure 1), and Addis Ababa roads authority, sub-city and Kebele officials and

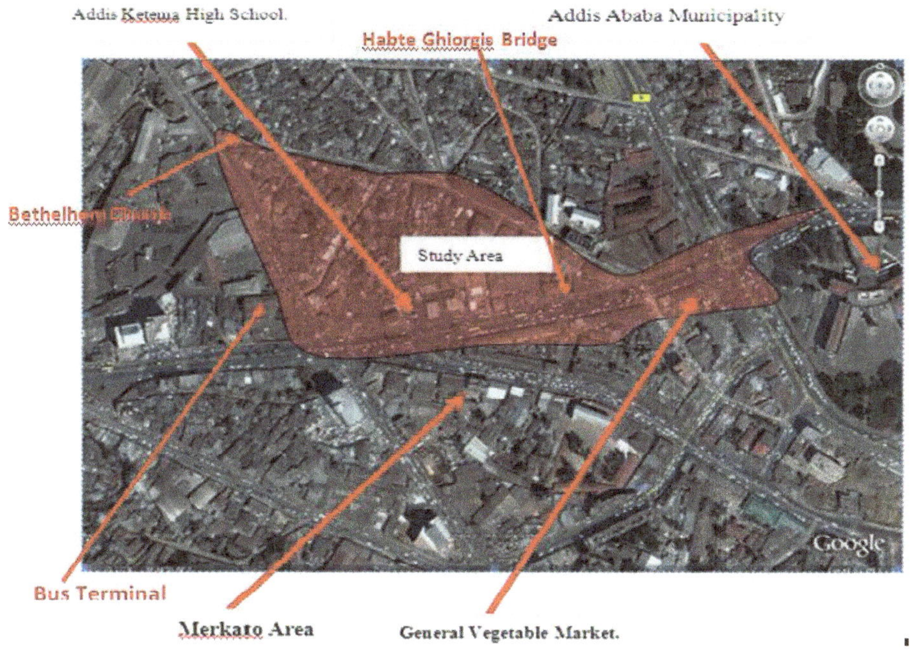

Figure 1. The actual study area (Source: Google earth and Field survey, 2009).

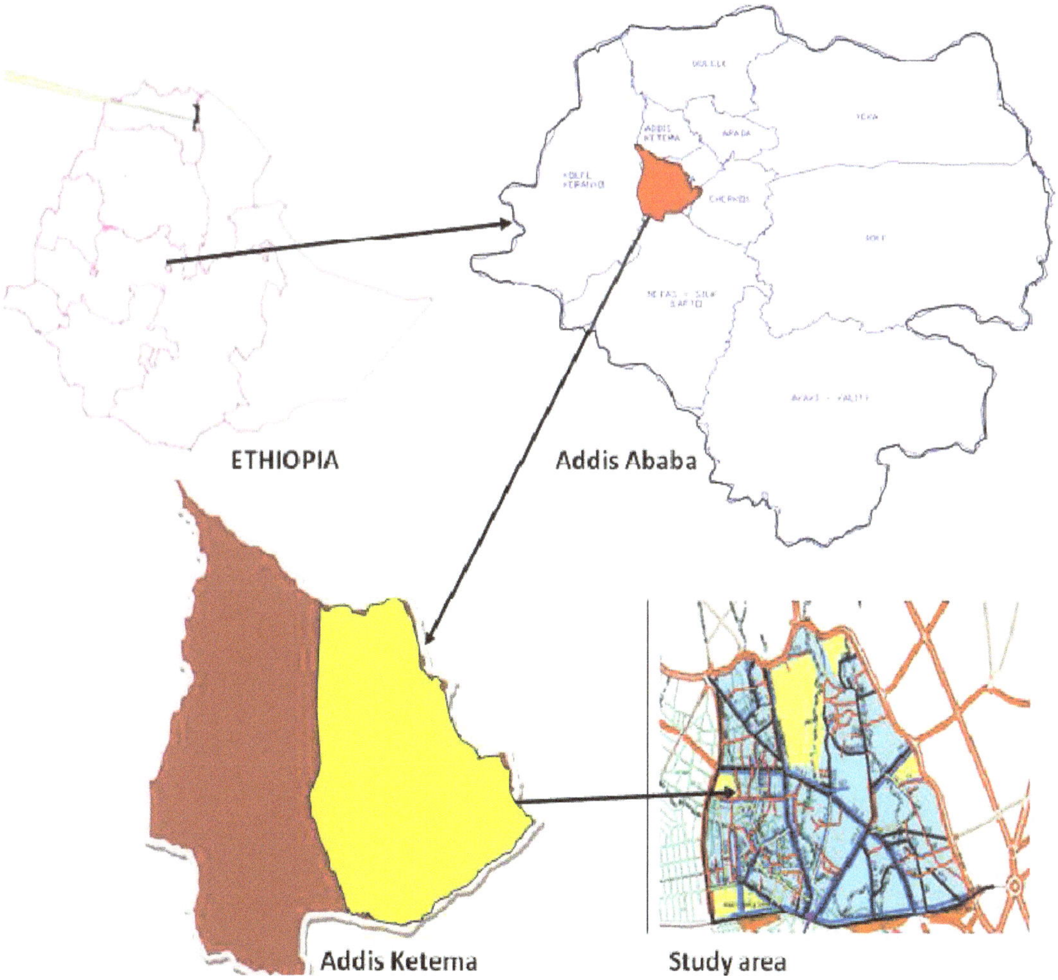

Figure 2. Location map of the study area (Source: GIS and AutoCAD analysis, 2009).

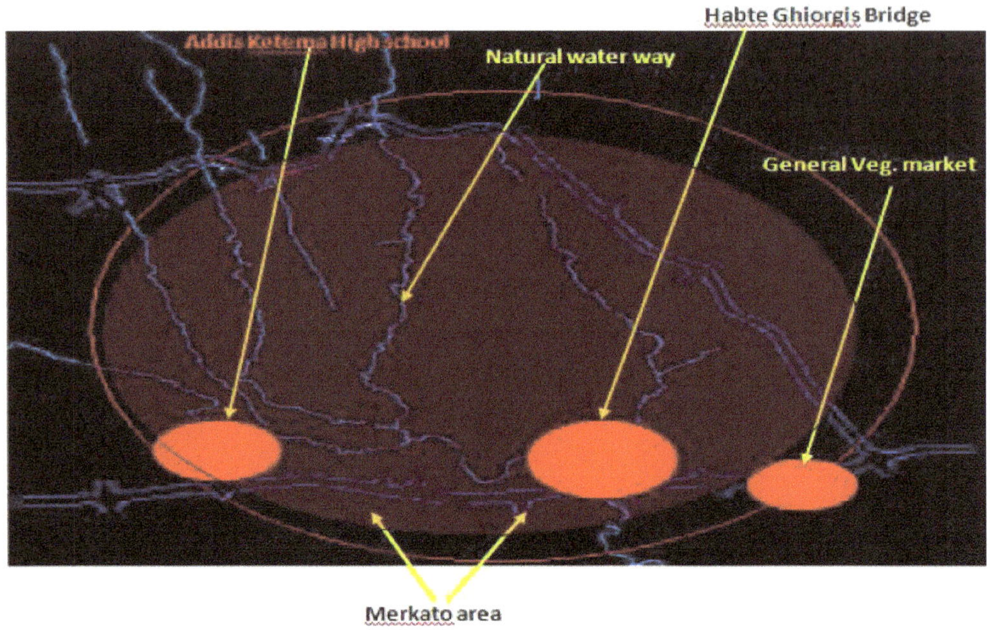

Figure 3. The major flood prone areas in the study area (Source: GIS analysis).

Table 1. Major flood prone areas in the study area.

Specific site/Kebele	Respondents in		Ranking of flood prone areas(high to low)	
	Number	percentage	Respondents' response	Field observation
Habte Ghiorgis area	12	35.29	1st	1st
Addis Ketema secondary school area	10	29.41	2nd	2nd
General vegetable market area	8	23.53	3rd	3rd
Neftegna locality(Kebele 31)	2	5.88	4th	4th
Addisu Michael area.	2	5.88	5th	-
Total	34	100	-	-

Source: Field survey April-May, 2009.

from researcher's direct observation The majority of this research work was drawn from researcher's field survey with the help of base map and checklists. The interviews and questionnaires were employed to reinforce the field survey data. In this research each and every road and urban storm water drainage infrastructures of the study area have fully been surveyed and observed.

Flood prone areas and causes of flooding in the study area

Major flood prone areas/sites in the study area

The major flood prone sites in the study area were identified through interviews from respondents, and questionnaires from kebele and sub-city officials along with field observations.

The purpose of identifying such flood prone areas was to investigate the root causes of the flooding problems and to give priority during the provision of urban storm water drainage facilities.

As depicted in Table 1 and Shown in Figure 4, a discussion with the community, the major flood prone area is Habte Ghiorgis (35.29%) followed by around Addis Ketema secondary school (29.4%) and General vegetable market (23.5%), which were evidenced from the respondents' response as well as field observation. The GIS Figure 3, shows the specific locations of these flood prone areas.

Causes of flooding

As it was observed during field survey the majority of the storm drains are blocked by solid wastes of various types and many residents illegally connect their sewerage system in to the existing drains. Besides, this idea is reinforced from respondents' response. Table 2 is an evidence to this truth.

Figure 4. Discussion with the respondents (Source: Field survey, 2009).

Table 2. The major causes of flooding.

No.	Major causes	Respondents	
		Number	Percentage
1	Blockage of drains by solid wastes	26	77
2	Sewer connection to drains	4	12
3	Inadequate drains	3	8
4	Absence of drains	1	3
	Total	34	100

Source: Residents and professionals interviews (April-May, 2009).

The implication of this analysis is that the Addis Ababa City Roads Authority, which is the sole urban storm water drainage and road provider in the sub-city, should give immediate attention for constructing closed/buried type urban storm water drainage lines in the provision of future road and urban storm water drainage infrastructure and a separate sewerage system. Besides, some corrective measures should be taken as per the actual design parameters/dimensions with a full consideration of the quantity of run-off generated in the respective catchment area.

Existing condition of road and urban storm water drainage net-work infrastructures

Existing condition and coverage of road infrastructure in the study area

The existing condition, coverage and hierarchy of roads in the study area were fully surveyed with the help of base map and check lists and the summary of the field survey is presented in Tables 3 and 4.

In this particular study, the least degraded road is stone surfaced and the highest one is asphalt. The implication of this is that: stone surfaced roads can give longer service life than asphalt and gravel roads. Thus, let alone comfort, stone surfaced roads are economical and asphalt with a high comfort has greater economic loss during construction as well as maintenance.

Existing condition and coverage of Urban storm water drainage net-work infrastructure in the study area

The flow of storm water run-off is one of the major problems in the study area as a result of urbanization and pavements of surfaces associated with blockage of urban storm water drainage net work management.

The summary of the condition and coverage of urban storm water drainage infrastructure is presented in Table 5.

From Table 5, the majority of drains is circular, which are closed/buried types. This implies that about half of the drains is open to the environment, which may cause

Table 3. Summary of road surface types and condition.

No.	Road type	Good(m)	Light (m)	Severe (m)	Total (m)
1	High way	2490	-	-	2490
1	Arterial	6345	-	-	6,345
2	Collector	3730	2923	3309	9,962
3	Local	4656	5678	2657	12,991
	Total	17,221	8,601	5,966	31,788 ~ 32km

Source: Field survey April-May, 2009.

Table 4. Detailed Length and condition of roads in the study area.

Road surface type	Total road		Counts	Road condition	Length (km)	Percent from total road
	Length (km)	Percent				
				Good	13.62	42.6
Asphalt	16.34	52	34	Light	0.533	1.70
				Severe	2.272	7.1
				Good	3.32	10.40
Stone surfaced	11.24	35	82	Light	6.20	19.4
				Severe	1.67	5.2
				Good	0.27	0.84
Gravel	4.2	13	19	Light	1.84	5.8
				Severe	2.02	6.3
Total	32	100	135		32	100

Source: Field survey (April-May, 2009).

Table 5. Summary of length and condition of urban drainage lines in the study area by type.

Drain shape	Construction material	Counts	USWD line length (km)	USWD condition	length (km)
				Good	0.72
Trapezoidal		13	3.82	Light	1.09
				Severe	1.58
			Percentage from total		17.4%
				Good	7.36
Circular/pipe	Masonry	58	10.50	Light	0.25
				Severe	2.98
			Percentage from total		47.7%
				Good	2.51
Rectangular		34	8.00	Light	4.27
				Severe	1.31
			Percentage from total		36.4%
Total		105	22.00		22.0

Source: Field survey (April, 2009).

many environmental problems including respiratory diseases. The other half is the open type it is because of this many of the drains are blocked by solid wastes.

All the circular/pipe drainage lines are buried/closed,

Table 6. Existing spatial integration between road and urban storm water drainage infrastructure.

The integration of	With			Total (km)	From the total drainage length (km) (%)
	Circular drainage line (km)	Rectangular drainage line (km) (%)	Trapezoidal drainage line (km)		
Asphalt road	10.59 (65.2%)	3.04 (18.7)	2.61 (16.1%)	16.24	74
Stone surface road	-	3.10 (86.4)	0.49	3.59	16
Gravel road	-	1.96 (87.1)	0.29	2.25	10
Total				22.08	100

whereas most of the rectangular and trapezoidal drains are open and are not environment friendly, which implies that these may encourage residents to connect and dump their sewerage systems and solid waste sources respectively resulting in unpleasant smells and are unaesthetic to the environment. As it was drawn from respondents during field survey, particularly open drainage lines are the major causes of respiratory diseases like Asthma, common cold and air pollution and other respiratory diseases.

The extent of integration of urban storm water drainage infrastructures in road projects provision

Provision of road and urban storm water drainage infrastructures are indispensable, particularly, in an urban center for safe and easy reachability from one area to another and to protect flood damage on infrastructure and utilities as a result of pavement. From the field survey and its analysis the integration of urban storm water drainage in road provision projects is summarized in Table 6.

From Table 6, for example the integration of asphalt road with circular, rectangular and trapezoidal drain types is 65.2, 18.7 and 16.1% respectively. From theses, the spatial integration between all roads and all urban storm water drainage infrastructures is found 42%. This implies that, it is only 42% of the roads have drains or for every kilometer of road there is only 420 m of drainage lines. The rest 58% of the road is without drainage facilities. Besides, the road density was analyzed and found 10%. The implication of these is that: the ever existing flooding problems will be continued, the gap between road andurban drainage integration gets widened, the environmental problems like soil erosion keeps continuing, the budget allocated for urban utilities will increase due to degradation of utilities and loss of effective land for other land use function increases.

The impacts of urban storm water drainage infrastructure integration on road performance and related environment issues

Deterioration of urban infrastructures and other urban utilities, bad smell and environmental pollution, urban land degradation, development of disease causing organisms were some of the findings resulted in inadequate integration between road and urban storm water drainage infrastructure.

Conclusions

1) Inadequate integration between road and urban storm water drainage lines followed by blockage of drains by solid wastes are the major causes of flooding in the study area.
2) Sewerage connection and solid waste dumping reduce the effective carrying capacity of drains.
3) To safely discharge the flood generated within the study area the urban storm water drainage facilities should be revised and designed.
4) Urban storm water drainage facilities should be contracted with roads for timely accomplishing and good work manship.
5) The budget allocated for the last and coming three years for road and urban storm water drainage provision has created a wider integration gap.

RECOMMENDATIONS

1) Proactive measures should be taken to reduce and manage flooding hazards (like clearing of drains before rain season begins).
2) Improvement on the integration of road and urban storm water drainage infrastructure.
3) Integrated solid waste management.
4) Urban storm water drainage infrastructure should be contracted with road infrastructure.
5) Encourage site infiltration through: Permeable pavements like porous concrete, coble stone, vegetated structures or grassing on road sides and vacant spaces/ gardens.
6) To ensure sustainable urban drainage management, there should be an integrated urban storm water drainage management.

REFERENCES

Addis Ababa city (AAC) administration (2006). Annual report, Addis

Ababa, Ethiopia.
Addis Ababa City Roads Authority (2008). Annual magazine on Roads and Drains construction and maintenance. Addis Ababa, Ethiopia.
Addis Ababa University (2003). Groundwater vulnerability mapping of the Addis Ababa water supply aquifers. Addis Ababa, Ethiopia.
American Society of Civil Engineers (2004). Existing sewer evaluation and rehabilitation. Manual and Report on Engineering Practice. Reston. VA., pp. 56-71.
CSIRO (1999). Best Practice Environmental Management Guidelines. Melbourne, Australia, pp. 19-31.
Ethiopian Central Statistics Agency (2008). Population and Housing census Addis Ababa, Ethiopia, p. 30.
Fabian P, Barry JA (2003). Urban storm water management planning: USA, pp. 57-89.
Federal Urban planning Institute (2008). Urban Storm Water Drainage Design Manual. Addis Ababa, Ethiopia, pp. 11-37.
GTZ-IS (2006). Urban Drainage manual series on infrastructure. Addis Ababa, Ethiopia, pp. 17-54.
Mays LW, Tung YK (1958). Hydro systems engineering. New York, USA., pp. 23-29.

13

Analytical and experimental study of velocity shear instability in the presence of inhomogeneous perpendicular D.C electric field

R. K. Tyagi[1], K. K. Srivastava[1] and R. S. Pandey[2*]

[1]Department of Mechanical Engineering, Birla Institute of Technology, Mesra Ranchi, Jharkhand, India.
[2]Department of Applied Physics, Amity School of Engineering and Technology, Amity University, Uatter Pradesh, Noida, India.

Analysis and experimental study of parallel flow velocity shear electrostatic ion cyclotron (EIC) instability has been done in plasma containing massive positive ions and electron by using the method of characteristics solution and kinetic theory in the presence of inhomogeneous direct current (DC) electric field perpendicular to ambient magnetic field. The effect of many parameters on growth rate and real frequency has been discussed by using the experimental data. Applications to possible laboratory plasmas and industries are also discussed.

Key words: Velocity shear instability with inhomogeneous D.C electric field.

INTRODUCTION

A parallel flow velocity shear which means a magnetic field (B) aligned flow with the velocity gradient perpendicular to B is known to be an origin of low frequency instabilities occurring in space plasmas. For example, ion acoustic instabilities observed in the ionosphere have recently been explained in terms of parallel flow velocity shear (Ichiki et al., 2005).

Many characteristics of EIC instability that have been investigated experimentally were in agreement with the local theory of Drummond and Resenbluth (1962) appropriate to uniform magnetized plasma. Their treatment however could not account for certain aspects of instability related to in homogeneities present in laboratory plasmas. Bakshi et al. (1983) found that if width of the current channel was reduced to just a few ion Larmor radii, the EIC instability was quenched. The effects of a transverse gradient in the plasma flow velocity parallel to a magnetic field on the excitation of EIC waves has also been analyzed by Ganguli et al. (2002). They showed that ion flow gradients (parallel velocity shear) can give rise to a new class of ion cyclotron waves via inverse cyclotron damping with a resulting multiple cyclotron harmonic spectrum.

The effect of parallel velocity shear on EIC wave excitation was studied experimentally by many workers. These experiments provided clear evidence. That parallel velocity shear does play a role in the excitation of EIC waves. Experimental evidence have also been persecuted by Agrimson et al. (2003) to show that the filament quenching of EIC instability may be rendered ineffective if, in addition to field aligned current, parallel velocity shear is present in ion flow. Observation shows that the EIC instability can be excited in narrow current channels if a sufficient amount of parallel velocity shear is also present (Agrimson et al., 2003).

An examination of the EIC instability in the light of recent theoretical work of Gaungli et al. (2002) and Gavrishchaka et al. (2000) suggest that an ion flow along the magnetic field with a flow gradient (shear) transverse to the magnetic field can have an additional destabilizing effect on EIC waves in the presence of magnetic field aligned electron drift (current). A related theoretical prediction by Gavrishchaka et al. (1998) that perpendicular shear in the flow of ions parallel to a magnetic field substantially reduces the electron drift necessary to excite ion acoustic-like waves was a subject of previous

*Corresponding author. E-mail: rspandey@amity.edu

experiment by Agrimson et al. (2001), since these predictions have important consequences for the inter prediction of low frequency plasma oscillations observations in the ionosphere.

Some experiments in Q-machines have been performed using Cesiveu plasma to test these in the laboratory. The purpose of these experiments was to investigate the possible influence of parallel velocity shear on the excitation of EIC waves. Experiment was performed in a single–ended Q machine using essentially the identical set up employed by many researches studying EIC waves. Addition evidence for the role of parallel velocity shear was obtained in the other experiment which employs a double ended Q machine Agrimson et al. (2002). Drift wave is observed to be destabilized by a magnetic-field-aligned ion flow velocity shear in the absence of field-aligned electron drift flow in laboratory experiments using a concentrically three segmented plasma source. The fluctuation amplitude increases with increasing a shear length, but the instability is found to be gradually stabilized where the shear strength exceeds a critical value (Kaneko et al., 2003). The stabilizing and destabilizing mechanisms are well explained by a plasma kinetic theory including the effect of radial density gradient (Kaneko et al., 2003).

Experimental demonstration of the certification of electrostatic ion cyclotron waves in plasma with a parallel ion flow shear but no magnetic field aligned current has already being discussed (Kim et al., 2004). Waves with frequencies near the ion cyclotron and multiple harmonics launched from an antenna are observed to grow in amplitude in a region of ion flow shear.

Particularly in the field of space plasma, the velocity shear of magnetic field-aligned flows has been studied to understand the physical properties of fluctuations occurring in space plasmas, which are observed by radars, rockets and satellites. The reason for predominant excitation of the ion acoustic instability in the Earth's ionosphere is explained by the fact that the parallel shear degrades the ion Landau damping rate of the ion acoustic mode to reduce this mode prominence (Gavrishchaka, 1998). This hypothesis was bench marked by several laboratory experiments. The reason for parallel shear modified instabilities has recently been extended to the drift wave instability (Teodorescu, 2002).

For simplicity it is assumed that there is no shear in the drift velocity. Generally speaking however, the velocity shear should be taken under consideration and may affect the system in two ways; on one half, it may lead to suppression or a complete stabilization of the perpendicular gradient instability; on the other hand, it may give rise to a traditional hydrodynamics instability of a shear flow. It has been focused that ion velocity shears can significantly lower the field aligned threshold current needed to trigger the instability, especially for wave vectors close to the perpendicular to the magnetic field. However the current density and shear requirements remain significantly higher than if collisions are neglected (Perron et al., 2009). The shear driven electrostatic ion cyclotron instability EIC has also been discussed using loss-cone distribution function by particle aspect analysis (Mishra et al., 2007).

In the present work, analytical and experimental study of velocity shear instability has been done in the presence of inhomogeneous perpendicular D.C electric field. The role of the electric field, magnetic field, shear scale length and temperature ratio has been discussed in detail on the growth rate of shear instability.

ANALYSIS

Spatially homogeneous anisotropic plasma subjected to an external magnetic field $\mathbf{B_0} = \mathbf{B_0}\hat{\mathbf{e}}_\mathbf{z}$ and an inhomogeneous DC electric field $\mathbf{E_0(x)} = \mathbf{E_0(x)}\hat{\mathbf{e}}_\mathbf{x}$ has been considered. In order to obtain the dispersion relation, the Vlasov–Maxwell equations are linearized for inhomogeneous plasma by small perturbations of E_1, B_1 and f_{s1}. These are perturbed quantities and are assumed to have harmonic dependence as exp i(**kr**-ωt). The linearized Vlasov equations obtained by separating the equilibrium and non equilibrium parts following the technique of Misra and Pandey (1995) and Pandey et al. (2001) in units of c=1 (c=velocity of light) are given as:

$$\mathrm{v}\frac{\partial \mathrm{f}_{s0}}{\partial \mathrm{r}} + \frac{\mathrm{e}_s}{\mathrm{m}_s}[\mathrm{E}_0(\mathrm{x}) + (\mathrm{v}\times \mathrm{B}_0)]\left(\frac{\partial \mathrm{f}_{s0}}{\partial \mathrm{v}}\right) = 0 \quad (1)$$

$$\frac{\partial \mathrm{f}_{s1}}{\partial \mathrm{t}} + \mathrm{v}\frac{\partial \mathrm{f}_{s1}}{\partial \mathrm{r}} + \left(\frac{\mathrm{F}}{\mathrm{m}_s}\right)\left(\frac{\partial \mathrm{f}_{s1}}{\partial \mathrm{v}}\right) = \mathrm{S}(\mathrm{r},\mathrm{v},\mathrm{t}) \quad (2)$$

where force is defined as F= mdv/dt

$$F = e_s\left[E_0(x) + (\mathrm{v}\times B_0)\right] \quad (3)$$

The particle trajectories are obtained by solving the equation of motion defined in Equation (3) and S(r,v,t) is defined as:

$$\mathrm{S}(\mathrm{r},\mathrm{v},\mathrm{t}) = \left(-\frac{\mathrm{e}_s}{\mathrm{m}_s}\right)[\mathrm{E}_1 + \mathrm{v}\times \mathrm{B}_1]\left(\frac{\partial \mathrm{f}_{s0}}{\partial \mathrm{v}}\right) \quad (4)$$

The method of characteristics solution is used to determine the perturbed distribution function f_{s1}. This is obtained from Equation (2) by

$$\mathrm{f}_{s1}(\mathrm{r},\mathrm{v},\mathrm{t}) = \int_0^\infty \mathrm{S}\left\{\mathrm{r}_0\left(\mathrm{r},\mathrm{v},\mathrm{t}'\right),\mathrm{v}_0\left(\mathrm{r},\mathrm{v},\mathrm{t}'\right),\mathrm{t}-\mathrm{t}'\right\}dt' \quad (5)$$

where s denotes species. We transformed the phase space coordinate system from (r,v,t) to (r_0,v_0,t-t') and $t = t - t'$. The particle trajectories obtained by solving Equation (3) for the given external field configuration by Misra and Tiwari (1977) are given as

$$x(t)=x_0+\frac{v_\perp}{\Omega_s}\left(1+\frac{\bar{E}'(x)}{4\Omega_s^2}\right)\left[\sin(\theta-\Omega_s t)-\sin\theta\right]$$

$$y(t)=y_0+\Delta'+\frac{v_\perp}{\Omega_s}\left(1+\frac{3}{4}\frac{\bar{E}'(x)}{4\Omega_s^2}\right)\left[\sin(\theta-\Omega_s t)-\cos\theta\right]$$

$$z=z_0+v_\parallel t \tag{6}$$

Where θ is the phase of $v_\perp$ at t=0, $E'(x), E''(x)$ are the derivatives of $E(x)$, and

$$\Delta=\frac{\bar{E}'(x)t}{\Omega_s^2}\left[1+\frac{E''(x)}{E(x)}\cdot\frac{1}{4}\left(\frac{v_\perp}{\Omega_s}\right)^2 \ldots\ldots\ldots\ldots\right],$$

$$\bar{E}(x)=\frac{e_s E(x)}{m_s},$$

$$E(x)=E_{0x}\left(1-\frac{x^2}{a^2}\right),\quad \Omega_s=\frac{e_s B_0}{m_s}.$$

Where a is thought to be comparable to the mean ion gyro-radius but larger than the debye length when $x^2/a^2<1$, $E(x)$ becomes a constant uniform field.

Using Equations (2), (4) and (6) and doing some lengthy algebraic simplifications and time integration, we obtained the perturbed distribution function from Equation (5) as:

$$f_{s1}(r,v,t)=\frac{ie_s}{m_s\omega}\sum_{m,n,p,g}J_n(\lambda_1)J_m(\lambda_1)J_p(\lambda_2)J_g(\lambda_2)e^{i(m-n)(\pi/2+\theta)}e^{i(g-p)(\pi/2+\theta)}\times$$

$$\left[E_{1x}U^*+V^*E_{1y}+W^*E_{1z}\right]\times\frac{1}{k_\parallel v_\parallel+n\Omega_s+p\Omega_s+k_\perp\Delta'-\omega} \tag{7}$$

Where;

$$U^*=Cv_\perp\frac{n}{\lambda_1}-Cv_\perp\frac{n}{\lambda_1}\frac{\bar{E}'(x)}{4\Omega_s}+k_\perp v_\perp\xi''\frac{n}{\lambda_1}+k_\perp v_\perp\xi''\frac{\bar{E}'(x)}{4\Omega_s^2}\frac{n}{\lambda_1}$$

$$V^*=iCv_\perp\frac{J'_n(\lambda_1)}{J_n(\lambda_1)}+\frac{3}{4}Cv_\perp\frac{\bar{E}'(x)n}{\Omega_s^2\lambda_1}-\Delta'C$$

$$W^*=\omega\frac{\partial f_{s0}}{\partial v_\parallel}+Dk_\perp v_\perp\frac{n}{\lambda_1}+\frac{3Dk_\perp v_\perp}{4}\frac{\bar{E}'(x)}{\Omega_s^2}\frac{n}{\lambda_1}$$

$$C=(\omega-k_\parallel v_\parallel)\left(\frac{-2f_{s0}}{\alpha_{\perp s}^2}\right)+k_\parallel\frac{\partial f_{s0}}{\partial v_\parallel}$$

$$D=v_\parallel\left(\frac{2f_{s0}}{\alpha_{\perp s}^2}\right)-\frac{\partial f_{s0}}{\partial v_\parallel}$$

Where the Bessel identity

$$e^{i\lambda_1\sin\theta}=\sum_{n=-\infty}^{\infty}J_n(\lambda_1)e^{in\theta}$$

$J_n(\lambda_1)=$ Bessel function

$$\lambda_1=\frac{k_\perp v_\perp}{\Omega_s},\quad \lambda_2=\frac{3}{4}\frac{k_\perp v_\perp\bar{E}'(x)}{\Omega_s^3},\quad \Delta'=\frac{\partial\Delta}{\partial t}$$

$$J'_n(\lambda_1)=\frac{J_n(\lambda_1)}{d\lambda_1}$$

The unperturbed bi-Maxwellian distribution function is written as

$$f_{s0}=f_{sm0}+v_{0y}\xi''$$

$$\xi''=\frac{1}{\Omega_s}\left[\frac{2(v_\parallel-v_{0z}(x))}{\alpha_{\parallel s}^2}\frac{\delta v_{0z}(x)}{\delta x}\right]f_{smo}$$

$$f_{mo}=\frac{n_0(x)}{(\pi)^{1/2}\alpha_{\perp s}^2\alpha_{\parallel s}}\exp\left[-\frac{(v_{0x}^2-v_{0y}^2)}{\alpha_{\perp s}^2}-\frac{(v_\parallel-v_{0z}(x))^2}{\alpha_{\parallel s}^2}\right] \tag{8}$$

where ξ'' is being constant of motion and $n_0(x)$ =plasma particle density

$$\alpha_{\perp s}=\left(\frac{2k_B T_{\perp s}}{m_s}\right)^{1/2},\ \alpha_{\parallel s}=\left(\frac{2k_B T_{\parallel s}}{m_s}\right)^{1/2}$$

Now simplifying m = n, g = p and using the definition of current density, conductivity and dielectric tensor, we get the dielectric tensor

$$\|\varepsilon(k_1\omega)\| = 1 - \frac{4\pi e_s^2}{m_s\omega^2}\int \frac{d^3v \sum J_p^2(\lambda_2) \| S_{ij} \|}{k_\| v_\| + n\Omega_s + p\Omega_s + k_\perp \Delta' - \omega} \tag{9}$$

Where;

$$\|S_{ij}\| = \begin{vmatrix} J_n^2 U^* \frac{n}{\lambda_1} v_\perp & J_n^2 V^* \frac{n}{\lambda_1} v_\perp & J_n^2 W^* \frac{n}{\lambda_1} v_\perp \\ -iJ_n J_n' U^* v_\perp & -iJ_n J_n' V^* v_\perp & -iJ_n J_n' W^* v_\perp \\ J_n^2 U^* v_\| & J_n^2 V^* v_\| & J_n^2 W^* v_\| \end{vmatrix}$$

Now we consider electrostatic ion-cyclotron instability

$$\| \varepsilon_\| \| = N^2 \tag{10}$$

Where N is refractive index

The required electrostatic dispersion relation can be obtained by using the approximation of Huba (1981) and from Equations (8-10).

$$D(k,\omega) = 1 + \frac{2\omega_{ps}^2}{k_\perp^2 \alpha_{\perp s}^2}\Gamma_n(\mu_s)\sum J_P^2(\lambda_2)\left[1 - \frac{E'(x)}{4\Omega_s^2}\right]\frac{k_\perp}{k_\|}$$

$$\left[\left(\frac{\bar{\omega}}{k_\|\alpha_{\|s}} - \frac{1}{2}\varepsilon_n\rho_s\frac{\alpha_{\perp s}}{\alpha_{\|s}}\right)Z(\xi) - A_s\frac{\alpha_{\perp s}^2}{\alpha_{\|s}^2}(1+\xi Z(\xi)) + A_T\frac{k_\|}{k_\perp}(1+\xi Z(\xi))\right] \tag{11}$$

Where Z(ξ) is plasma dispersion function

$$\xi = \frac{\bar{\omega} - (n+p)\Omega_s - k_\perp\Delta'}{k_\|\alpha_\|}$$

$$A_s = \frac{1}{\Omega_s}\frac{\delta v_{oz}(x)}{\delta x}$$

$$\varepsilon_n = \frac{\delta \ln n_0(x)}{\delta x}$$

$$A_T = \frac{\alpha_{\perp s}^2}{\alpha_{\|s}^2} - 1$$

$$\bar{\omega} = \omega - k_\| v_{oz}(x)$$

$$\mu_s = \frac{k_\perp^2\rho_i^2}{2}, \lambda_{Ds}^2 = \frac{\alpha_{\perp s}^2}{2\omega_{ps}^2}, \omega_{ps}^2 = \text{Plasma frequency}$$

$\Gamma_n(\mu_s) = \exp-(\mu_s)I_n(\mu_s)$, $I_n(\mu_s)$ =Modified Bessel function of order n.

In order to get dispersion relation for electrons and ions, approximations for electrons are assumed as $k_\perp\rho_e << 1$ and for ions no such assumptions is done thus above equation for p=1 and species s= i, e becomes:

$$D(k,\omega) = 1 + \frac{1}{k_\perp^2\lambda_{De}^2}\eta_e\frac{T_{\perp e}}{T_{\|e}} + \frac{1}{k_\perp^2\lambda_{Di}^2}\eta_i\left[\frac{T_{\perp i}}{T_{\|i}}\Gamma_n(\mu_i)\frac{k_\perp}{k_\|}\left[\left(\frac{\bar{\omega}}{k\alpha_{\|i}}\times\frac{T_{\perp i}}{T_{\|i}} - \frac{1}{2}\varepsilon_n\rho_s\frac{\alpha_{\perp i}}{\alpha_{\|i}}\right.\right.\right.$$

$$\left.\left.\left. + \frac{n\Omega_i + k_\perp\Delta'}{k_\|\alpha_{\|i}}\left(1 - \frac{T_{\perp i}}{T_{\|i}}\right)\right)Z(\xi_i) - A_i\frac{T_{\perp i}}{T_{\|i}}(1 + \xi Z(\xi_i))\right]\right] \tag{12}$$

After substituting the asymptotic expansion of $Z(\xi_i) = -\frac{1}{\xi_i} - \frac{1}{2\xi_i^3}$ for ion i.e. $|\xi_i|$<<1 Fried and Conte (1961) and $n_{0i} = n_{0e}$ multiplying throughout by $\frac{k_\perp^2\lambda_{Di}^2}{\eta_i}$ the above equation becomes:

$$0 = \frac{\lambda_{Di}^2}{\lambda_{De}^2}\frac{\eta_e}{\eta_i}\frac{T_{\perp e}}{T_{\|e}} + \frac{T_{\perp i}}{T_{\|i}} - \Gamma_n(\mu_i)\frac{T_{\perp i}}{T_{\|i}} + \frac{\Gamma_n(\mu_i)k_\perp}{2k_\|}\varepsilon_n\rho_i\frac{\alpha_{\perp s}}{\alpha_{\|s}}\frac{k_\|\alpha_{\|i}}{\bar{\omega} - n\Omega_i + k_\perp\Delta'} -$$

$$\frac{\Gamma_n(\mu_i)k_\perp}{k_\|}\frac{n\Omega_i + k_\|\alpha_{\|i}}{\bar{\omega} - n\Omega_i - k_\perp\Delta'} - \frac{\Gamma_n(\mu_i)k_\perp}{2(\bar{\omega} - n\Omega_i - k_\perp\Delta')^2}\cdot\frac{T_{\perp i}}{T_{\|i}}(k_\|\alpha_{\|i})^2\left(1 - \frac{k_\perp}{k_\|}A_i\right) \tag{13}$$

Where

$$\eta_i = 1 - \frac{\overline{E_i}'(x)}{4\Omega_i^2}$$

$$\eta_e = 1 - \frac{\overline{E'_e}(x)}{4\Omega_e^2}$$

Multiplying Equation (13) throughout by $\left(\frac{\bar{\omega} - n\Omega_i - k_\perp\Delta'}{k_\|\alpha_{\|i}}\right)^2$ we obtain a quadratic equationion as;

$$\frac{\overline{\omega}'}{\Omega_i} = -\frac{b_1}{2a_1}\left[1 \pm \sqrt{\left(1-\frac{4a_1c_1}{b_1^2}\right)}\right] \quad (14)$$

Where;

$$a_1 = a_2\left(\frac{\Omega_i}{k_{\|}\alpha_{\|}}\right)^2,$$

$$a_2 = \frac{\eta_e}{\eta_i}\frac{T_{\perp i}}{T_{\| i}} + \frac{T_{\perp i}}{T_{\| i}} - \Gamma_n(\mu_i)\frac{T_{\perp i}}{T_{\| i}}$$

$$b_1 = \frac{\Omega_i}{k_{\|}\alpha_{\| i}}b_2 - \frac{2k_{\perp}\Delta'}{k_{\|}^2\alpha_{\| i}^2}a_2\Omega_i$$

$$b_2 = \frac{\Gamma_n(\mu_i)k_{\perp}}{2k_{\|}}\varepsilon_n\rho_i\frac{\alpha_{\perp i}}{\alpha_{\| i}} - \frac{\Gamma_n(\mu_i)k_{\perp}}{2k_{\|}} - \frac{\Gamma_n(\mu_i)k_{\perp}n\Omega_i}{2k_{\|}^2\alpha_{\| i}}$$

$$c_1 = \frac{\Gamma_n(\mu_i)T_{\perp i}}{2T_{\| i}}\left(1-\frac{k_{\perp}}{k_{\|}}A_i\right) - \frac{b_2k_{\perp}\Delta'}{k_{\|}\alpha_{\| i}} + \frac{k_{\perp}^2\Delta'^2}{k_{\|}^2\alpha_{\| i}^2}$$

$$\overline{\omega}' = \overline{\omega} - n\Omega_i$$

$$P = \frac{x^2}{a}$$

$$E(x) = E_{0x}\left(1-\frac{x^2}{a^2}\right)$$

The solution of Equation (13) is

$$\overline{\omega}' = -\frac{b_1}{2a_1}\left[1 \pm \left(1-\frac{4a_1c_1}{b_1^2}\right)\right] \quad (15)$$

From this expression dimensionless growth rate has been calculated by computer technique when $b_1^2 < 4a_1c_1$.

RESULTS AND DISCUSSION

Here, we show the solution of Equation (15) using parameters may be representative of the laboratory by Kim and Merlino (2007) and Rosenberg and Merlino (2008). We consider plasma in which the heavy positive ions are produced due to ionization of K^+ and light electron are produced from $SF6^-$. It is further assumed that the plasma is immersed in magnetic field strengths, are varying like 0.24 to 0.32 T and homogeneous DC electric field strength 4 to 12 V/m which is perpendicular to magnetic field. In this case, the positive ions gyro radius ρ~ 0.095 cm and having a temperature anisotropy $A_T = \frac{T_{\perp i}}{T_{\| i}} - 1$ = 0.5 to 1.5 with density gradient $\varepsilon_n\rho_i$ =0.2 and inhomogeneity in DC electric field P/a= 0.5,0.55. In this case, we would except that the heavy positive ions electrostatic ion cyclotron instability could become excited by parallel velocity shear scale length like $A_i = 0.5$ to 0.55.

In Figure 1, variation of growth rate $\frac{\gamma}{\Omega_i}$ versus $k_{\perp}\rho_i$ for different values of velocity shear scale length A_i has been shown for other fixed plasma parameters. The growth rate increase with increasing the value of shear scale length A_i band width did not shift for lower and higher valued of $k_{\perp}\rho_i$. The maximum peak value of growth rate is 4.00×10^{-3} at $k_{\perp}\rho_i$=1.75 for shear length A_i=0.55 and in homogeneity in DC electric field P/a=0.5. The mechanism for instability of this mode is due to coupling of regions of positive and negative wave energy. This coupling occurs if velocity shear is non-uniform and hence velocity shear is the source of instability.

Figure 2 shows the variation of growth rate $\frac{\gamma}{\Omega_i}$ versus $k_{\perp}\rho_i$ for various value of electron ion temperature ratio $\frac{T_e}{T_i}$. In this Figure 2, the growth rate increases by increasing the values of electron ion temperature ratio due to in-homogeneity in electron ion temperature that dependence on the applied voltage of electrodes. The maximum peak values of growth rate is 3.86×10^{-3} at the $k_{\perp}\rho_i$=1.75 with inhomogeneous DC electric field. As velocity shear term is proportional to electron ion temperature ratio $\frac{T_e}{T_i}$.

Figure 3a and 3b shows the variation of real frequency $\frac{\omega_r}{\Omega_i}$ and growth rate $\frac{\gamma}{\Omega_i}$ versus $k_{\perp}\rho_i$ for different values of magnetic field strength B_0 with other fixed parameters listed in the figure caption. The growth rate and real frequency decrease with increasing the magnetic field strength. Due to change in magnetic field, gyro frequency has been changed and also modified Doppler shift

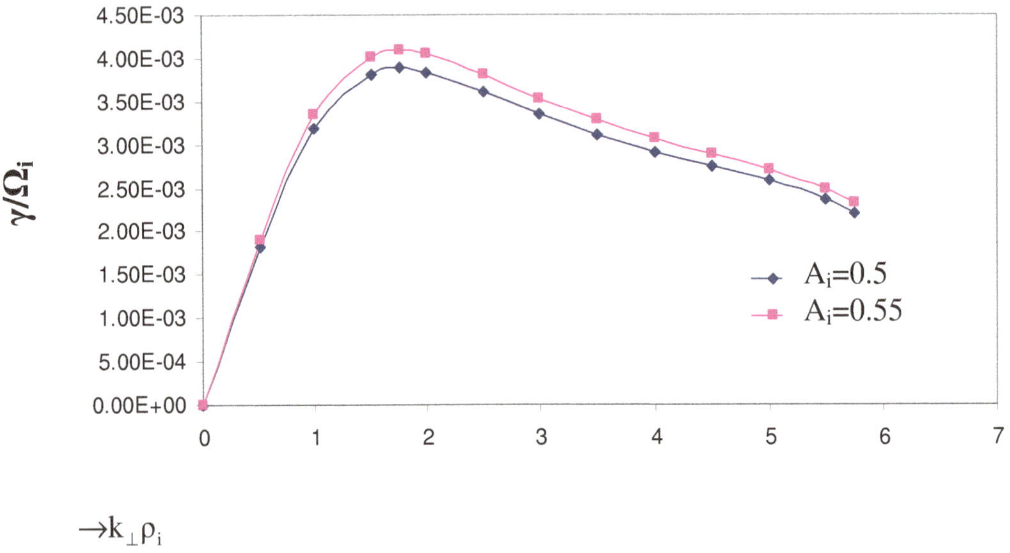

Figure 1. Variation of growth rate γ/Ω_i verses $k_{\perp}\rho_i$ for different values of A_i and other parameters are $B_0 = 0.24$ T, $T_e/T_i = 2$, $E_0 = 8$ V/m, $x/a = 0.5$, $\theta_1 = 88.5^0$, $A_T = 1.5$, $\varepsilon_n\rho_i = 0.2$.

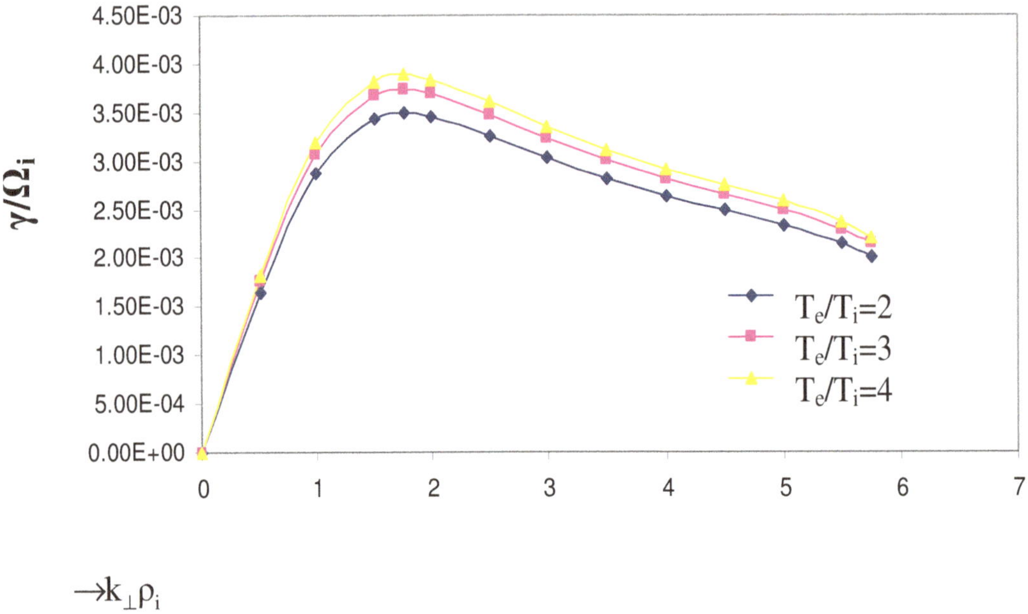

Figure 2. Variation of growth rate γ/Ω_i verses $k_{\perp}\rho_i$ for different values of T_e/T_i and other parameters are $A_i = 0.5$, $B_0 = 0.24$ T, $E_0 = 8$ V/m, $x/a = 0.5$, $\theta = 88.5^0$, $A_T = 1.5$, $\varepsilon_n\rho_i = 0.2$

frequency. The homogeneous magnetic field couples positive and negative energy waves and changes the growth of the wave. The magnetic field strength is a useful parameter for required velocity of EIC wave. Hence this is useful result for designing a machine for cold spray, metal cutting operations and micromachining.

Figure 4a shows the variation of growth rate $\frac{\gamma}{\Omega_i}$ versus $k_{\perp}\rho_i$ for different values of magnitude of inhomogeneous DC electric field. The growth rate decreases with increasing the value of magnitude of inhomogeneous

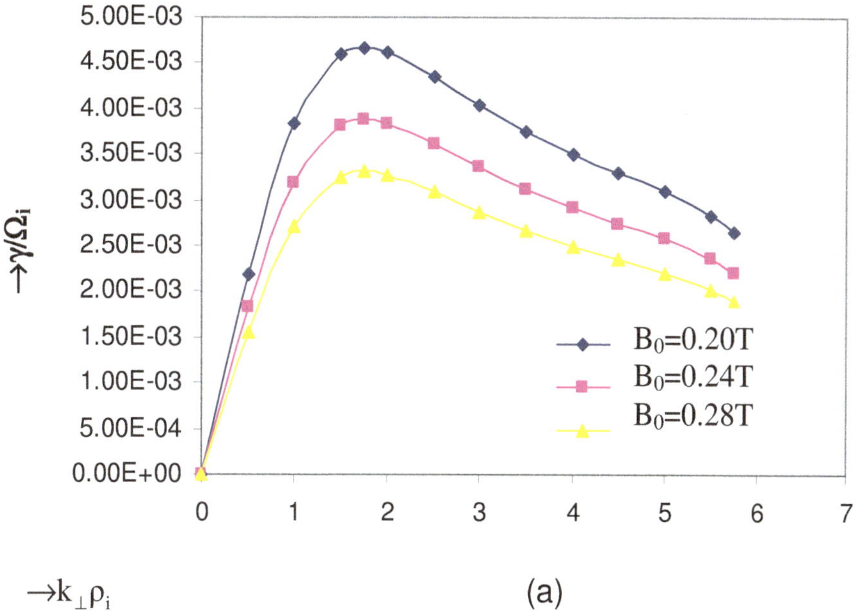

(a)

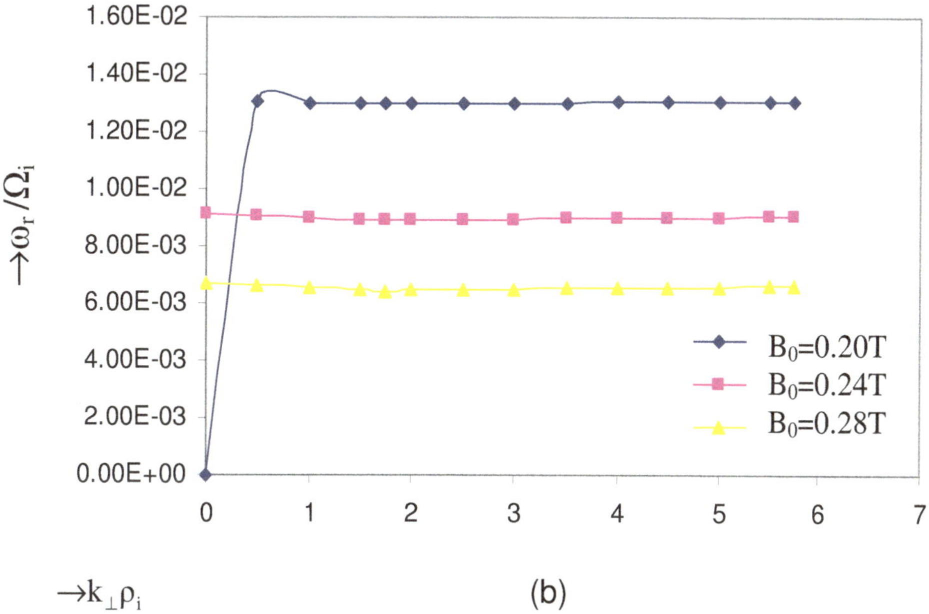

(b)

Figure 3. (a) Variation of growth rate γ/Ω_i verses $k_\perp \rho_i$ for different values of B_0 and other parameters are A_i = 0.5, T_e/T_i =2, E_0 = 8 V/m, x/a = 0.5, Θ = 88.5°, A_T = 1.5, $\varepsilon_n\rho_i$ = 0.2. (b) Variation of real frequency ω_r/Ω_i verses $k_\perp \rho_i$ for different values of B_0 and other parameters are A_i = 0.5, T_e/T_i =2, E_0 = 8 V/m, x/a =0.5, Θ = 88.5°, A_T = 1.5, $\varepsilon_n\rho_i$ = 0.2.

DC electric field like 4 to 12 V/m. It means the magnitude of inhomogeneous DC electric field shows the stabilizing effect on the growth rate. Figure 4b shows the variation of real frequency $\frac{\omega_r}{\Omega_i}$ versus $k_\perp \rho_i$ for different values of real frequency. It is clear from the Figure 4a that the real frequency increases with increasing the value of magnitude of inhomogeneous electric field. The above discussion shows that the magnitude of inhomogeneous DC electric field has a destabilizing effect on the real frequency. In general, the non-resonant particles support the oscillatory motion of the wave while the resonant particles participate in energy exchange with the wave.

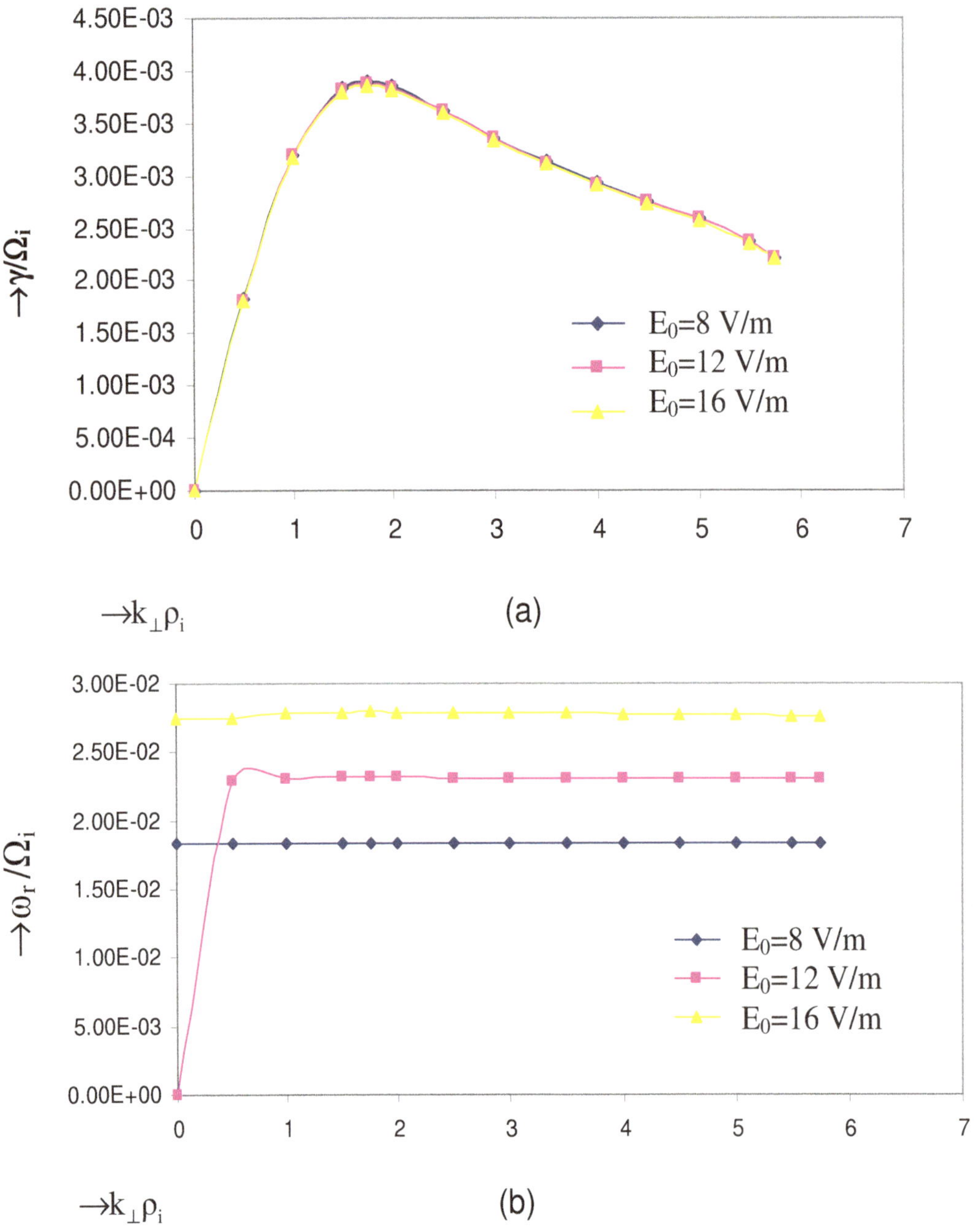

Figure 4. (a) Variation of growth rate γ/Ω_i verses $k_\perp \rho_i$ for different values of E_0 and other parameters are A_i =0.5, B_0=0.24 T, T_e/T_i =2, x/a =0.5, $\theta = 88.5^0$, A_T =1.5, $\varepsilon_n\rho_i$=0.2. (b) Variation of real frequency ω_r/Ω_i verses $k_\perp \rho_i$ for different values of E_0 and other parameters are A_i =0.5, B_0=0.24 T, T_e/T_i =2, $\theta = 88.5^0$, x/a =0.5,A_T =1.5, $\varepsilon_n\rho_i$=0.2.

The velocity of EIC wave is 1000 m/s for the value of magnitude of inhomogeneous DC electric field 8 V/m and it is 2500 m/s for 20 V/m with other fixed parameters listed in figure caption. Figure 5 shows the variation of growth rate versus inhomogeneity in DC electric field x/a with others parameters listed in figure caption. The maximum value of dimensionless growth rate is 3.895×10^{-3} at $k_\perp\rho_i$=1.75 obtained for x/a =0.3. This is the effective parameter for group velocity of the generation and control modulation of the wave.

The parameters like density gradient $\varepsilon_n\rho_i$ and

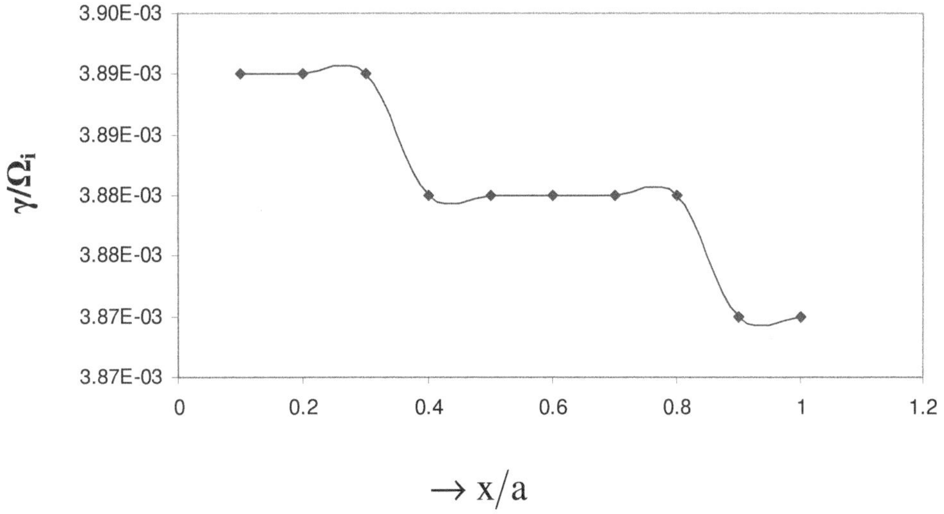

Figure 5. Variation of growth rate γ/Ω_i verses x/a for parameters like $k_\perp\rho_i$ = 1.75, A_i = 0.5, B_0 =0.24T, T_e/T_i = 2, E_0 = 8 V/m, A_T = 1.5, $\varepsilon_n\rho_i$ = 0.2.

temperature anisotropy $A_T = \frac{T_{\perp i}}{T_{\|i}} - 1$ have less effect of growth rate for the experimental data but the growth rate increases slightly with increasing the value of $\varepsilon_n\rho_i$ and A_T.

CONCLUSION

In this paper the effect of magnetic field, electric field electron ion temperature ratio, temperature anisotropy in ions, shear scale length, density gradient and inhomogeneity in DC electric field x/a on the growth rate and real frequency have been studied separately.

REFERENCES

Agrimson E, Kim SH, Angelo ND, Merlino RL (2003). Effect of parallel velocity shear on the electron static ion-cyclotron instability in filamentary current, Phys. Plasmas, 10: 3850.

Agrimson E, Angelo ND, Merlino RL (2001). Excitation of Ion-Acoustic-Like Waves by Subcritical Currents in a Plasma Having Equal Electron and Ion Temperatures, Phys. Rev. Lett., 86: 5282-5285.

Agrimson EP, Angelo ND, Merlino RL (2002). Effect of Parallel Velocity Shear on the Excitation of electrostatic ion cyclotron waves, Phys. Lett., 293: 260.

Bakshi P, Ganguli G, Palmadesso P (1983). Finite width currents magnetic shear and current driven ion-cyclotron instability Phys. Fluids. 26: 1808-1983.

Drummond WE, Rosenbluth MN (1962). Anomalous diffusion arising from microinstabilities in a plasma, Phys. Fluids, 5: 1507.

Fried BD, Conte SD (1961). The Plasma dispersion function, San Diego, Calif.

Ganguli, G, Slinker S, Gavrishchaka V, Scales W (2002). Low frequency oscillation in a plasma with spatially variance field-aligned flow, Phys. Plasmas, 9: 2321.

Gavrishchaka V, Ganguli GI, Scales WA (2000). Multiscale coherent structures and broadband waves due to parallel inhomogeneous flows Phys. Rev., 85: 4285-4288.

Gavrishchaka VV, Ganguli SB, Ganguli GI (1998). Origin of low frequency oscillations in the ionosphere, Phys. Rev. Lett., 80: 728.

Huba JD (1981). The Kelvin-Helmholtz instability in inhomogeneous plasmas, J. Geophys. Res., 86: 3653-3656.

Ichiki R, Kaneko T, Hatakeyama R (2005). Eindhoven, Netherlands. pp. 18-22.

Kaneko T, Tsunoyarna H, Hatakeyama R (2003). Drift-Wave Instability Excited by Field-Aligned Ion Flow Velocity Shear in the Absence of Electron Current, Phys. Rev., 90: 1205001.

Kim SU, Agrimson E, Miller MJ, Angelo ND, Merlino RL (2004). Amplification of electrostatic ion-cyclotron waves in a plasma with magnetic-field-aligned ion flow shear and no electron current, Phys. Plasmas, 11: 4501-4505.

Kim SH, Merlino RL (2007). Electron attachment to C_7F_{14} and SF_6 in a thermally ionized potassium plasma, Phys. Rev., 76 : 35401.

Mishra R, Verma P, Tiwari MS (2007). Effect of general loss loss-cone distribution on shear- driven electrostatic ion cyclotron instability, J. Planetary Space Sci., 9: 4.

Misra KD, Pandey RS (1995). Generation of Whistler Emissions by Injection of Hot Electrons in the Presence of a Perpendicular Ac Electric Field, J. Geophys. Res., 100: 19405-19411.

Misra KD, Tiwari MS (1977). Particle aspect analysis of drift instability in the presence of non- uniform electric field, Phys. Scripta, 16: 142.

Pandey RS, Misra KD, Tripathi AK (2001). Kelvin –Helmholtz instabilityin an Anisotropic Magneto-plasma in the presence of Inhomogeneous Perpendicular Electric Field and Parallel Flow Velocity Shear Indian J. Radio Space Phys., 30: 113.

Pandey RS, Misra KD, Tripathi AK (2003). Generation of electrostatics Ion-Cyclotron like Wave by Parallel Flow Velocity Shear in the presence of Inhomogeneous Electric Field in an Anisotropic Magneto-plasma, Indian J. Radio Space Phys., 30: 75-80.

Perron PJ, Noel G, St-Maurice JP (2009). Velocity shear and current driven instability in a collisional F-region, Ann. Geophys. 27: 381-394.

Rosenberg M, Merlino RL (2008). Instability of higher harmonic electrostatic ion cyclotron waves in a negative ion plasma, J. Plasmas Phys., pp. 1-14.

Teodorescu C, Reynolds EW, Koepke ME (2002). Observation of inverse Ion-cyclotron damping induced by parallel velocity shear, Phys. Rev. Lett., 88: 185003.

14

A study on the engineering properties of sandcrete blocks produced with rice husk ash blended cement

G. L. Oyekan[1]* and O. M. Kamiyo[2]

[1]Department of Civil and Environmental Engineering, University of Lagos, Nigeria.
[2]Department of Mechanical Engineering, University of Lagos, Nigeria.

Sandcrete blocks have been in used in many nations of the world including Nigeria, playing a major role in the building industry. The material constituents, their mix, presence of admixtures and the manufacturing process are important factors that determine the properties of sandcrete blocks. This paper investigates the effects of a partial replacement of cement with rice husk ash (RHA) on some engineering properties of hollow sandcrete blocks with 1:6 cement-sand mix ratios. Single block size 225 x 225 x 450 mm, is produced with a vibrating machine. Results show that the addition of RHA produces blocks of lower density. Particularly, the density of the blocks decreases as the RHA content increases. The compressive strength of the block is also not enhanced. Results also reveal that the thermal and hygrothermal properties of the blocks are significantly affected.

Key words: Sandcrete blocks, rice husk ash, compressive strength, hygrothermal properties, thermal properties.

INTRODUCTION

Hollow sandcrete blocks containing a mixture of sand, cement and water are used extensively in many countries of the world especially in Africa. In many parts of Nigeria, sandcrete block is the major cost component of the most common buildings. The high and increasing cost of constituent materials of sandcrete blocks has contributed to the non-realization of adequate housing for both urban and rural dwellers. Hence, availability of alternatives to these materials for construction is very desirable in both short and long terms as a stimulant for socio-economic development. In particular, materials that can complement cement in the short run, and especially if cheaper, will be of great interest.

Over the past decade, the presence of mineral admixtures in construction materials has been observed to impart significant improvement on the strength, durability and workability of cementitious products (Mental, 1994; Falade, 1997; Oyekan, 2001). In areas prone to flood, hygrothermal properties of the buildings' construction materials are of importance. Also, energy requirements for residential and commercial buildings are known to be influenced by building design and by the materials used. In both temperate and tropical regions, thermal properties of building materials are of significant importance to the determination of the heating or cooling load within the building and hence the capacity of the mechanical equipment required in handling the load. This is necessary to provide a given level of thermal comfort within the building and over the annual climatic cycle. Substitution of any of these admixtures is aimed at enhancing at least one of the properties of the block. Rice husk is a residue produced in significant quantity on a global basis. While it is utilized as fuel in some regions, it is regarded as a waste in others thereby causing pollution; due to problem with disposal.

Hence, it's profitable use in an environmentally friendly manner, will be a great solution to what would otherwise be a pollutant. When burnt under controlled conditions, the rice husk ash (RHA) is highly pozzolanic and very suitable for use in lime-pozzolana mixes and for Portland cement replacement (Smith, 1984; Chandrasekhar et al., 2003; Yogenda et al., 1988; Anwar et al., 2000; Paya et al., 2000; Nair et al., 2006; Goncalves and Bergmann, 2007; Rodriguez et al., 2008). Effect of RHA blended cement on the strength and permeability properties of concrete has been investigated by Ganesan et al. (2008). On sandcrete block, Cisse and Laquerbe (2000)

*Corresponding author. E-mail: gloyekan@yahoo.com.

observed that the mechanical resistance of sandcrete blocks obtained when unground ash is added increased in performance over the classic mortar blocks. Their studies on Senegalese RHA also revealed that the use of unground RHA enabled production of lightweight sandcrete block with insulating properties at a reduced cost. The ash pozzolanic reactivity was responsible for the enhanced strength obtained. Okpala (1993) partially substituted cement with RHA in the percentage range of 30–60% at intervals of 10% while considering the effect on some properties of the block.

His results revealed that a sandcrete mix of 1:6 (cement/sand ratio) required up to 40% cement replacement and a mix of 1:8 ratio required up to 30%, are adequate for sandcrete block production in Nigeria. However, it is worthy of note that due to the high cost of procuring the rice husk required for producing large number of blocks needed for an average-size building, and in the light of the diminishing agricultural activities in Nigeria, replacing cement with such high volume of RHA could be economically counterproductive for local sandcrete block manufacturers thereby defeating the main purpose of the substitution which is to reduce the unit cost of the block.

Also, the hygrothermal properties and some vital thermal properties are not considered in the investigation. Therefore, to encourage continual production and for better characterization of the sandcrete blocks, this paper investigates experimentally the effect of partially replacing cement with the Nigerian rice husk ash on some structural, thermal and hygrothermal properties of the blocks. The replacement is by volume and it is in the range of 5–20% at intervals of 5%.

MATERIALS AND METHODS

The material constituents, their mix, presence of admixtures and manufacturing process are important factors that determine the properties of sandcrete blocks. The materials used and method of manufacture employed in this investigation are thus presented.

Materials of the sandcrete blocks

The sandcrete blocks are made of sand, cement and water.

Sand

Sharp river quartzite sand that is free of clay, loam, dirt and any organic or chemical matter is used. It is sieved with the 3.35 mm zone of British standard (BS) test sieves. The sand has a specific gravity of 2.66 and an average moisture content of 0.90%. The coefficient of uniformity of the sand is 2.95.

Cement

The cement used is the ordinary portland cement (OPC) from the West African Portland Cement Company, Ogun State, Nigeria with properties conforming to BS 12 (British Standards Institution, 1971).

Water

Fresh, colourless, odourless and tasteless potable water that is free from organic matter of any kind.

Rice husk ash

Rice husks used were collected from a threshing site at Ifo, Ogun State, Nigeria. Open field burning method is employed. The husks are burnt into ash and sifted in order to remove all particles that do not pass the 45 μm test sieve (British Standards Institution, 1997). Table 1 shows the result of the chemical analysis of the RHA and of the Portland cement. From the chemical analysis of RHA, the sum of SiO_2, Al_2O_3 and Fe_2O_3 is 79%. This satisfies the minimum percentage requirement for pozzolana when these constituents are added, which is 70% for ASTM C618 (American Society for Testing and Materials, 1978).The moisture content is less than 1.5% specified by BS 3892. It therefore implies that the RHA used in the present study is a good pozzolana since it meets the requirements of BS 3892 and ASTM C618 for pozzolana. The specific gravity of the RHA is 2.17 and its density is 2170 kg/m^3. The specific gravity value is almost the same as 2.15 obtained by Sampaio et al. (2003) but less than 2.36 reported by Cook et al. (1977) and higher than 1.54 obtained by Okpala (1993).These values indicate that the specific gravity of RHA is location and harvest-time dependent.

Grading of aggregates

The grading of an aggregate defines the properties of different sizes in the aggregate. This grading has a considerable effect on the workability and stability of the mix. Wet sieving analysis which is in accordance with BS 1377 (British Standards Institution, 1990) is used. The particle size distribution curve of the sand used in this study is shown in Figure 1.

Manufacture of sandcrete blocks

The blocks (all hollow) are manufactured with the use of a vibrating machine. The standard mix proportion of 1:6 cement-sand ratio; that is, one part by volume of cement to six parts by volume of coarse sand, is used in this investigation. The size of the block produced is 225 x 225 x 450 mm with one-third of the volume void so as to produce the type of hollow sandcrete blocks commonly used for construction of buildings in Nigeria. Four percentages of cement substitution with RHA (that is, 5, 10, 15 and 20%) by volume are used. A specific gravity of 2.17 for RHA in comparison with 3.14 for OPC implies that replacement of cement with RHA by weight would result in a much greater volume of the cementitious material in the mix. Consequently, the replacement of cement with RHA is by volume, as well as the batching of cement and sand.

In the manufacture of the blocks, hand mixing is employed and the materials are turned over a number of times until an even colour and consistency are attained. Water is added through a fire hose and it is further turned over to secure adhesion. It is then rammed into the machine moulds, compacted and smoothed off with a steel face tool. After removal from the machine moulds, the blocks are left on pallets under cover in separate rows, one block high and with a space between 2 blocks for the curing period. They are kept wet during this period by watering daily. The laboratory condition is 27 ± 2°C dry-bulb temperatures, 50 ± 5% relative humidity. Testing for compressive strength is then carried out at ages 7, 14, 21 and 28 days. On the average, four specimens are tested at each age and each percentage replacement of cement with RHA.

Block units immersed in liquid absorb the liquid due to their porous nature. The volume, the rate and the dominant method of

Table 1. Chemical analysis of RHA and cement (OPC).

Parameter	Rice husk ash (%)	Cement (%)
Silica (SiO_2)	76.0	21.0
Aluminium oxide (Al_2O_3)	3.0	5.22
Ferrous oxide (Fe_2O_3)	Not detected	4.75
Calcium oxide (CaO)	6.0	64.73
Magnesium oxide (MgO)	1.3	2.01
Sodium oxide (Na_2O)	1:18	0.19
Potassium oxide (K_2O)	0.10	0.42
Barium oxide (BaO)	0.24	-
Lead oxide (PbO)	Not detected	-
Sulphite (SO_3^{2-})	Not detected	1.48
Chloride	Not detected	-
Moisture	0.27	-
Ash	11.28	-

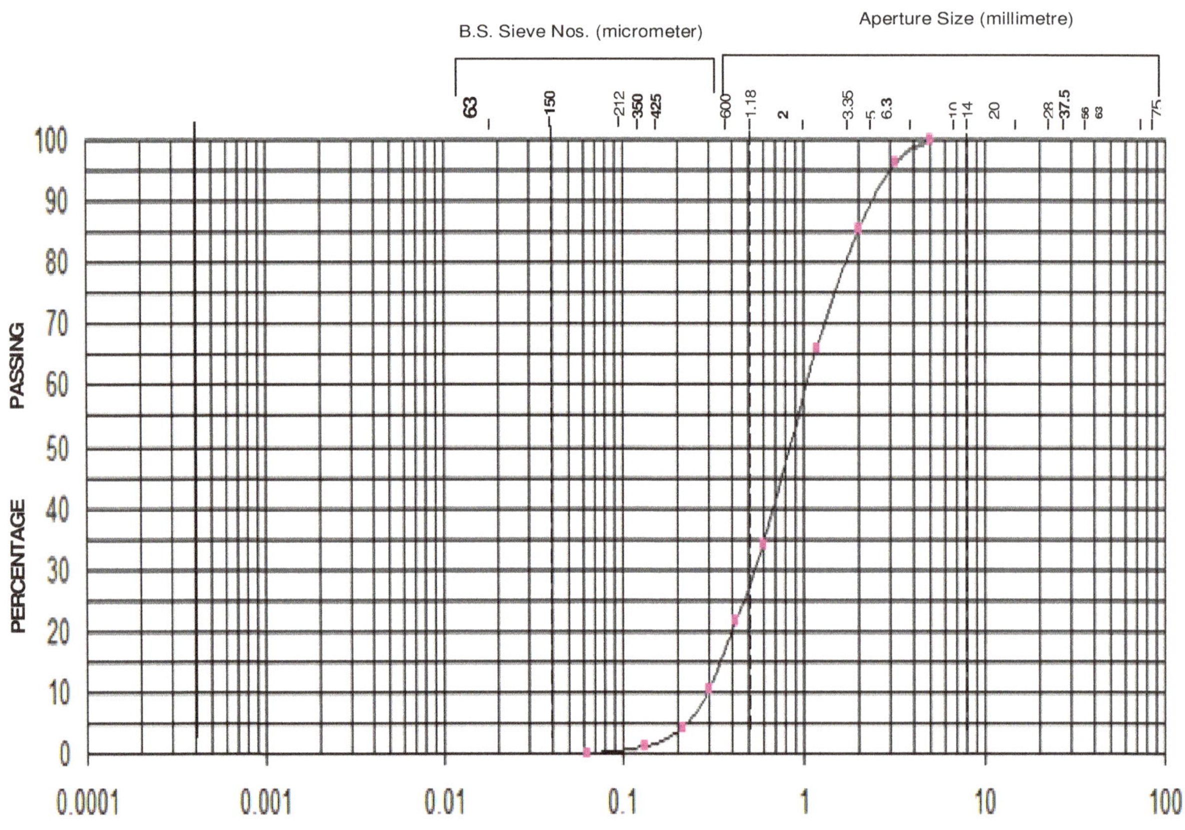

Figure 1. Particle size distribution curve of sand.

absorbing the liquid depends on the interstitial arrangement of the particles of the constituent materials at macro level. It is therefore necessary to investigate the effects of replacing cement with RHA on the hygrothermal properties of the block. Some of the properties are determined as follows:

Porosity

Presence of admixtures may increase, decrease or maintain the porosity of the main material depending on the aggregate sizes. When exposed to persistent flooding, a highly porous block could absorb much water, consequently become weakened and eventually fail. The volume of liquid absorbed by a porous medium is an indication of its pore volume and it is a good approximate measure of its porosity. Hence, porosity υ is obtained with the relation.

$$\upsilon = \frac{V_f}{V} x\, 100\ \% \qquad \text{......(1)}$$

Permeability

The term "permeability" is often loosely used to cover a number of different properties. In this paper, it is defined as the property of a porous medium which characterizes the ease with which a fluid will pass through it under atmospheric pressure. Darcy's law for fluid flow in a permeable medium expresses permeability in terms of measurable quantities and states that the steady state rate of flow is directly proportional to the hydraulic gradient. Thus, for uni-axial penetration employed, permeability, K can be expressed as

$$K = \frac{\upsilon H^2}{2th} \qquad \text{.......(2)}$$

Sorptivity

In using blocks for external walls in tropical humid climate, water-resistance ability of the blocks must be considered in order to minimize penetration of moisture or rain water into the interior of the building. Many times, block work is used in the construction of channels for drainage. Blocks to be used for such purposes must have low sorptivity value. Sorptivity is a measure of the capacity of a porous medium to absorb liquid by capillarity. The absorption of water under capillary action is directly proportional to the square-root of time (Hall, 1989).

$$A'' = S\sqrt{t} \qquad \text{....... (3)}$$

Various test methods are used to determine hygrothermal properties of a material. However, in most cases, the test method chosen for a particular property is always the one appropriate to the predominant transport mechanism acting on the block. After evaluation of various methods, capillary rise method is employed in this investigation due to its simplicity and accuracy. Basically, a sample of sandcrete block is placed with one edge fully in contact with the water surface. Through that area, water is absorbed. The height of capillary rise is then measured at increasing time intervals. The fineness of the capillary pores in sandcrete blocks promotes absorption of water by capillary attraction; hence, a measure of the rate of absorption provides a useful indication of the pore structure. If water is absorbed rapidly, it shows that the pores are either large or straight; if the absorption rate is slow, then the pores are small or not easily accessible.

Thermal properties

Thermal properties of most cementitious materials are found to change with the presence of admixtures (Cisse and Laguerbe, 2000; Okpala, 1993). The change is found to depend on the admixture's grain structure or interstitial arrangement within the main material and other micro structural parameters including the volumetric fraction of each constituent, the shape of the particles, and the size distribution of the particles. In predicting the thermal performance of buildings, it is necessary to consider the dynamic effects of this variation. The thermal properties investigated in this study are:

Thermal conductivity (k)

Thermal conductivity is a measure of the quantity of heat that flows through a material per unit time. Thermal conductivity of most materials is found to change with the presence of impurities or admixtures. From the Fourier's steady-state heat conduction equation, thermal conductivity is determined as

$$k = \frac{Q}{A}\frac{\Delta x}{\Delta T} \qquad \text{......... (4)}$$

Specific heat capacity (c)

It is a measure of the thermal storage capacity of the material. The specific heat capacity of a sandcrete block indicates the relative amount of heat energy the wall built with it is capable of storing per unit mass. Walls with high specific heat capacity can store more energy, have a larger thermal lag, and thus, generally be more effective for thermal storage and peak load shifting. This time lag effect contributes to shifting demand to off-peak periods and improves overall thermal efficiency. Specific heat capacity of the sandcrete block is determined from the classical heat capacity equation

$$c = \frac{Q}{m\theta} \qquad \text{..(5)}$$

Thermal diffusivity (α)

Thermal diffusivity is a measure of the material's ability to undergo a temperature change. It describes the heat transfer capability of a material relative to its heat storage ability. Materials with a low thermal diffusivity have a slow rate of heat transfer relative to heat storage. Thermal diffusivity is obtained through the relation

$$\alpha = \frac{k}{\rho c} \qquad \text{...(6)}$$

Thermal effusivity (β)

Thermal effusivity represents the capacity of a material to absorb and release heat. The value of the thermal effusivity is useful in calculating the heat-accumulation capacity of materials. Materials with high thermal effusivity cannot hold heat long enough because heat will quickly dissipate from its surface as soon as surrounding temperature drops. On the other hand, materials with low thermal effusivity (but with high thermal inertia) will hold heat much longer. Thermal effusivity is calculated as

$$\beta = \sqrt{k\rho c} \qquad \text{.....(7)}$$

RESULTS AND DISCUSSION

Necessary precautions were taken at every stage and for each procedure to minimize error as much as possible. The results are presented in graphical form in most cases.

Effect of RHA substitution on compressive strength

Figures 2a and b show the variation of compressive

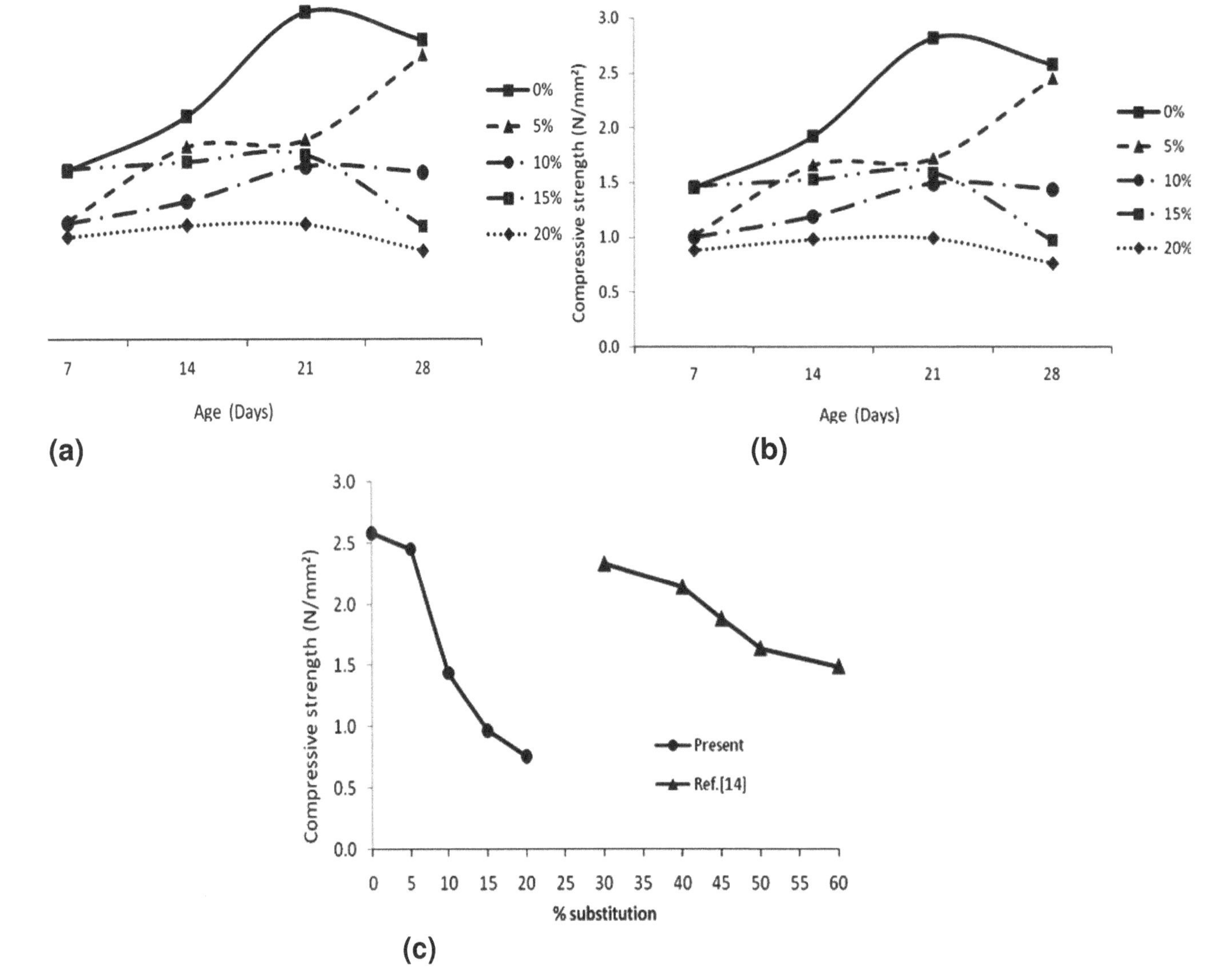

Figure 2. Plot of compressive strength against (a) percentage RHA substitution; (b) age of sandcrete blocks; (c) percentage RHA substitution.

strength with percentage RHA content for the mix proportion used. The graphs show that the blocks decrease in strength as the RHA percentage content in the mix increases. It shows that RHA does not appreciably enhance the compressive strength of the conventional sandcrete block. The reduction in strength could be attributed to the fact that the partial replacement of cement with the RHA caused a reduction in the quantity of cement in the mix available for the hydration process and hence a reduction in the formation of the stable strength- producing cementitious compounds. The difference in the chemical composition of RHA and of cement it replaced is also a major factor. As shown in Table 1, the major constituent of RHA is silica (SiO_2) while that of cement is quicklime (CaO). On mixing with water, the silica produced in solution reacts with calcium to produce calcium fly ash (C_3H) which is further hydrated in the reaction with water to produce tobermorite (CSH) gel. The formation of CSH gel is responsible for slower rate of hydration of cement which commences after about 14 days and extends up to about 150 days after which the fly ash particles would almost be completely disintegrated.

Figure 2c shows the plot of the compressive strength against percentage RHA substitution for the 0-20% range considered in the present study and the plot for the 30-60% range investigated by Okpala (2003) for the 28-day-old blocks. Rather than the gradual decrease in the compressive strength observed in Okpala (1993), the decrease is sharp in the present study. At 20% RHA substitution, the compressive strength has reduced drastically to 29.5% in the strength of the control block. But in Okpala (1993), at 60% RHA substitution, the strength is just 39% of that of the control block. This means that the strength of the blocks produced with RHA used by Okpala within the 5-20% range would be higher

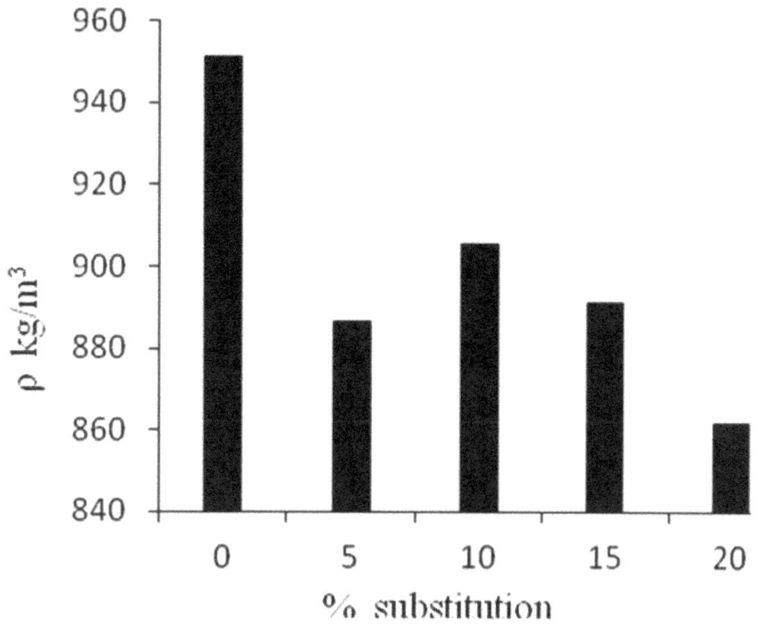

Figure 3. Plot of density against percentage RHA substitution.

than that obtained in the present study. This may be due to differences in the physical properties of the RHA used. These properties depend on the source location, harvest time, rice husks processing and burning procedure used. The differences in block size and block manufacturing process could also be some of the possible reasons.

Effect on density

The plot of density variation with the percentage increase in the RHA substitution after 28 days is shown in Figure 3. The plot indicates decrease in density with percentage increase in RHA substitution. Other data indicate that the density decreases with curing age. This means that partially substituting cement with RHA produces lighter sandcrete blocks.

Effect on porosity

The variation of porosity with percentage substitution of the RHA is presented in Figure 4. The results show that all the blocks with the admixture are more porous than that of the control (0%). Figure 4 shows a gradual increase in the value of porosity when only 5% RHA is added. But further addition of the admixture rapidly increases the porosity value. This increase in porosity may be as a result of trapped air bubbles that are interconnected. It can be concluded that sandcrete blocks with RHA blended cement is more porous, absorbs more liquid and hence could fail faster than those without admixture. This makes such type of block not suitable for areas prone to flood or with year-round rainfall like Niger Delta areas.

Effect on permeability

Figure 5 shows the variation of permeability with the percentage increase in RHA content. It is observed that partially replacing cement with RHA increases the permeability of the sandcrete block. The value of the permeability steadily increases with percentage increase in RHA content. The value is almost doubled for every 5% addition of RHA. This means that the inclusion of the admixture opens up the block in a way that encourages upflow of fluid. This rapid flow of water indicates that the pores are either large or straight.

Effect on sorptivity

It is observed in Figure 6a that the value of sorptivity of the sandcrete blocks gradually increases with the percentage content of RHA. This implies that blocks made with cement partially replaced with RHA are not suitable for drainage channels construction but could be useful for partitioning of building spaces. The results of the present study were compared with the experimental data of Sampaio et al. (2003) on sandcrete blocks produced with Portuguese RHA (Figure 6b). The plot shows that while the sorptivity value of the blocks produced with cement and blended with Nigerian RHA increases, that of the blocks produced with the Portuguese RHA blended cement decreases as the

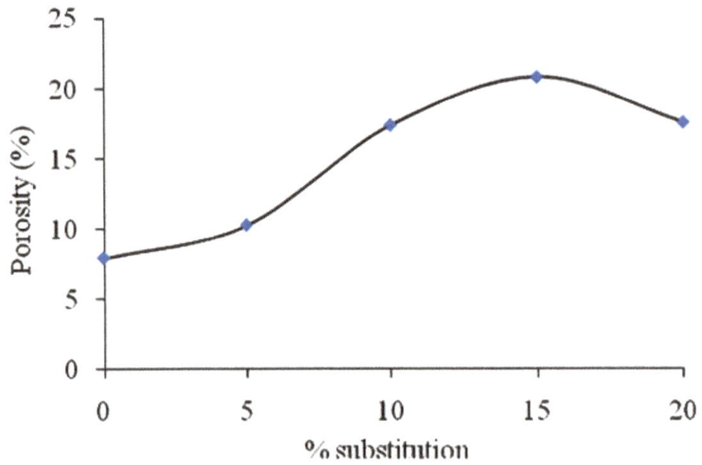

Figure 4. Plot of porosity against percentage RHA substitution.

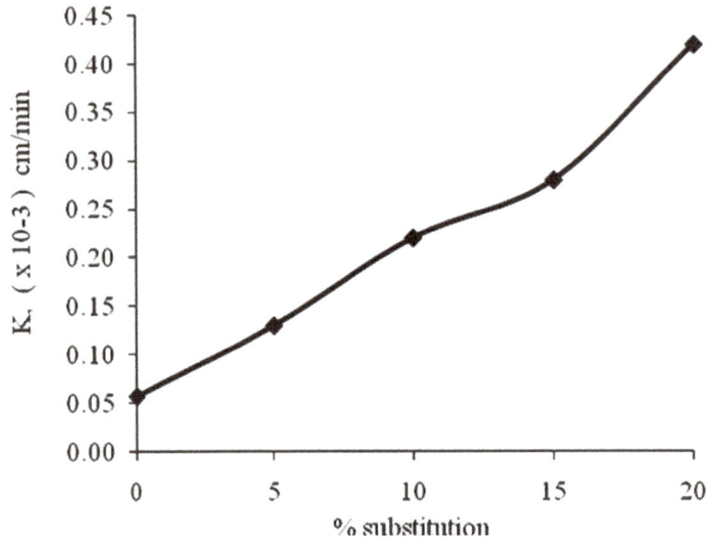

Figure 5. Plot of permeability against percentage RHA substitution.

percentage RHA content increases. The implication of this is that the substitution of cement with the Nigerian RHA forms a material structure that encourages liquid absorption as the RHA content increases whereas the blocks produced with the cement partially replaced with the Portuguese RHA forms a more compact structure that reduces absorption of liquid.

Effect of temperature change on the hygrothermal properties

As the temperature of the block increased above the room temperature, it was observed that the block absorbed water at a faster rate. This means that the higher the temperature around a building, the greater the rate of liquid absorption.

Effect on thermal conductivity

The thermal conductivity of the sandcrete blocks is determined using the Guarded Hot Plate Box conforming to the requirements of ASTM C177 (American Society for Testing and Materials, 2004). The effect of the substitution of RHA on the thermal conductivity of the sandcrete blocks is presented in Figure 7a. The value of the thermal conductivity increases with the percentage RHA content. The higher k value obtained when cement was partially substituted with RHA may be as a result of the products of reaction of RHA with cement in the mix during the hydration process. The products that tend to increase the k value of the block despite its increased porosity are subject to further investigation. For a hot climate, higher thermal conductivity block is undesirable for construction as it increases the cooling/heating load

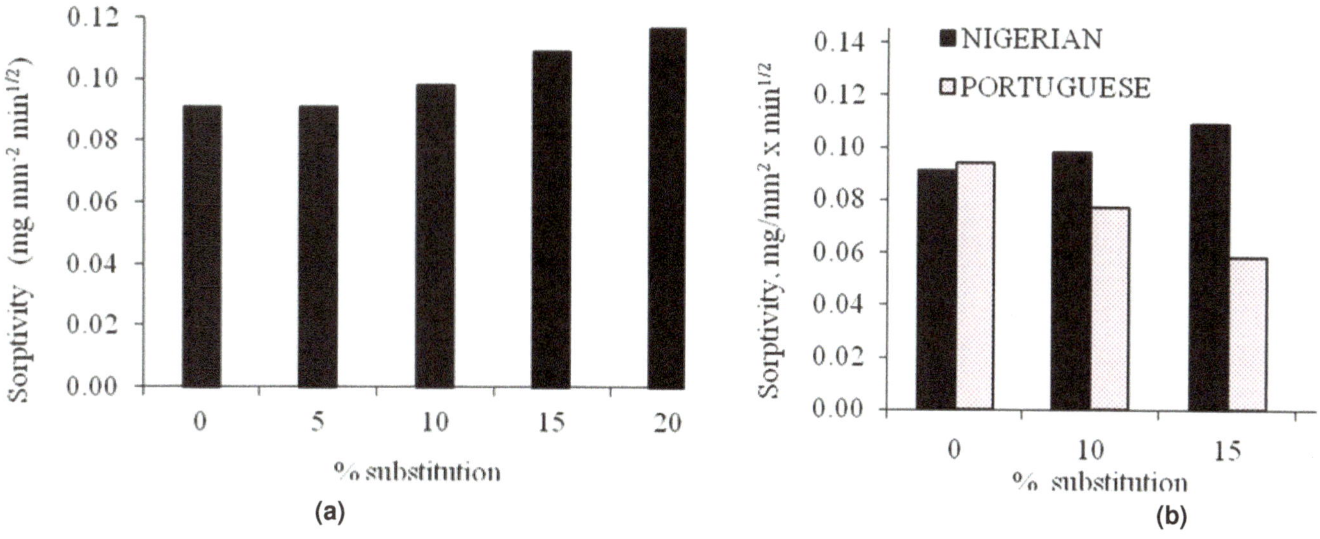

Figure 6. Plot of sorptivity against (a) percentage RHA substitution;(b) percentage RHA substitution for Nigeria and Portugal.

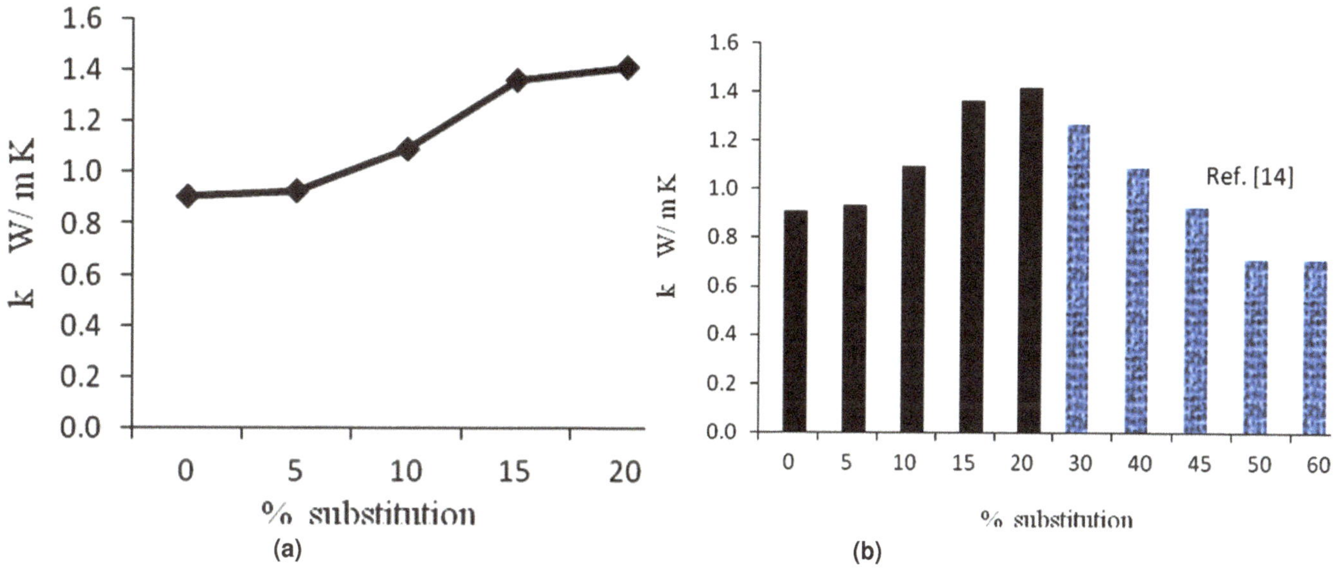

Figure 7. Plot of thermal conductivity against percentage RHA substitution for (a) 0-20; (b) 0-60.

in the building space.

The plot of the 0-20% RHA substituted range investigated in this study and the 30-60% range of Okpala (1993) in Figure 7b, shows an interesting variation. The value of the thermal conductivity is found to increase with percentage RHA substitution up to a maximum at 20% (or possibly 25%) and then decreases thereafter as the percentage RHA increases. The initial increase in k value with increase in the percentage RHA content may be as a result of higher k values of the products of reaction between RHA and cement. However, as more RHA is added to the mix and the quantity of cement reduces, the inert nature of the excess RHA probably caused the thermal conductivity to reduce. The complex structure of the block material may be a contributing factor. Also, being a non-metallic solid, heat is transferred via lattice vibrations which is characterised by much variability.

Effect on specific heat capacity

The determination of the specific heat capacity of the sandcrete blocks is carried out using the adiabatic calorimetric technique. The plot of the specific heat capacity against percentage RHA substitution in Figure 8, indicates that all the blocks with the admixture have slightly lower values (except the 5% block) than that of the control block; hence, lower heat energy storing capacity and lower thermal mass. In tropical environment, these blocks will lose heat gained during the day faster, thereby

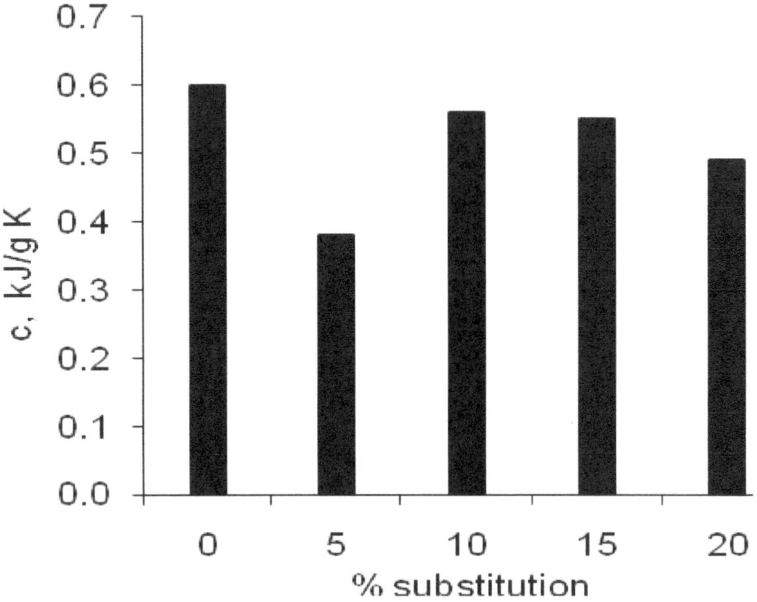

Figure 8. Plot of specific heat capacity against percentage RHA substitution.

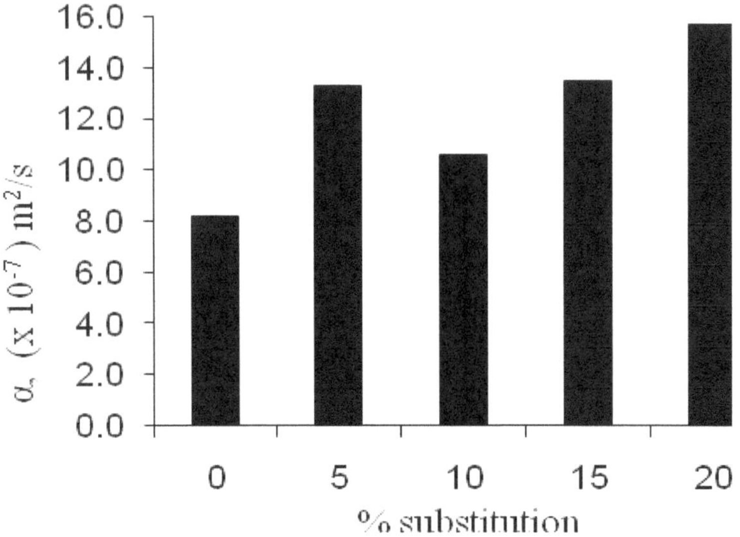

Figure 9. Plot of thermal diffusivity against percentage RHA substitution.

making the building space comfortable for early sleep.

Effect on thermal diffusivity

In a manner slightly consistent with the thermal conductivity variation, Figure 9 shows that the values of the thermal diffusivity of all the blocks is higher than that of the control. The reducing values of both density and the specific heat capacity with RHA content caused the thermal diffusivity value to increase with the percentage RHA substitution. In effect, sandcrete blocks made with RHA-blended cement will undergo a faster temperature change or allow more rapid heat flow through them than those without the admixture.

Effect on thermal effusivity

The thermal effusivity value in Figure 10 fluctuates as the percentage RHA content increases. The β values of most of the blocks containing RHA are more than those of the control block. The β value of the 15% RHA block, which is the largest, is 17.5% more than that of the control block. In practical terms, the higher thermal effusivity of the blocks made with RHA-blended cement increases its ability to conduct heat away from the building space faster thereby reducing the air-conditioning load, and

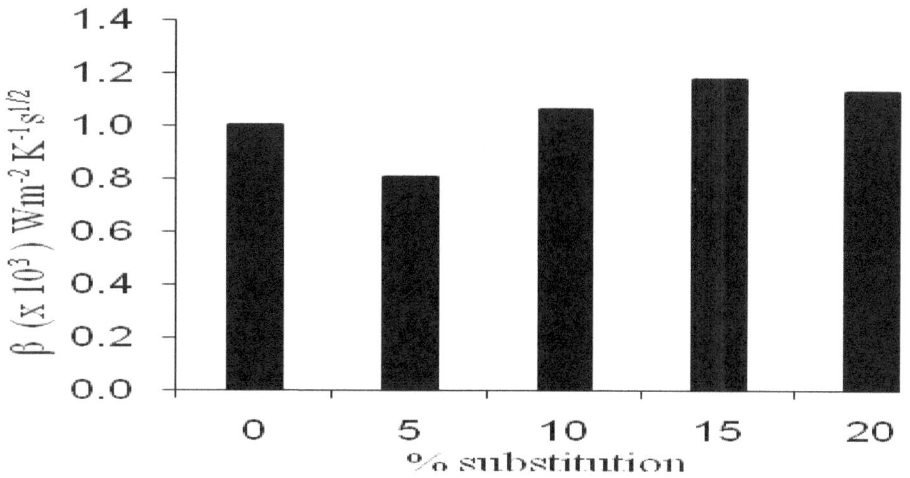

Figure 10. Plot of thermal effusivity against percentage RHA substitution.

consequently increasing the period of thermal comfort.

It is worthy of note that, despite much effort in minimizing possible areas of errors, the observed thermal test results may have some influences which might likely be caused by two factors: differences in heat transfer pathways and differences in sample homogeneity. There may be inhomogeneous intrasample effects, contact resistance effects and differences in the heat transfer process through the sample which could affect the test results. Though the particle sizes of RHA and OPC are the same, for a given volume of sample, changes in the particle size distribution of both materials in the mix with sand may lead to changes in the number of particle-particle and particle-gas interfaces as well as the interstitial voids and the volume of those voids. A wider variety of heat transfer pathways are available to the penetrating heat wave through a sample with broader particle size distribution. This also could, by extension, be applied to the hygrothermal properties due to the phenomenon of flowing water's 'path of least resistance'. Nevertheless, since the test results agree to some extent with the internationally-accepted standard values available in the literature, it would appear that these factors did not have any significant effects.

Conclusions

The main conclusions derived from this investigation are as follows:

1. As the percentage RHA content in the mix increases, the compressive strength of the sandcrete block decreases. Hence, RHA does not enhance the compressive strength of the conventional sandcrete block with 1:6 cement-sand mix.
2. The addition of RHA into the cement-sand matrix produces sandcrete blocks of lighter weight. The density of the blocks decreases as the percentage RHA content in the mix increases.
3. A sandcrete block with RHA-blended cement is more porous than a pure sand-cement block. Its permeability increases with RHA content and ambient temperature.
4. The products of the reaction between the rice husk ash and the ordinary Portland cement tend to initially slightly increase the thermal conductivity of the block. However, as the percentage RHA content increases beyond 20%, the thermal conductivity reduces with increasing RHA content.
5. The sandcrete blocks made with RHA-blended cement have lower heat storage capacity and lower thermal mass than those without the admixture.
6. Due to the increasing values of thermal diffusivity with increasing RHA content, blocks containing RHA tend to undergo faster temperature change as heat flows through them more rapidly.
7. The increased thermal effusivity of the sandcrete block with RHA content is an advantage over pure sandcrete block as it improves its ability to conduct heat away from the building space faster thereby reducing the air-conditioning load and hence increasing the period of human thermal comfort.

The practical significance of this study is that the strength, liquid-absorption and heat transfer characteristics of the sandcrete blocks presented will be of a great value to building professionals engaged in the design and analysis of building structures. Designers would be able to properly size the heating/cooling equipment necessary to provide a given level of thermal comfort within the building and over the annual climatic cycle. Consequently, there is reduction in the size of the air-conditioning system required to cool or heat the space, reduction in the thickness of the thermal insulator and extension of the period of human comfort without reliance on mechanical air-conditioning. The above qualities reduce the annual energy cost in addition to other energy conservation and environmental effects.

ACKNOWLEDGEMENT

This work was funded by the Central Research Committee of the University of Lagos, Nigeria. Their assistance is hereby gratefully acknowledged.

Abbreviations: BS, British standard; **OPC,** ordinary Portland cement; **RHA,** rice husk ash; **A,** area of material perpendicular to heat flow (m^2); **A″,** cumulative infiltration; **c,** specific heat capacity (J/g K); **H,** height of liquid rise after time t (m); **h,** hydraulic head (m); **k,** thermal conductivity (W/m K); **K,** permeability (cm/min); m, mass of the sample (kg); **P,** percentage substitution of admixture (%); **Q,** heat flux (W); **S,** sorptivity ($m/s^{1/2}$); **ΔT,** temperature change (K); **T,** time taken for liquid to rise (s); **V,** volume of material sample (m^3); V_f**,** volume of water absorbed (m^3); **Δx,** sample thickness (m); **α,** thermal diffusivity (m^2/s); **β,** thermal effusivity ($W/m^2\ K\ s^{1/2}$); **θ,** change in temperature (K); **ρ,** density (kg/m^3); **υ,** porosity of the material (%).

REFERENCES

American Society for Testing and Materials (1978). Specifications for pozzolanas. ASTM International, USA, ASTM C618.

Anwar M, Miyagawa T, Gaweesh M (2000). Using rice husk ash as a cement replacement material in concrete. Wast Manage. Series, 1: 671-684.

American Society for Testing and Materials (2004). Standard test method for steady-state heat flux measurements and thermal transmission properties by means of the guarded-hot-plate apparatus. ASTM International, USA. ASTM C177.

British Standards Institution (1971). Ordinary and rapid-hardening Portland cement. London. BS, 12: 2.

British Standards Institution (1990). Methods of testing for soils for Civil Engineering purposes. London. pp. 1377.

British Standards Institution (1997). Specifications for pulverised fuel ash for use with Portland cement in concrete. London, BS 3892: I.

Chandrasekhar S, Satyanarayana KG, Pramada PN, Raghavan P, Gupta TN (2003). Processing, properties and applications of reactive silica from rice husk – an overview. Mater. Sci., 38(15): 3159-3168.

Cisse IK, Laguerbe M (2000). Mechanical characterization of filler sandcretes with rice husk ash additions; study applied to Senegal. Cem. Concr. Res. 30(1):13–18.

Cook DJ, Pama, RP, Paul BK (1977). Rice husk ash-lime-cement mixes for use in masonry units. Build Envir., 12: 281-288.

Falade F (1997). The use of ground broken bottles as partial replacement of cement in concrete. In: Proceedings of Fourth International Conference on Structural Engineering Analysis and Modelling, Ghana. p. 473–486.

Ganesan K, Rajagopal K, Thangavel K (2008). Rice husk ash blended cement: assessment of optimal level of replacement for strength and permeability properties of concrete. Constr. Build. Mater., 22(8):1675-1683.

Goncalves MRF, Bergmann CP (2007). Thermal Insulators made with rice husk ashes: Production and correlation between properties and microstructure. Constr. Build. Mater., 21(12): 2059-2065.

Hall C (1989). Water sorptivity of mortars and concretes. Rev Mag. Concr. Res., 41(147): 51-61.

Mental PQ (1994). Mineral admixtures for concrete – an overview of recent developments. In: Advances in Cement and Concrete. Proceedings of Engineering Foundation Conference, University of New Hampshire, Durham. p. 243-256.

Nair DG, Jagadish KS, Fraaij A (2006). Reactive pozzolanas from rice husk ash: An alternative to cement for rural housing. Cem Concr. Res., 36(6): 1062-1071.

Okpala DC (1993). Some engineering properties of sandcrete blocks containing rice husk ash. Build. Environ., 28(3): 235-241.

Oyekan GL (2001). Effect of granite fines on the compressive strength of sandcrete blocks, In: Proceedings of Conference on Construction Technology, Sabah, Malaysia. p. 14-17.

Paya J, Monzo J, Borrachero MV, Peris-Mora E, Ordonez LM (2000). Studies on crystalline rice husk ashes and the activation of their pozzolanic properties. Wast Manage. Series, 1: 493-503.

Rodriguez de Sensale G, Ribeiro AB, Goncalves A (2008). Effects of RHA on autogenous shrinkage of Portland cement pastes. Cem. Concr. Compos., 30(10): 892-897.

Sampaio J, Coutinho SJ, Sampaio MN (2003). Portuguese rice husk ash as a partial cement replacement In: Proceedings of 1st Inter-American Conference on Non-Conventional Materials and Technologies on the Eco-construction and Infrastructure. Joao Pessoa, Brazil.

Smith R (1984). Rice husk ash cement: progress in development and applications. Intermediate Technology Publications Ltd.

Yogenda MR, Jagadish KS (1988). Pozzolanic properties of rice husk ash, burnt clay and red mud. Build Environ., 23: 303–308.

Analysis of smart crack detection methodologies in various structures

Dayal R. Parhi* **and Sasanka Choudhury**

Department of Mechanical Engineering, National Institute of Technology, Rourkela, India.

This paper describes a comprehensive review of various technical papers in the domain of crack detection in Beam-Like Structure. Sensibility analysis of experimentally measured frequencies as a criterion for crack detection has been extensively used in the last decades due to its simplicity. However determination of crack parameters (like depth and location) is not straight forward. The various techniques discussed on the basis of dynamic analysis of Crack. The techniques mainly of fuzzy logic neural network, fuzzy system, hybrid neuro genetic algorithm, artificial neural network, artificial intelligence.

Key words: Neural network, fuzzy logic, genetic algorithm, artificial intelligence, crack detection.

INTRODUCTION

Since last few decades engineers and scientists are working on various techniques for detection of crack in the beamlike structure. The recent trends in crack detection in beam like structures are generally done by using fuzzy logic, neural network, artificial intelligence...etc. Some of the methods done by Researchers mentioned below: Das et al. (2009) have performed analytical studies on fuzzy inference system for detection of crack location and crack depth of a cracked cantilever beam structure using six input parameters to the fuzzy member ship functions. The six input parameters are percentage deviation of first three natural frequencies and first three mode shapes of the cantilever beam. The two output parameters of the fuzzy inference system are relative crack depth and relative crack location Experimental setup has been developed for verifying the robustness of the developed fuzzy inference system. The developed fuzzy inference system can predict the location and depth of the crack in a close proximity to the real results.

Taghi et al. (2008) have proposed a method in which damage in a cracked structure was analyzed using genetic algorithm technique. For modeling the cracked-beam structure an analytical model of a cracked cantilever beam was utilized and natural frequencies were obtained through numerical methods. A genetic algorithm is utilized to monitor the possible changes in the natural frequencies of the structure. The identification of the crack location and depth in the cantilever beam was formulated as an optimization problem.

(Bakhary et al., 2007) applied artificial neural network (ANN) for damage detection. In his investigation an ANN model was created by applying Rosenblueth's point estimate method verified by Monte Carlo simulation, the statistics of the stiffness parameters were estimated. The probability of damage existence (PDE) was then calculated based on the probability density function of the existence of undamaged and damaged states. The developed approach was applied to detect simulated damage in a numerical steel portal frame model and also in a laboratory tested concrete slab. The effects of using different severity levels and noise levels on the damage detection results are discussed.

Maity et al. (2004) have presented a method called damage assessment in structures from changes in static parameter using neural network. The basic strategy applied in this study was to train a neural network to recognize the behavior of the undamaged structure as well as of the structure with various possible damaged states. When this trained network was subjected to the measured response; it was able to detect any existing damage. The idea was applied on a simple cantilever beam. Strain and displacement were used as possible candidates for damage identification by a back

*Corresponding author. E-mail: dayalparhi@yahoo.com.

propagation neural network and the superiority of strain over displacement for identification of damage has been observed.

Saridakis et al. (2008) applied neural networks, genetic algorithms and fuzzy logic for the identification of cracks in shafts by using coupled response measurements. In this research the dynamic behavior of a shaft with two transverse cracks characterized by three measures: position, depth and relative angle. Both cracks were considered to lie along arbitrary angular positions with respect to the longitudinal axis of the shaft and at some distance from the clamped end. A local compliance matrix of two degrees of freedom (bending in both the horizontal and the vertical planes) was used to model each crack.

Huijian Li et al. (2005) has presented a method for crack damage detection. In which he has used a combination of global (changes in natural frequencies) and local (strain mode shapes) vibration-based analysis data as input in artificial neural networks (ANNs) for location and severity prediction of crack damage in beam-like structures and also used finite element analysis to obtain the dynamic characteristics of intact and damaged cantilever steel beams for the first three natural modes.

Panigrahi et al. (2009) has firstly formulate of an objective function for the genetic search optimization procedure along with the residual force method are presented for the identification of macroscopic structural damage in an uniform strength beam. Two cases have been investigated here. In the first case the width is varied keeping the strength of beam uniform throughout and in the second case both width and depth are varied to represent a special case of uniform strength beam. The developed model requires experimentally determined data as input and detects the location and extent of the damage in the beam. Here, experimental data are simulated numerically by using finite element models of structures with inclusion of random noise on the vibration characteristics.

Suh et al. (2000) have presented a method to identify the location and depth of a crack on a structure by using hybrid neuro-genetic technique. Feed-forward multi-layer neural networks trained by back-propagation are used to learn the input (the location and depth of a crack)–output (the structural eigen frequencies) relation of the structural system. With this trained neural network, genetic algorithm is used to identify the crack location and depth minimizing the difference from the measured frequencies.

Huai Chou et al. (2001) stated that, the problem is initially formulated as an optimization problem, which is then solved by using genetic algorithm (GA). Static measurements of displacements at few degrees of freedom (DOFs) are used to identify the changes of the characteristic properties of structural members such as Young's modulus and cross-sectional area, which are indicated by the difference of measured and computed responses. In order to avoid structural analyses in fitness evaluation, the displacements at unmeasured DOFs are also determined by GA. The proposed method is able to detect the approximate location of the damage.

CRACK DETECTION USING ARTIFICIAL INTELLIGENCE TECHNIQUES

Crack detection in various classical methods as well as finite element method has been discussed in various papers. In this paper different types of Artificial Intelligence Techniques adapted by researchers in the domain of crack detection in damaged structures are reviewed briefly.

Neural network method for fault diagnosis of cracked cantilever beam

Das et al. (2009) have performed an analytical studies by on fuzzy inference system for detection of crack location and crack depth of a cracked cantilever beam structure using six input parameters to the fuzzy member ship functions. The six input parameters are percentage deviation of first three natural frequencies and first three mode shapes of the cantilever beam. The two output parameters of the fuzzy inference system are relative crack depth and relative crack location Experimental setup has been developed for verifying the robustness of the developed fuzzy inference system. The developed fuzzy inference system can predict the location and depth of the crack in a close proximity to the real results.

Dynamics of structures with crack has been studied for last four decades intensively. Natural frequencies and modes shapes undergo variation due to presence of crack. The deviations of natural frequencies and modes shapes mainly dependant on location and intensity of the crack. Scientists are focusing their thoughts to find out the damage location and its intensity. The investigations reported in this regards are jotted below. Measurement of flexural vibrations of a rectangular cross-section cantilever beam having a transverse surface crack extending uniformly along the width of the beam and analytical results are used to relate the measured vibration modes to the crack location and depth (Rizos et al., 1990). From the measured amplitudes at two points of the structure vibrating at one of its natural modes, the respective vibration frequency and an analytical solution of the dynamic response, the crack location can be found and depth can be estimated with satisfactory accuracy. The method of crack localization and sizing in a beam can be obtained from free and forced response measure-ments (Karthikeyan et al., 2007).

A method using singular value decomposition is developed to handle the ill-conditioned system equations that occur in the experimental investigation by using the measured natural frequencies of the modified structure. An extensive study has been done on diagnosis of fracture damage in structure (Akgun et al., 1983). The concept of 'fracture hinge' is developed analytically and the same is applied to a cracked section for detecting fracture damage in simple structures. It is experimentally verified that the structural effect of a cracked section can be represented by an equivalent spring loaded hinge.

Artificial neural networks (ANN) can be used as an alternative effective tool for solving the inverse problems because of the pattern-matching capability (Sahoo et al., 2007). The results of ANN are quite encouraging and prove the robustness of the proposed damage assessment algorithm. The fault detection model is developed by using two different artificial neural network approaches, namely feed forward network with back propagation algorithm and binary adaptive resonance network (ART1). The performance of the developed back propagation and ART1 model is tested for a total of seven categories of faults in the centrifugal pumping system. The modal frequency parameters for the flexural vibration of a cantilever beam having a transverse surface crack are analytically computed for various crack locations and depths using a fracture mechanics based crack model (Suresh et al., 2004). The computed modal frequencies are used to train a neural network to identify both the crack location and depth. This modular neural network architecture can be used as a non-destructive procedure for health monitoring of structures. A damage detection algorithm has been developed using a combination of global (changes in natural frequencies) and local (curvature mode shapes) vibration based analysis data as input in artificial neural networks (ANNs) for location and severity prediction of damage in beam-like structures (Sahin et al., 2003). The trained feed-forward back propagation ANNs using the data obtained from the experimental damage case for quantification and localization of the damage is tested. Das et al. (2009) has proposed a new methodology for prediction of crack intensity and its location using neural network technique.

Genetic algorithm method for fault detection in beam-like structures

Taghi et al. (2008) have proposed a method in which damage in a cracked structure was analyzed using genetic algorithm technique. For modeling the cracked-beam structure an analytical model of a cracked cantilever beam was utilized and natural frequencies were obtained through numerical methods. A genetic algorithm is utilized to monitor the possible changes in the natural frequencies of the structure. The identification of the crack location and depth in the cantilever beam was formulated as an optimization problem.

Beams are common structures used to carry and transfer high loads in machines and civil structures, and cracks are the main cause of structural failure. Sudden failure during high load operation could be disastrous, thus early crack detection is important. Non-destructive inspection techniques are generally used to investigate the critical changes in the structural parameters, so that an unexpected failure can be predicted. Damage can be detected, quantified and localized by on-line damage assessment techniques, through measuring vibration parameters, which indicate the global conditions of the structures. In general a crack causes a reduction in the stiffness and an increase in the damping of the structure.

These changes of physical properties cause a reduction in the natural frequencies and a deviation of the mode shape. Therefore, it is possible to predict the location and the depth of a crack by measuring changes in the vibration parameters. Changes in the natural frequencies are used more often than deviation of mode shapes, since frequencies can be measured more easily than mode shapes, and they are less seriously affected by experimental errors (Morassi et al., 2001). Taghi et al. (2008) have proposed a GA-based procedure for estimating crack location and depth in an aluminium beam is described and some guidelines for selecting the GA parameters are presented. The damage effect is modelled as a torsional spring (Rizos et al., 1990), and to develop the equations of the motion of the cracked beam the Hamilton's principle is used. Then the eigenvalue problem is solved to obtain the natural frequencies of the beam. Another way for evaluating the natural frequencies of a cracked beam is based on the use of the finite elements method (Rizos et al., 1990), but it is less accurate than a continuous model which is used in this paper. The identification of crack location and depth in a cantilever beam is formulated as an optimization problem, and the binary and continuous genetic algorithms (BGA, CGA) are used to find the optimal location and depth by minimizing the cost function which is based on the difference of measured and calculated frequencies. In comparison with traditional GAs a GA with small population size combined with large mutation rate is used, so the great exploration of the search space is achieved with a small number of cost function evaluations. We iterate the GA from five different initial points (variables) and choose the best answer and in this way the rate of success in finding the global minimum instead of a local one is increased. In order to compare CGA to BGA we have implemented a set of test points. Both GAs have equal population size and maximum number of iterations. The variables in BGA are represented by a total of 21 bits, while in CGA there is no need to convert the variable values. The obtained results demonstrate higher accuracy of the CGA over the BGA.
The various process used in GA as follows:

1. Selecting the variables and cost function
2. The gene size

3. Initial population
3. Cost evaluation
4. Selection
5. Reproduction
6. Mutation
7. Re-evaluating the costs and iterating the algorithm.
8. Stopping criteria (Based upon fitness function)

Artificial neural network method for fault detection using with consideration of uncertainties

Bakhary et al. (2007) applied Artificial Neural Network for damage detection. In his investigation an ANN model was created by applying Rosenblueth's point estimate method verified by Monte Carlo simulation, the statistics of the stiffness parameters were estimated. The probability of damage existence (PDE) was then calculated based on the probability density function of the existence of undamaged and damaged states. The developed approach was applied to detect simulated damage in a numerical steel portal frame model and also in a laboratory tested concrete slab. The effects of using different severity levels and noise levels on the damage detection results are discussed.

Artificial Neural Networks (ANN) have been utilized by many researchers to identify damage location and severity from various types of input and output variables, as they provide an efficient tool for pattern recognition. Wu et al. (1992) explored the use of an ANN to detect member damage in a 3-storey frame. Pandey et al. (1995) provided a more detailed treatment of ANN architecture in their study identifying damage in a 21- bar bridge truss. Zhao et al. (1998) applied a counter-propagation neural network to detect damage and support movement in a continuous beam. Zapico et al. (2001) developed a procedure for damage assessment of steel structures based on ANN. Most of the studies concluded that ANNs are capable of providing correct damage identification, especially when the structural damage and the associated changes in vibration properties are simulated numerically and are error free. However, in practice uncertainties in the FE model parameters and modelling errors are inevitable. The existence of modelling error in the FE model due to the inaccuracy of physical parameters, non-ideal boundary conditions, finite element discretization and nonlinear structural properties may result in the vibration parameters generated from such a FE model not exactly representing the relationship between the modal parameters and the damage parameters of the real structure. On the other hand, the existence of measurement error in the measured data that is normally used as testing data in an ANN model is also unavoidable.

Since the efficiency of an ANN prediction relies on the accuracy of both components, the existence of these uncertainties may result in false and inaccurate ANN predictions. Therefore, the impact of uncertainties on the reliability of ANN models for structural damage detection needs to be analysed. The objective of this paper is to study the influence of uncertainty on damage identify-cation using a combination of frequency and mode shape as the input variables. To consider the uncertainties in the FE modelling and the measurement data, an approach introduced by Papadopoulos et al. (1998) is applied. Using this method, the probability of damage existence (PDE) can be estimated by comparing the probability distribution of the undamaged and damaged models. To consider the effect of FE modelling error, a statistical ANN model is trained with vibration data generated from the FE model, but smeared with random variations. To include the effect of noise in the measurement data, the testing data used as input to the statistical ANN model for damage identification are also smeared with random noises.

The probability moments of the undamaged and damaged states of the structural parameters are estimated using the point estimation method and verified by Monte Carlo simulation. The Monte Carlo simulation data is also used to determine the type of probability distribution function of the structural parameters of both the undamaged and damaged states. The PDEs are determined from the probability distribution for each structural member.

Damage assessment in structure from changes in static parameter using neural networks

Maity et al. (2004) have been presented a method called damage assessment in structures from changes in static parameter using neural network. The basic strategy applied in this study was to train a neural network to recognize the behavior of the undamaged structure as well as of the structure with various possible damaged states. When this trained network was subjected to the measured response; it was able to detect any existing damage. The idea was applied on a simple cantilever beam. Strain and displacement were used as possible candidates for damage identification by a back propagation neural network and the superiority of strain over displacement for identification of damage has been observed.

The objective of the above method is to locate and assess the damage occurring at any position in a cantilever beam by back-propagation neural network considering displacement and strain as input parameter to the network. The approach here consists of three sub-processes. Firstly, by varying the model parameters of the structure, their corresponding response to the system is calculated through the finite element method. Secondly, a neural network is iteratively trained using a number of training patterns. Here, structural responses are given as input to the neural network, while parameters to be identified are shown to the network as desired data. Finally some structural responses measured are given to the well-trained network, which immediately outputs the appropriate value of parameters for untrained patterns. The model parameter taken here is the EI value of the structural member and the structural responses are displacement and strain for a comparison of the performance of the damage assessment algorithm.

Vibration based damage detection in a uniform strength beam using genetic algorithm

Panigrahi et al. (2009) firstly formulated an objective function for the genetic search optimization procedure along with the residual force method presented for the identification of macroscopic structural damage in an uniform strength beam. Two cases have been investigated here. In the first case the width is varied keeping the strength of beam uniform throughout and in the second case both width and depth are varied to represent a special case of uniform strength beam. The developed model requires experimentally determined data as input and detects the location and extent of the damage in the beam. Here, experimental data are simulated numerically by using finite element models of structures with inclusion of random noise on the vibration characteristics.

In this paper the authors used roulette wheel selection criteria for damage identification of uniform simple supported beam as well as for cantilever beam. Fault classification has been done for cylindrical shells with auto associative neural network along with GA in reference (Marawala et al., 2006). As compared with the traditional optimization and search algorithms, GA search from a population of points in the region of the whole solution space, rather than a single point, and can obtain the global optimum. Other advantages of using GA are that it is a self-start method with no special requirement on the initial value of unknown parameters, other than defining a search range, and also it does not need information such as gradients or derivatives of the function to be minimised.

Moreover, GA has the advantage of easy computer implementation. These properties make GA successful and powerful in the field of structural optimization (Rajasekaran et al., 2003). Genetic algorithm (GA) has been established which may be used intelligently to identify and quantify the damage in a uniform strength beam. Rao et al. (2004a, b) have used this procedure for uniform cantilever beam, truss structures and portal frames. Panigrahi et al. (2007) addressed the problem of damage identification in a cantilever beam with uniform thickness only by changing the selection methods in GA. Here GA along with residual force vector method has been used for damage identification of a uniform strength beam with variation of depth along its length and another case with both width and depth varying. In the present paper, first the concept of residual force vector is introduced to specify an objective function for an optimization procedure, which is then solved by using G.A. The aim is to formulate an objective function in terms of parameters related to the physical properties and state of the structure. The objective function must be formulated in such a way that the minimum or null value is obtained when evaluated with true parameters of the structure. Here the parameters used are the damage factors which are nothing but the reduction in stiffness factor. GA is employed to determine the values of these parameters by following an iteration process. In this study a method known as steady-state selection is selected for reproduction purpose in GA which requires less number of iterations (Michalewicz, 1994).

The main idea of the selection is that bigger part of the chromosome should survive to next generation. When the objective function is optimized, values of the parameters indicate the state of the structure. Two cases have been investigated in this study. In the first case, a uniform strength beam with inclusion of slope function in width has been discussed. The second case demonstrates this method with inclusion of slope functions for both the depth and width of the beam. For simulating the experimental measure-ment, the vibration characteristics viz. natural frequencies and the mode shapes were perturbed randomly. A computer program using MatLab is employed to find out the location and extent of the damage.

Crack identification by the method of hybrid neuro-genetic technique

Suh et al. (2000) have presented a method to identify the location and depth of a crack on a structure by using hybrid neuro-genetic technique. Feed-forward multi-layer neural networks trained by back-propagation are used to learn the input (the location and depth of a crack)–output (the structural eigen frequencies) relation of the structural system. With this trained neural network, genetic algorithm is used to identify the crack location and depth minimizing the difference from the measured frequencies.

The use of neural networks in detecting the damage has been developed for several years, because of their ability to cope with the analysis of the structural damage without the necessity for intensive computation. Recently, neural networks are expected to be a necessity for intensive computation.

Recently, neural networks are expected to be a potential approach to detect the damage of the structure (Furukawa et al., 1995). In these researches, both the modal frequencies and the modal shapes are needed for the training of neural network to detect the structural damage, since the frequency information alone is not sufficient to train the neural network for the inverse problem of the crack identification. To identify the location and depth of a crack in a structure with only frequency information, a method is presented in this paper which uses hybrid neuro-genetic technique.

Feed-forward multi-layer neural networks trained by back-propagation are used to learn the input (the location and depth of a crack)-output (the structural eigen frequencies) relation of the structural system. With this trained neural network, genetic algorithm is used to identify the crack location and depth minimizing the difference from the measured frequencies. This approach needs only the modal frequencies for use of hybrid neuro-genetic techniques.

FUNDAMENTAL STRUCTURE OF THE GENETIC ALGORITHM

There is preparation phase and application phase. In the preparation phase, firstly, the learning data of various sets of crack parameters and the corresponding response of the structure, which is the eigen frequency in this study, are prepared by the computational structure analysis. Figure 1 shows the various processes involved in genetic algorithm.

t=0;
Initialize P(t);
do
Crossover P(t);
Mutate P(t);
Evaluate P(t);
Select P(t);
t=t+1;
While terminal condition is not satisfied

Figure 1. Process of Genetic Algorithm

Fuzzy-genetic algorithm and Fuzzy inference method

In this Various fuzzy inference and fuzzy-genetic algorithm methods followed for crack identification are outlined.

Pawar et al. (2007) have used a genetic fuzzy system to identify the crack depth location in a composite matrix cracking model. As described by him the genetic fuzzy system combines the uncertainty characteristics of fuzzy logic with the learning ability of genetic algorithm and observed that the success rate of the genetic fuzzy system in the presence of noise is dependent on crack density. He has found that the genetic fuzzy system shows excellent damage detection ability even in the presence of noise data.

Das et al. (2008) have proposed a fuzzy inference system for detection of crack location and crack depth of a cracked cantilever beam using the percentage deviation of first three natural frequencies and first three mode shapes by the fuzzy member ship functions. The two output parameters of the fuzzy inference system are relative crack depth and relative crack location. The vibration signatures are used to derive the fuzzy rules and experimental setup has been developed for verifying the robustness of the developed fuzzy inference system.

Wada et al. (1991) have proposed a fuzzy control method with triangular type membership functions using an image processing unit to control the level of granules inside a hopper. He has stated that the image processing unit can be used as a detecting element and with the use of fuzzy reasoning methods good process responses are obtained.

Fox et al. (1977) have studied the use of fuzzy logic in medical diagnosis and raised a broad range of issues in connection to the role of information-processing techniques in the development of medical computing.

Lo et al. (2007) have developed an intelligent technique based on fuzzy-genetic algorithm (FGA) for automatically detecting faults on HVAC system. The proposed automatic fault detection system (AFD) monitors the HVAC system state continuously by fuzzy system and the optimization capability of genetic algorithms allows the generation of optimal fuzzy rules. As stated simulation studies are conducted to verify the proposed automatic fault detection (AFD) system for the single zone air handler system.

Parhi et al. (2008) have addressed the fault detection of cracked beam using AI technique. In this work fuzzy controller has been used to find out the relative crack depth and relative crack location of the cracked beam. The input parameters to the fuzzy controller are percentage deviation of first three relative natural frequencies and first three relative mode shapes differences. The fuzzy rules are set up with the help of natural frequencies, mode shapes, crack depths and crack location. The output from fuzzy controller has been experimentally validated in this work.

Zimmermann et al. (1978) has applied fuzzy linear programming approach for solving linear vector maximum problem. The solutions are obtained by fuzzy linear programming. These are found to be efficient solutions then the numerous models suggested solving the vector maximum problem.

DISCUSSION

The neural network technique has been discussed for detection of the relative crack location and relative crack

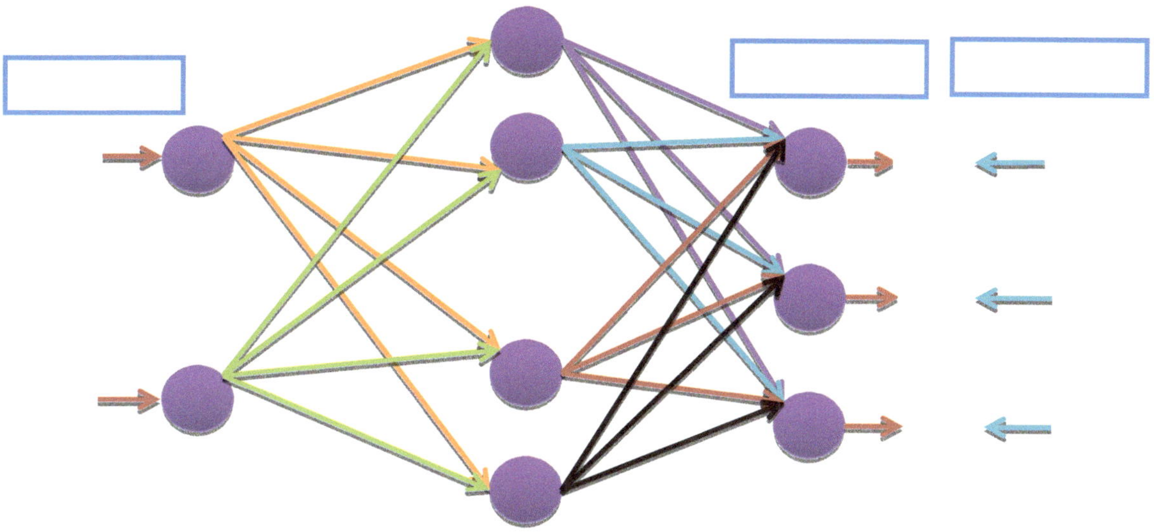

Figure 2. Three layer Neural Network utilized in this study.

depth. The neural network has got six input parameters and two output parameters. The structure of the neural network has been shown in Figure 2. The inputs to the neural network are as follows;

Relative first natural frequency = "fnf"; relative second natural frequency = "snf"; relative third natural frequency = "tnf"; relative first mode shape difference = "fmd"; relative second mode shape difference = "smd" and relative third mode shape difference = "tmd".The outputs from the neural network are as follows; Relative crack location = "rcl" and Relative crack depth = "rcd".

The back propagation neural network has got ten layers (that is, input layer, output layer and eight hidden layers). The neurons associated with the input and output layers are six and two respectively. The neurons associated in the eight hidden layers are twelve, thirty-six, fifty, one hundred fifty, three hundred, one hundred fifty, fifty and eight respectively. The input layer neurons represent relative deviation of first three natural frequencies and first three relative mode shape difference. The output layer neurons represent relative crack location and relative crack depth. A genetic algorithm approach for detecting cracks in beam-like structures has been analyzed. The search process proposed in this method utilizes binary and continuous genetic algorithms to find the crack location and depth whose natural frequencies have maximum similarity with the input natural frequencies. Also a new cost function based on natural frequencies was presented. Some guidelines for selecting GA parameters are provided. In comparison with traditional GAs, we use a GA with small population size combined with a large mutation rate, so the great exploration of the search space with small number of cost function evaluation is achieved. A statistical ANN method gives a good value in that accounts for the inevitable FE modeling error and measurement noise for structural condition identification. Rosenblueth's point estimation method is used to derive the statistical ANN model and to identify the structural condition. Both the modeling error and measurement noise are assumed to have normal distribution and zero means. The accuracy of the statistical approach was proved using Monte Carlo simulation. Using this method, the probability of damage existence can be estimated. The numerical and experimental results demonstrated that, compared with the normal ANN approach, the statistical ANN approach gives more reliable identification of structural damage. Condition monitoring tools like expert system, acoustic emission, shock pulse method and extended neuro-fuzzy schemes are proposed for a real time machinery condition monitoring. The uses of various Artificial Intelligence (AI) methods for fault detection are discussed briefly. The genetic fuzzy system combines the uncertain characteristics of fuzzy logic with the learning ability of genetic algorithm which shows an excellent capability for damage detection even in the presence of noise data. The fuzzy technique can be used for fault detection by feeding the percentage deviation of first three natural frequencies and relative mode shapes as input parameters to the fuzzy controller. Fuzzy control method with triangular type membership functions and image processing unit can be used in a manufacturing environment for optimize the input and output parameters. Probabilistic neural network, artificial neural network, fuzzy proportional integral (PI), genetic algorithm, neuro fuzzy, evolution based fuzzy neural diagnostic can be used for fault diagnosis in mechanical and electrical machines. A computer code is developed in which structural response due to damage is carried out. The response data are fed into the network to determine the damage. It is observed that neural networks can successfully identify and calculate the amount of damage for both single and multiple element damage cases. The main advantage of a neural network is that response measurement is required only at a limited number of points.

CONCLUSION

In this paper different techniques and methodologies used for identification of crack in the domain of dynamic vibration of cracked structures have been comprehensively reviewed and their applications for damage detection have been described briefly.

Based on the current analysis, it has been noticed that other than conventional methods like classical, finite element method different Artificial Intelligence (AI) methods like genetic algorithm, neural network, and fuzzy inference technique are successfully used for calculation of damage severity that is the location and depth of the crack of cracked structures.

1. The finite element method has been used to find out the natural frequencies, and mode shapes of the cracked beam which are being used to locate the crack location and size.
2. The fuzzy controller has been used to find out the relative crack depth and relative crack location of the cracked beam.
3. The genetic controller has been used to find out the relative crack depth and relative crack location of the cantilever beam.
4. Some of the researchers have also applied neural network for crack identification in damaged vibrating structures like artificial neural network, hybrid neuro genetic technique.
5. In some of the analysis the fuzzy technique, neural network and genetic algorithm are applied in combination to detect depth and location of the damage.

REFERENCES

Akgun M, Ju FD, Paez TL (1983). Fracture Diagnosis in Beam Frame Structures using Circuit Analogy, Recent Adv. Eng. Mechanics Their Impact Civil Eng. Pract., 2: 767-769.

Bakhary N, Hao H, Deeks AJ (2007). Damage detection using artificial neural network with consideration of uncertainties, Eng. Struct., 29(11): 2806-2815.

Chou JH, Ghaboussi J (2001). Genetic algorithm in structural damage detection, Comput. Struct., 79(14): 1335-1353.

Das H, Parhi DR (2008). Online fuzzy logic crack detection of a cantilever beam, Int. J. Knowledge-based Intelligent Eng. Syst., 12(2): 157-171.

Das H, Parhi DR (2009). Application of Neural network for fault diagnosis of cracked cantilever beam, World Congress on Nature & Biologically Inspired Computing.

Fox J (1977). Some observations on fuzzy diagnosis and medical computing, Int. J. Bio-Medical Comput., 8(4): 269-275.

Furukawa T, Yagawa G (1995). Parameter identification using an evolutionary algorithm and its performance under measurement errors", International Conference on Computational Engineering Science Computational Mechanics 195 Springer-Verlag, pp. 122-127.

Hagiwara I, Shi Q, Inoue A (1997). Development of inverse analysis methods applicable to modification calculation model by using experimental data, Proceedings of the International Symposium on Optimization and Innovative Design, Tokyo, Japan, JSME.

Huijian L, Changjun H, Jialin J, Wang H, Hao C (2005). Crack damage detection in beam-like structures using RBF neural networks with experimental validation, Int. J. Innovative, Comput. Inf. Control, 1: 625-634.

Karthikeyan MA, Tiwari RA, Talukdar SB (2007). Crack localization and sizing in a beam based on the free and forced response measurement, Mechanical Syst. Signal Process., 21(3): 1362-1385.

Kubo S (1993). Inverse Problem, Baifu-kan (in Japanese).

Lo CH, Chan PT, Wong YK, Rad AB, Cheung KL (2007). Fuzzy-genetic algorithm for automatic fault detection in HVAC systems, Appl. Soft Comput., 7(2): 554-560.

Maity D, Saha A (2004). Damage assessment in structure from changes in static parameter using neural networks, Publisher Springer India, 29(3).

Marawala T, Chakraverty S (2006). Fault classification in structures with incomplete measured data using auto associative neural networks and genetic algorithm, Curr. Sci., 90(4): 542–548.

Michalewicz Z (1994). Genetic algorithms + data structure = evolution programs, 2nd edition. Spring, Berlin.

Morassi A (2001). Identification of a crack in a rod based on changes in a pair of natural frequencies, J. Sound Vibration, 242(4): 577–596.

Pandey PC, Barai SV (1995). Multilayer perceptron in damage detection of bridge structures, Comput. Struct., 54(4): 597–608.

Panigrahi SK, Chakraverty S and Mishra BK (2009). Vibration based damage detection in a uniform strength beam using genetic algorithm, Springer Publication, pp. 44.

Panigrahi SK, Chakraverty S, Mishra BK (2007). Identification of damage in structural members using genetic algorithm, National conference on emerging technology and developments in civil engineering, (ETDCE-07), Govt Eng College, Amaravati, pp. 71–78.

Papadopoulos L, Garcia E (1998). Structural damage identification: A probabilistic approach, AIAA J., 39 (11): 2137–2145.

Parhi DR, Das H (2008). Smart crack detection of a beam using fuzzy logic controller, Int. J. Comput. Intelligence Theory Pract., 3(1): 9-21.

Pawar PM, Reddy KV, Ganguli R (2007). Damage Detection in Beams using Spatial Fourier Analysis and Neural Networks, J. Intelligent Mater. Syst. Struct., 18(4): 347-360.

Rao MA, Srinivas J, Murthy BSN (2004). Damage detection in vibrating bodies using genetic algorithms, Comput. Struct., 82: 963–968.

Rao DVPR (2004). Identification of damage in structures using genetic algorithm, IE (I) J MC, pp. 154–160.

Rajasekaran S, Pai GAV (2003). Neural networks, fuzzy logic, and genetic algorithms, synthesis and applications, Prentice-Hall of India Pvt. Ltd., New Delhi.

Rizos PF, Aspragathos N, Dimarogonas AD (1990). Identification of crack location and magnitude in a cantilever beam from the vibration modes", J. Sound Vibration, 138(3): 381-388.

Sahin M and Shenoi RA (2003). Quantification and localisation of damage in beam-like structures by using artificial neural networks with experimental validation", Eng. Structu., 25(14): 1785-1802.

Sahoo B, Maity D (2007). Damage assessment of structures using hybrid neuro-genetic algorithm, Appl. Soft Comput., 7(1):89-104.

Saridakis KM (2008). Applying neural networks, genetic algorithms and fuzzy logic for the identification of cracks in shafts by using coupled response measurements, Comput. Struct., 86(11-12): 1318-1338.

Shi Q, Hagiwara I (1999). Structural damage detection and identification using learning vector quantization neural network, Proceedings of the ASME Design Conference on Engineering Technology.

Suh MW, Shim MB, Kim MY (2000). Crack identification using hybrid neuro-genetic technique, J. Sound Vibration, 238(4): 617-635.

Suresh S, Omkar SN, Ganguli R (2004). Identification of crack location and depth in a cantilever beam using a modular neural network approach, Smart Mater. Struct., 13: 907-915.

Taghi M, Baghmisheh V (2008). Crack detection in beam-like structures using genetic algorithms, Appl. Soft Comput., 8(2):1150-1160.

Tsou, Shen MH (1993). Structural damage detection and identification

using neural network, AIAA-93-1708-CP, 3551-3560.
Wada K, Hayano N, Oka H (1991). Application of the fuzzy control method for level control of a hopper, Adv. Powder Technol., 2(3): 163-172.
Wu X, Ghaboussi J, Garrett JH (1992). Use of neural networks in detection of structural damage, Comput. Struct., 42(4): 649–59.
Zapico JL, Worden K, Molina FJ (2001). Vibration-based damage assessment in steel frames using neural networks, J. Smart Mater. Struct., 10: 553–559.
Zhao J, Ivan JN, DeWolf JT (1998). Structural damage detection using artificial neural network, J. Infrastructure Syst., 4(3): 93–101.
Zimmermann HJ (1978). Fuzzy programming and linear programming with several objective functions, Fuzzy Sets Syst., 1(1): 45-55.

Design and construction of fish smoking kiln

ADAMU, I. G.[1], KABRI, H. U.[2], HUSSAINI, I. D.[1]* and MADA, A. D.[1]

[1]Department of Agricultural Engineering Technology, Adamawa State College of Agriculture, P. M. B. 2088 Ganye, Nigeria.
[2]Department of Agricultural and Environmental Engineering, Modibbo Adama University of Technology, P. M. B. 2076 Yola, Nigeria.

The design and construction of fish smoking kiln was undertaken with the aim of improving the existing fish smoking kiln. The kiln consists of metal frame made of angle iron on which the entire component was assembled. The handle controls the speed of the ventilation fan through chain and sprocket. The charcoal housing is made of perforated metal sheet to allow for proper ventilation. The kiln is operated manually by turning the handle which drives the fan to augment natural ventilation. The machine performance was evaluated using fish samples (African mud Fish) smoked to an average moisture content of 11.46% within an average time span of 5 h, and the average final weight of the dried fish was 0.9827 Kg. It was found that the fish can be kept for at least two months before showing the sign of spoilage.

Key words: Fish, kiln, moisture content, smoke charcoal.

INTRODUCTION

Fish is one of the most diverse groups of animals known to man with more than 20,500 species in existence. According to FAO (1985), the World fish production from capture fisheries and aquaculture has reached 121 tones. A decline in fish availability will have a detrimental effect on the nutritional status of the citizenry particularly in places where fish contributes significantly to the protein intake of the people such as Nigeria where fish constitutes 40% of animal intake Olafunde (1989). Efforts should therefore, be geared towards increased fish production through improved resource management matched with effective post-harvest handling, preservation and processing to prevent spoilage.

Most of the fish harvested in the tropics is used for direct consumption but a great deal is processed into fish meal for use in compounding feeds (Eyo, 2001). Also according to Eyo (2001), a significant quantity is lost as a result of the absence of adequate technology to prevent post harvest losses. An estimated 20 to 50% of the fish produced in the remote and hinter lands of many tropical countries perish before these get to the consumers (Olafunde, 1989). Also, according to Eyo (2001), is due to poor handling, preservation and processing practices adopted by the artisanal fishermen, fish farmers and fish entrepreneurs.

In most developed countries, canning or freezing methods are used for fish preservation to improve the shelve life of fish. However, in underdeveloped and developing countries, traditional smoking kiln is still practiced, especially, in the tropics due to the sophisticated nature, erratic power supply and costs of the equipments. Consequently, there is the need to develop a more efficient and affordable equipment to enhance fish processing and reduce drudgery associated with traditional method.

RESEARCH METHODOLOGY

The smoking kiln was designed and constructed with the aim of fulfilling the following objectives; safe operation, general maintenance and servicing, low initial and operating costs, efficient and effective performance among others.

Design calculations

Volume of the kiln: the capacity of the kiln is directly related to the

*Corresponding author. Email: hussainidanjuma@gmail.com.

volume of the fish tray (vf) and is calculated as:

$$Vf = l \times b \times h \quad (1)$$

Where l = length of the tray = 49 cm
b = width of the tray = 29 cm
h = height of the tray = 5 cm

$$\therefore Volume = 49 \times 29 \times 5 = 7105\, cm^3$$

Volume of the Charcoal Tray (Vc): is calculated as;

$$Vc = L \times b \times h \quad (2)$$

Where: L = Length of Charcoal Tray = 49 cm
b = width of Charcoal Tray = 29 cm
h = height of charcoal tray = 10 cm

$$\therefore\ Vc = 49 \times 29 \times 10 = 14210\ cm^3$$

Diameter of kiln handle

Moments (m) of force = force (f) x length (height) (3)

Where f = maximum of a man (500 N) and that of a woman (300 N)
h = height of handle = 0.16 m
f = 300 N (force of a woman)

$$m = 300 \times 0.16 = 48\ Nm$$

The diameter of the kiln handle can be calculated as

$$(\delta) = \frac{m}{w} \quad (4)$$

Where: $\delta = working\ \ stress = \frac{\delta ult}{K}$ (5)

δult = 552 Mpa
K = Safety factor = 2

$$\delta = \frac{\delta ult}{K} = \frac{552}{2} = 276\ \ Mpa$$

M = moment of force
W = Resistance moment = $0.1d^3$
Where d = diameter of the kiln handle (mm)

$$\delta = \frac{m}{0.1d^3}$$

$$276 = \frac{48}{0.1d^3}$$

$$d = \sqrt[3]{\frac{48 \times 10^3}{0.1 \times 276}}$$

$$d = 12\ mm$$

According to Khurmi and Gupta (2003), the speed ratio may be determined from the expression below:

Speed ratio (V.R)

$$V.R. = \frac{N_1}{N_2} = \frac{T_2}{T_1} \quad (6)$$

Where N_1= and N_2 = number of revolutions of the smaller and larger sprockets (rpm) respectively
T_1 = number teeth on larger sprocket = 48
T_2 = number teeth on smaller sprocket = 20

$$V.R. = \frac{48}{20} = 2.4$$

Pitch Circle diameter of the smaller and larger sprockets (d_1 and d_2)
Pitch circle diameter of the larger sprocket (d_2)

$$d_2 = \frac{P}{\sin\left(180^0 / T_2\right)} \quad (7)$$

Where P = pitch = 1.3 (from Table 1)
d_2 = 20 cm
Pitch circle for the smaller sprocket (d_1)

$$d_1 = \frac{P}{Sin\left(180^0 / T_1\right)} \quad (8)$$

d_1 = 8 cm

Determination of speed of chain

$$p = [(wb \times v)/k] \times Ks \quad (9)$$

Where;
P = power developed by man = 0.1 Kw (Onwuala et al., 2006).
v = velocity of the chain in m/s
K = safety factor
Ks = Service factor
Wb = breaking load = 300 N

$Ks = K_1.K_2.K_3$ (10)

Where;
K_1 = Load factor = 1 (for constant load)
K_2 = Lubrication factor = 1.5 (for periodic lubrication)
K_3 = rating factor = 1.25 (for 16 hours per day)
K_s = 1.875
Now

$v = (P.K.K_s)/W_b$ (11)

$$V = 0.1 \times 10^3 \times 2 \times 1.875 / 300$$

$$V = 1.25\ \ m/s$$

Determination of number of revolution of the larger and smaller sprockets

$$v = \frac{\pi DN}{60} \quad (12)$$

Where

v = Velocity = 1.25 m/s
D = Pitch circle diameter (mm)
N = Number of revolutions (rev)

From Equation (12) number of revolutions of the larger sprocket (N_2) is given as

$$N_2 = \frac{V \times 60}{\pi D}$$

Where
D = 20 cm = 0.2 m
V = 1.25 m/s
N_2 = 119 revolutions

Similarly from equation (12) number of revolutions of the smaller sprocket (N_1)

$$N_1 = \frac{V \times 60}{\pi D}$$

Where
D = 0.08 m
N_1 = 298 revolutions

Torque on the shaft

$$T = \frac{P \times 60}{2\pi N_2} \qquad (13)$$

T = Torque (Nm)
P = Power (0.1 Kw)
N_2 = Number of revolutions of the larger sprockets = 119

$$\Rightarrow T = \frac{0.1 \times 60}{2\pi \times 119} = 8.03\ Nm$$

Fan speed
Fan speed is given by the speed of gear multiplied by the velocity ratio

Fan speed = N_2 X V.R (14)
= 119 X 2.4 = 286 rpm
Fan air density (ρ) is given as

ρ = 1.325B/T (15)

Where
B = atmospheric pressure = 760 mmHg
T = absolute temperature of air at room temperature
= 35 + 273^{0}K = 308^{0}K
ρ = 13.25 x 760/308 = 3.30 Kg/m^3

Fan capacity Q is given as

$$Q = \frac{Q_x P_x}{P} \qquad (16)$$

Where
Q_x = Volume flow rate (m^3/s)
Q = Fan capacity

$$\rho_x = fan\ air\ density\ at\ plane\ X\ (Kg/m^3)$$

$$Q_x = 1097 A_x \sqrt{\frac{P_{vx}}{\rho_x}}. \qquad (17)$$

Where
P_{vx} = velocity pressure at plane X
A_x = Area of the larger sprocket (m^2)

$$A_x = \frac{\pi d_2^{\ 2}}{4} = \frac{3.142 \times 0.2^2}{4}$$

$$A_x = 0.031\ m^2$$

Where d_2 = pitch circle diameter of the larger sprocket
Velocity pressure (Pvx) = force/Area
Where
F = 300 N

$$\Rightarrow P_{vx} = \frac{300}{0.03} = 9677\ N/m^2$$

Now $Q = \dfrac{Q_x \rho_x}{\rho}$; $\rho_x = \rho = 3.30\ Kg/m^3$

Q = Q_x = 1836 m^3/s

Heat transfer by conduction is obtained as:

$$q = \frac{KA(T_1 - T_2)}{L} \qquad (18)$$

Where
K = Stainless steel conductivity = 30 Wm/K
A = Area of the fish tray = L x b = 0.142 m^2
L = distance between the fish tray and the charcoal tray = 0.1 m
T_1 = Temperature of the inside = 40°C (assumed)
T_2 = Temperature of the outside = 30°C (assumed)

$$\Rightarrow q = \frac{30 \times 0.142 \times 10}{0.1} = 426\ \ J/m^2$$

Materials and construction

A mild steel metal sheet was used for the construction of charcoal tray and fish tray and general cover. While angle bar iron was used for the construction of the frame which formed the main support and the shaft was made using steel rods.

Construction of components

The important components of smoking kiln are

Frame: made up of angle iron of 25 x 25 mm. The angle irons were cut and welded together to form the mounting support for the kiln with 570 mm x 300 mm x 580 mm as the length, width and height respectively (Figure 1).

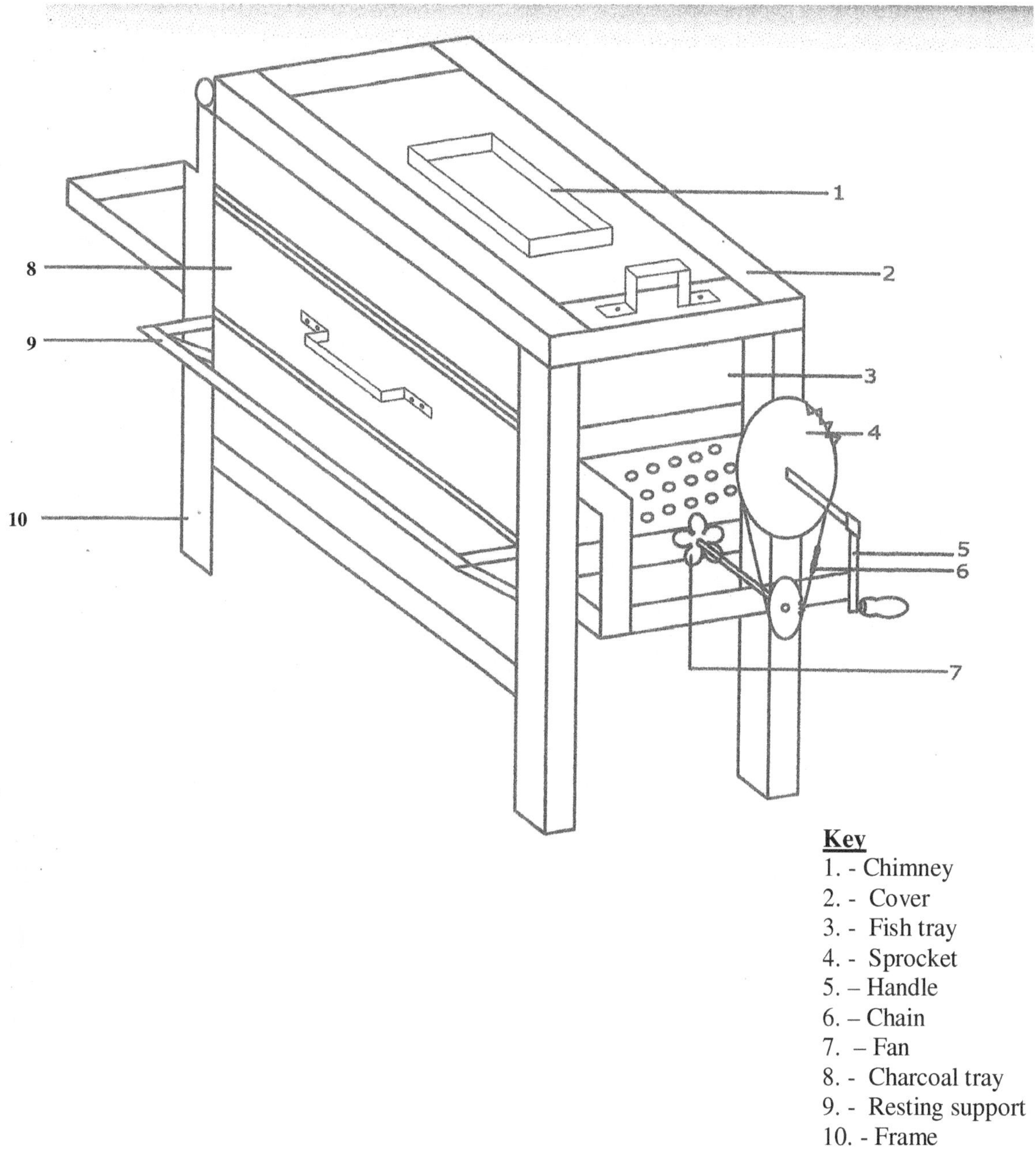

Figure 1. The Assembled smoking kiln

Fish Tray: This is the container into which the fish is loaded and smoked. It is constructed from metal sheet and wire mesh with length 4980 mm, width 290 mm and height 50 mm.

Charcoal Tray: This is the container for burning the charcoal to provide energy for smoking the fish. It is constructed from metal sheets of 1 mm thick. It has the dimension of 490 mm x 290 mm x 100 mm.

Rotating Handle: The rotating handle was constructed from steel rod. This handle controls the speed of the fan which provides air for effective burning of the charcoal.
Axial Fan: The axial fan of 20 mm diameter with five blades and thickness of 1 mm was chosen. The fan was screwed to the shaft attached to the smaller sprocket and directly facing the charcoal tray which is situated 10 mm below the fish tray.

Machine assembly

The various components of the machine were either fastened using bolts or nuts or welded. The completed assembly of the kiln is shown in Figure 1.

Operational techniques

Fish to be smoked are fed into the fish tray and covered. The

Table 1. Performance evaluation results.

S/No	Initial weight (Kg)	Time taken to smoke the fish (h)	Final weight (Kg)	Final moisture content (%)
1.	1.2	5.0	1.0727	10.60
2.	1.3	5.0	1.1527	11.33
3.	0.82	4.0	0.7127	13.08
Mean	1.11	4.66	0.9827	11.46

charcoal is lit and the burning charcoals supplies the heat which dries the fish and smokes that envelope the fish serves as a preservative. The handle is rotated from time to time to drive the fan which supplies air to facilitate the burning. The ashes dropout through the perforation at the base of the tray. To control the heat transfer to the fish, the charcoal tray can be pulled out to rest on support.

Performance evaluation

Three sets of fish samples (African Mudfish) were obtained from Jimeta main market; the samples were thoroughly washed and salted, then allowed to dry for an hour under the sun.

The charcoal in the charcoal tray was first ignited with the help of kerosene, the ignited charcoal was allowed to burn for 10 to 15 min to allow the kerosene odour to be exhausted. More charcoal was added to the burning charcoal.

The first set of the sun dried fish was then weighed (Initial weight) and put into the burning tray. As the charcoal was burning, the fan is operated to supply a fairly constant heat. During the smoking, the sides of the fish facing the burning charcoal are changed routinely by pulling out the burning tray and then turning the fishes upside down then pushing the tray back. Similarly the weight of the fish is determined intermittently whenever the position of the fish is to be changed and the corresponding moisture content is determined. This is continued until the final weight and hence the final moisture content is determined. The final moisture content that is, safe moisture content (10 to 15%) (Ikenweiwe et al., 2010) is calculated using Equation (19) when there is virtually no further reduction in the moisture content. The time taken for the smoking was the total time taken including the time for the intermittent determination of moisture content and that of changing the position of the fish. The result is given on Table 1. After smoking and determination of final moisture content, the fish is allowed to cool and kept in plastic bucket, well covered carton and polythene bag to determine its storage life. These containers were selected because they are among the common storage media used for handling fish in Adamawa State. The determination of the storage life was done between the months of August to October, 2010.

$$mc_{db}(\%) = \frac{Initialweight - Finalweight}{Finalweight} \times 100 \qquad (19)$$

Where

Mc_{db} = Moisture content dry basis in percentage

Initial weight = Weight of sun dried fish (Kg)

Final weight = Weight of smoked fish (Kg)

RESULTS AND DISCUSSION

The assembled machine was rigid and operates smoothly with little effort needed to operate the fan.

The smoking kiln was tested with three sets of fish in the month of August, 2010; the result of the test is given in Table 1.

Table 1 above shows that within the time frame of averagely five (5) hours, the moisture content of each set of the fish was reduced to a safe moisture content this is similar to the result obtained in Ikenweiwe et al. (2010). The time taken for smoking as stated earlier depends on the weights/sizes of fish, the initial moisture content, the fat content of the fish and the uniformity and intensity of heat supplied. The use of this machine is found to be efficient when compared to the local method of smoking fish that takes almost 24 h in addition to the other labour stresses.

During the storage period, the fish were checked on weekly basis. At the end of the tenth week it was found that the fish in the carton maintains its normal colour, odour and taste but has some problems of insect and fungal attack (Table 2). While the fish in the plastic bucket and polythene bag were found to be in their normal characteristics up to the eighth week. Though in the ninth and tenth week there were no problems of insect attack, but traces of fungal attack (the formation of mold) was observed. This implies that storing in cartons need to be accompanied by the use less harmful insecticides that is, insecticides that can be exhausted within the storage period. While fish to be stored in plastic buckets and polythene bags, must be well dried. In general stored fish need to be checked at least once every two weeks, though frequency of opening may give room for absorption of moisture and penetration of insects. The storability of fish could also depend on the atmospheric weather condition.

The machine was found to be affordable by the local farmers as shown in Table 3. At the end of testing the machine it was cleaned, movable parts were oiled and the machine was well covered and kept under shed.

Conclusion

A smoking kiln that is easy to operate, cheap, portable and easily maintained was designed and constructed. The kiln was tested using one of the most available types of fish in the area. The smoking kiln performed efficiently by drying the fish to safe moisture content (10 to 15%) within five hours. The storability of the fish was found to be about two month. However, the duration of storage

Table 2. Characteristics of the fish stored under different conditions.

Week	Storage media		
	Carton	Plastic Bucket	Polythene bag
1st week	Normal colour, normal odour and normal taste	Normal colour, normal odour and normal taste	Normal colour, normal odour and normal taste
2nd week	Normal colour, normal odour and normal taste	Normal colour, normal odour and normal taste	Normal colour, normal odour and normal taste
3rd week	Normal colour, normal odour and normal taste	Normal colour, normal odour and normal taste	Normal colour, normal odour and normal taste
4th week	Normal colour, normal odour and normal taste	Normal colour, normal odour and normal taste	Normal colour, normal odour and normal taste
5th week	Normal colour, normal odour and normal taste	Normal colour, normal odour and normal taste	Normal colour, normal odour and normal taste
6th week	Normal colour, normal odour and normal taste	Normal colour, normal odour and normal taste	Normal colour, normal odour and normal taste
7th week	Normal colour, normal odour and normal taste	Normal colour, normal odour and normal taste	Normal colour, normal odour and normal taste
8th week	Normal colour, normal odour and normal taste	Normal colour, normal odour and normal taste	Normal colour, normal odour and normal taste
9th week	Normal colour, normal odour and normal taste	Normal colour, normal odour and normal taste	Normal colour, normal odour and normal taste
10th week	Normal with traces of insect and fungal attack	There are traces of fungal attack	There are traces of fungal attack

Table 3. Cost analysis.

Item	Cost (₦)
Metal sheet	1000
Angle bar iron	1000
wire mesh	200
Steel rod	1000
Axial fan	1000
Shaft	500
Sprockets	500
Charcoal	500
Paint	400
Labour	1000
10% Miscellaneous	710
Total	7810

could be increased in the dry season because of lower atmospheric humidity. It can also be concluded that fish that should be stored in cartons should be accompanied by the application of less harmful insecticides (those that are exhausted within the storage period). The overall production cost is eight thousand Naira (₦8000). This can be affordable by the local fish farmers. The machine can also be used for roasting meat and corn like maize. However the machine requires attention during operation to regulate or control the smoking process to avoid damage of the fish.

It is therefore recommended that the fan should be driven by d. c. motors for uniform blowing of air, constant heat energy supply and to further reduce drudgery to minimum.

REFERENCES

Eyo AA (2001). Fish Processing Technology in the Tropics. National Institute of Fresh water Fisheries Research (NIFER) Publications, New Bussa, Nigeria. University of Ilorin Press. pp. 66-70.

FAO (1985). World catch and Trade of Fisheries Productions in 1984. Info Fish Marketing Digest. No. 25.

Ikenweiwe NB, Bolaji BO, Bolaji GA (2010). Fabrication and Performance Assessment of a Locally Developed Fish Smoking Kiln. Ozean J. Appl. Sci. 3(4):363-369.

Khurmi RS, Gupta JA (2003). A Textbook of Machine Design. 24th Edition Published by Rajendra Ravindra Printers Limited pp. 618-700.

Olafunde AA (1989). The Approaches to the Study of Fisheries Biology in Nigeria Inland Waters.

Onwualu AP, Akubuo CO, Ahanaku IE (2006). Fundamentals of Engineering for Agriculture. Published by Immaculate Publications Ltd pp. 137-144.

17

Smart crack detection of a cracked cantilever beam using fuzzy logic technology with hybrid membership functions

Dayal R. Parhi* and Sasanka Choudhury

Department of Mechanical Engineering, National Institute of Technology, Rourkela, India.

Damage detection methods have been considerably increased over the past few decades. A crack in a structural member introduces local flexibility that would affect vibration response of the structure. This property may be used to detect existence of a crack together with its location and depth in the structural member. The presence of damage leads to changes in some of the lower natural frequencies and mode shapes. Damage detection is one of the important aspects in structural engineering both for safety reasons and because of economic benefits that can result. This technique has been used in the present investigation for crack detection. Here the crack is transverse surface crack. The crack is analyzed using fuzzy logic system and finite element analysis. The fuzzy controller uses the hybrid membership functions (combination of triangular, trapezoidal and Gaussian) as input and trapezoidal membership functions as output. The input parameters to the fuzzy controller are the first three natural frequencies. The output parameters of the fuzzy controller are the relative crack depth and relative crack location. Finite element analysis has been done for modeling the cracked cantilever beam. By using Several fuzzy rules the results obtained for crack depth and crack location in the Matlab Simulink environment and have been compared with the results obtained from finite element analysis. It is observed that the fuzzy controller can predict the depths and locations accurately close to the finite element analysis. Finally a deviation of the results obtained by comparing the finite element analysis results and fuzzy controller result.

Key words: Damage, vibration, natural frequency, fuzzy logic, membership function, fuzzy controller.

INTRODUCTION

Many researchers have been carried out in an attempt to find methods for non-destructive crack detection in structural members. Vibration-based methods have been proved as a fast and inexpensive means for crack identification. A crack in a structure induces a local flexibility which affects the dynamic behavior of the whole structure to a considerable degree. It results in reduction of natural frequencies and changes in mode shapes. An analysis of these changes makes it possible to determine the position and depth of cracks. Most of the researches used in their studies are open crack models, that is, they assume that a crack remains always open during vibration. The assumption of an open crack leads to a constant shift of natural frequencies of vibration. Various studies investigated over the last decades, however, indicate that a real fatigue crack opens and closes during vibration. It exhibits non-linear behavior due to the variation of the stiffness which occurs during the response cycle. As a result, a breathing crack gives rise to natural frequencies falling between those corresponding to the open and closed states. Therefore, if an always open crack is assumed, the decrease in experimental natural frequencies will lead to an under-estimation of the crack depth. Harish and Parhi, 2009 have performed analytical studies on fuzzy inference system for detection of crack location and crack depth of a cracked cantilever beam structure using six input parameters to the fuzzy membership functions. The six input parameters are percentage deviation of first three natural frequencies and first three mode shapes of the cantilever beam. The two output parameters of the fuzzy inference system are relative crack depth and relative

*Corresponding author. E- mail: dayalparhi@yahoo.com.

crack location. Experimental setup has been developed for verifying the robustness of the developed fuzzy inference system. The developed fuzzy inference system can predict the location and depth of the crack in a close proximity to the real results. (Mohammad and Vakil, 2008) have proposed a method in which damage in a cracked structure was analyzed using genetic algorithm technique. For modeling the cracked-beam structure an analytical model of a cracked cantilever beam was utilized and natural frequencies were obtained through numerical methods. A genetic algorithm is utilized to monitor the possible changes in the natural frequencies of the structure. The identification of the crack location and depth in the cantilever beam was formulated as an optimization problem. Norhisham et al. (2007) applied Artificial Neural Network for damage detection. In his investigation an ANN model was created by applying Rosenblueth's point estimate method verified by Monte Carlo simulation, the statistics of the stiffness parameters were estimated. The probability of damage existence (PDE) was then calculated based on the probability density function of the existence of undamaged and damaged states. The developed approach was applied to detect simulated damage in a numerical steel portal frame model and also in a laboratory tested concrete slab. The effects of using different severity levels and noise levels on the damage detection results are discussed. Saridakis (2008) applied neural networks, genetic algorithms and fuzzy logic for the identification of cracks in shafts by using coupled response measurements. In this research the dynamic behavior of a shaft with two transverse cracks characterized by three measures: position, depth and relative angle. Both cracks were considered to lie along arbitrary angular positions with respect to the longitudinal axis of the shaft and at some distance from the clamped end. A local compliance matrix of two degrees of freedom (bending in both the horizontal and the vertical planes) was used to model each crack. Ganguli (2001) has developed a fuzzy logic system (FLS) for ground based health monitoring of a helicopter rotor blade. Structural damage is modeled as a loss of stiffness at the damaged location that can result from delamination. The fuzzy system is trained by a batch least squares algorithm based on desired input–output data so that the trained fuzzy system can behave like the training data. Parhi and Amiya (2009) have presented comprehensive review of methodologies in the domain of dynamic vibration of cracked structures using energy methods, finite element methods, fuzzy inference techniques, neural networks, neuro-fuzzy adaptive techniques and genetic algorithms for identifying the intensity and location of cracks.

FINITE ELEMENT FORMULATION

Theory

The beam with a transverse edge crack is clamped at left end, free at right end and has uniform structure with a constant rectangular cross-section of 800 × 50 × 6 mm as shown in Figure 1. The Euler-Bernoulli beam model is assumed for the finite element formulation. The crack in this particular case is assumed to be an open crack and the damping is not being considered in this theory. Both single and double edged crack are considered for the formulation.

Governing equation of free vibration

The free bending vibration of an Euler-Bernoulli beam of a constant rectangular cross section is given by the following differential equation as given in:

$$EI\frac{d^4y}{dx^4} - m\omega_i^2 y = 0 \tag{1}$$

where 'm' is the mass of the beam per unit length (kg/m), 'ωi' is the natural frequency of the ith mode (rad/s), E is the modulus of elasticity (N/m2) and I is the moment of inertia (m4).

By defining $\lambda^4 = \frac{m\omega_i^2}{EI}$ equation is rearranged as a fourth-order differential equation as follows:

$$\frac{d^4y}{dx^4} - \lambda^4 y = 0 \tag{2}$$

The general solution to the equation is:

$$y = A\cos\lambda_i x + B\sin\lambda_i x + C\cosh\lambda_i x + D\sinh\lambda_i x \tag{3}$$

where A, B, C, D are constants and 'λi' is a frequency parameter. Adopting Hermitian shape functions, the stiffness matrix of the two-noded beam element without a crack is obtained using the standard integration based on the variation in flexural rigidity.
The element stiffness matrix of the un cracked beam is given as:

$$\left[K^e\right] = \int [B(x)]^T EI[B(x)]dx \tag{4}$$

$$[B(x)] = \{H_1(x)H_2(x)H_3(x)H_4(x)\} \tag{5}$$

where are the Hermitian shape functions defined as:

$$H_1(x) = 1 - \frac{3x^2}{l^2} + \frac{2x^3}{l^3} \tag{6a}$$

$$H_2(x) = x - \frac{2x^2}{l} + \frac{x^3}{l^2} \tag{6b}$$

$$H_3(x) = \frac{3x^2}{l^2} - \frac{2x^3}{l^3} \tag{6c}$$

$$H_4(x) = -\frac{x^2}{l} + \frac{x^3}{l^2} \tag{6d}$$

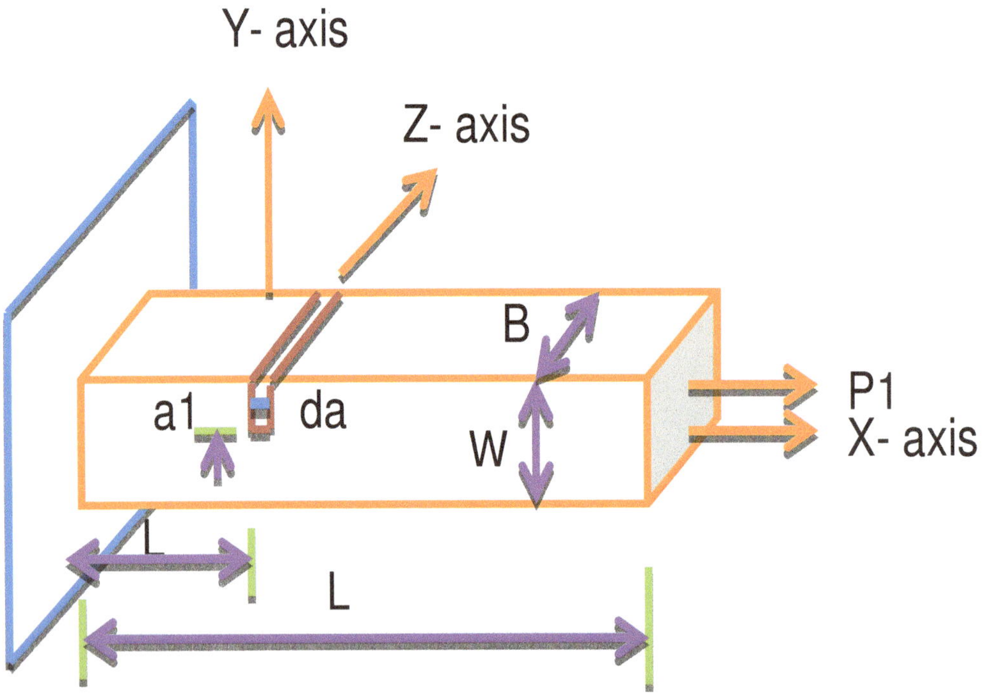

Figure 1. Geometry of Cantilever beam.

Assuming the beam rigidity EI is constant and is given by EI0 within the element, and then the element stiffness is Equation (6):

$$[K^e]=\frac{EI_0}{l^3}\begin{bmatrix} 12 & 6l & -12 & 6l \\ 6l & 4l^2 & -6l & 2l^2 \\ -12 & -6l & 12 & -6l \\ 6l & 2l^2 & -6l & 4l^2 \end{bmatrix} \tag{7}$$

$$[K_c^e]=[K^e]-[K_c] \tag{8}$$

Here, $\left|K_c^e\right|=$ Stiffness matrix of the cracked element, $\left|K^e\right|=$ Element stiffness matrix, $[K_c]=$ Reduction in stiffness matrix due to the crack.

According to Peng et al. (2007), the matrix [Kc] is:

$$[K_c]=\begin{bmatrix} K_{11} & K_{12} & -K_{11} & K_{14} \\ K_{12} & K_{22} & -K_{12} & K_{24} \\ -K_{11} & -K_{12} & K_{11} & -K_{14} \\ K_{14} & K_{24} & -K_{14} & K_{44} \end{bmatrix} \tag{9}$$

where:

$$K_{11}=\frac{12E(I_0-I_c)}{L^4}\left[\frac{2l_c^3}{L^2}+3l_c\left(\frac{2L_1}{L^2}-1\right)^2\right] \tag{9a}$$

$$K_{12}=\frac{12E(I_0-I_c)}{L^3}\left[\frac{l_c^3}{L^2}+l_c\left(2-\frac{7L_1}{L}+\frac{6L_1^2}{L^2}\right)^2\right] \tag{9b}$$

$$K_{14}=\frac{12E(I_0-I_c)}{L^3}\left[\frac{l_c^3}{L^2}+l_c\left(2-\frac{5L_1}{L}+\frac{6L_1^2}{L^2}\right)^2\right] \tag{9c}$$

$$K_{22}=\frac{12E(I_0-I_c)}{L^3}\left[\frac{3l_c^3}{L^2}+2l_c\left(\frac{3L_1}{L}-2\right)^2\right] \tag{9d}$$

$$K_{24}=\frac{12E(I_0-I_c)}{L^2}\left[\frac{3l_c^3}{L^2}+2l_c\left(2-\frac{9L_1}{L}+\frac{9L_1^2}{L^2}\right)^2\right] \tag{9e}$$

$$K_{44}=\frac{12E(I_0-I_c)}{L^2}\left[\frac{3l_c^3}{L^2}+2l_c\left(\frac{3L_1}{L}-1\right)\right] \tag{9f}$$

Here, lc=1.5W, L=Total length of the beam, L1=Distance between the left node and crack, $I_0=\frac{BW^3}{12}=$ Moment of inertia of the beam cross section, $I_0=\frac{B(W-a)^3}{12}=$ Moment of inertia of the beam with crack. It is supposed that the crack does not affect the mass distribution of the beam. Therefore, the consistent mass matrix of the beam element can be formulated directly as:

$$[M^e] = \int_0^l \rho A [H(x)]^T [H(x)] dx \qquad (10)$$

$$[M^e] = \frac{\rho A l}{420}\begin{bmatrix} 156 & 22l & 54 & -13l \\ 22l & 4l^2 & 13l & -3l^2 \\ 54 & 13l & 156 & -22l \\ -13l & -3l^2 & -22l & 4l^2 \end{bmatrix} \qquad (11)$$

The natural frequency then can be calculated from the relation:

$$[-\omega^2 [M] + [K]]\{q\} = 0 \qquad (12)$$

where: q=displacement vector of the beam.

ANALYSIS OF FUZZY LOGIC SYSTEM FOR CRACK DETECTION

Fuzzy logic is a tool for Embedding Human structured knowledge (Experience, Expertise and Heuristic). P. L. Zadeh says: “Fuzzy logic may be viewed as a bridge over the excessively wide gap between the precision of classical crisp logic and the imprecision of both the real world and its human interpretation”. Fuzzy logic attempts to model the way of reasoning that goes in the human brain. Almost all of human experience is stored in the form of the If-Then rules. Human reasoning is pervasively approximate, non-quantitative, linguistic and dispose-tional. Fuzzy logic can be explained in the following steps.

Fuzzy set

A fuzzy set, as the name implies, is a set without a crisp boundary. That is the transition from “belongs to a set” to not belong to a set is gradual and this smooth transition is characterized by membership functions that give fuzzy sets flexibility in modeling commonly used linguistic expressions. A membership function assigns to each element in the set under consideration a membership grade, which is a value in the interval [0, 1].

Membership function

The basic structure of a fuzzy interface system consists of three components: a rule base, which contains a selection of fuzzy rules, a database which defines the membership functions used in the fuzzy rules and a reasoning mechanism, which performs the interface procedure. The membership function μA(x) describes the membership of the elements x of the base set X in fuzzy set A, where by μA(x) a large class of function can be taken. Reasonable functions are often piecewise linear function, such as triangular or trapezoidal functions. The value for the membership function can be taken in the interval [0, 1]. When the functions are nonlinear the Gaussian membership function will be taken for the smooth operation.

Fuzzy logic

In Crisp logic, the truth values acquired by proposition or predicates are 2-valued, namely True, False which may be treated numerically equivalent to (0, 1). However in fuzzy logic, truth values are multivalued such as absolutely false, partly true, absolutely false, and very true and so on and are numerically equivalent to 0-1.

Fuzzy linguistic variables

Just like an algebraic variable takes numbers as values, a linguistic variable takes words or sentences as values. The set of values that it can take is called its term set. Each value in the term set is a fuzzy variable defined over a base variable. The base variable defines the universe of discourse for all the fuzzy variables in short. In short the hierarchy is as follows: Linguistic variable → Fuzzy variable → Base variable.

Fuzzy if-then rule

A fuzzy if-then rule (also known as fuzzy rule, fuzzy implication or fuzzy conditional statement) assumes the form “if x is A then y is B”. Where A and B are linguistic values defined by fuzzy sets on universes of discourse x and y respectively. Often “x is A” is called the antecedent or premise, while “y is B” is called the consequence or conclusion (Some of the linguistic terms used are shown in Table 1).

Fuzzy mechanism used for crack detection

The fuzzy controller (as shown in Figure 2) has been developed where there are 3 inputs and 2 outputs parameter. The natural linguistic representations for the input are as follows:

Relative first natural frequency = “FNF”
Relative second natural frequency = “SNF”
Relative third natural frequency = “TNF”
The natural linguistic term used for the outputs are
Relative crack depth = “RCD”
Relative crack length= “RCL”

Based on the above fuzzy subset the fuzzy rules are defined in a general form as follows:

If (FNF is FNFi and SNF is SNFj and TNF is TNFk) then

Table 1. Linguistic term used for membership functions.

Name of the membership function	Linguistic terms	Description and range of the linguistic terms
L1F1,L1F2,L1F3	fnf_{1to3}	Low ranges of relative natural frequency for first mode of vibration in ascending order respectively.
M1F1,M1F2,M1F3	$fnf_{4,6}$	Medium ranges of relative natural frequency for first mode of vibration in ascending order respectively.
H1F1,H1F2,H1F3	$fnf_{7,9}$	Higher ranges of relative natural frequency for first mode of vibration in ascending order respectively
L2F1.L2F2,L2F3	snf_{1to3}	Low ranges of relative natural frequency for second mode of vibration in ascending order respectively.
M2F1,M2F2,M2F3	$snf_{4,6}$	Medium ranges of relative natural frequency for second mode of vibration in ascending order respectively.
H2F1,H2F2,H2F3	snf_{7to9}	Higher ranges of relative natural frequency for first mode of vibration in ascending order respectively
L3F1,L3F2,L3F3	tnf_{1to3}	Low ranges of relative natural frequency for second mode of vibration in ascending order respectively
M3F1,M3F2,M3F3	$tnf_{4,6}$	Medium ranges of relative natural frequency for second mode of vibration in ascending order respectively
H1F1,H1F2,H1F3	tnf_{7to9}	Higher ranges of relative natural frequency for first mode of vibration in ascending order respectively
SD1,SD2,SD3	rcd_{1to3}	Small ranges of relative crack depth in ascending order respectively.
MD1,MD2,MD3	rcd_{4to6}	Medium ranges of relative crack depth in ascending order respectively
LD1,LD2,LD3	rcd_{7to9}	Larger ranges of relative crack depth in ascending order respectively.
SL1,SL2,SL3	rcl_{1to3}	Small ranges of relative crack location in ascending order respectively.
ML1,ML2,ML3	rcl_{4to6}	Medium ranges of relative crack location in ascending order respectively.
BL1,BL2,BL3	rcl_{7to9}	Bigger ranges of relative crack location in ascending order.

(CD is CDijk and CL is CLijk)

Where i= 1to 9, j=1 to 9, k=1 to 9 (13)

Because of "FNF", "SNF", "TNF" have 9 membership functions each.

From the above expression (13), two set of rules can be written:

If (FNF is FNFi and SNF is SNFj and TNF is TNFk) then CD is CDijk (14a)

If (FNF is FNFi and SNF is SNFj and TNF is TNFk) then CL is CLijk (14b)

According to the usual Fuzzy logic control method (Harish and Parhi, 2008), a factor Wijk is defined for the

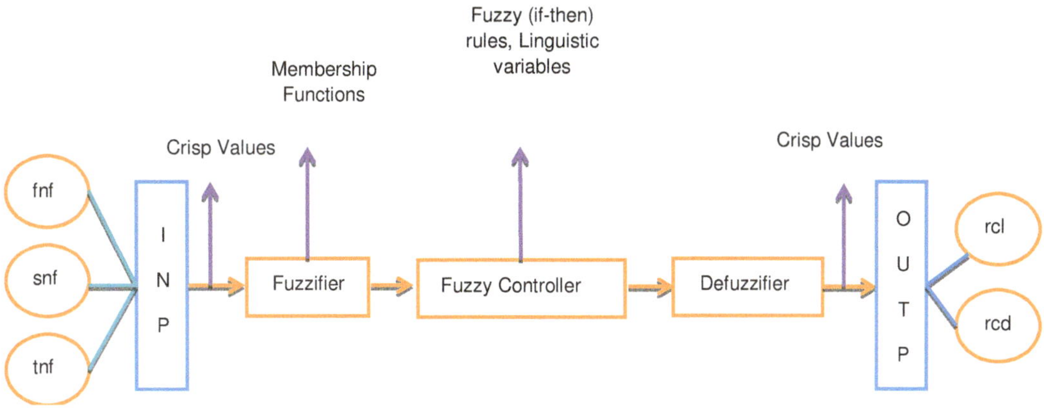

Figure 2. Fuzzy controller.

rules as follows:

Wijk=μfnfi (freqi) Λ μsnfj (freqj) Λ μtnfi (freqk)

where freqi, freqj and freqk are the first, second and third natural frequency of the cantilever beam with crack respectively; by applyng composition rule of interference (Harish and Parhi, 2008) the membership values of the relative crack location and relative crack depth (location)CL:

μrclijk (location) = Wijk Λ μrclijk (location) length CL

As:
μrclijk (depth) = Wijk Λ μrclijk (depth) depth CD

The overall conclusion by combining the output of all the fuzzy can be written as follows:

μrclijk (location) = μrcl111 (location) V.....V μrclijk (location)

V.V μrcl9 9 9 (location) (15a)

μrclijk (location) = μrcl111 (depth) V......V μrclijk (depth)

V....V μrcl9 9 9 (depth) (15b)

The crisp values of relative crack location and relative crack depth are computed using the center of gravity method (Das et al., 2008) as:

$$\text{Relative crack location} = rcl = \frac{\int \text{location}.\mu_{rcl}(\text{location}).d(\text{location})}{\int \mu_{rcl}(\text{location}).d(\text{location})} \quad (16a)$$

$$\text{Relative crack depth} = rcd = \frac{\int \text{depth}.\mu_{rcd}(\text{depth}).d(\text{depth})}{\int \mu_{rcd}(\text{depth}).d(\text{depth})} \quad (16b)$$

WHY WE USE FUZZY LOGIC

(i) Provides an easy to use interface for applying modern fuzzy logic techniques.
(ii) Easily integrated into Model-Based design through the use of the Simulink blocks.
(iii) Provides the ability to use fuzzy logic when appropriate with other control techniques.
(iv) Provides the ability to generate code for various uses.
(v) Supplies a fuzzy inference engine that can execute the fuzzy system as a stand-alone application.

DISCUSSION AND CONCLUSION

In this paper a cantilever beam with a single crack has been taken into consideration. The change in local flexibility due to the presence of the crack is used to calculate the change in the natural frequencies of the cantilever beam. For this theoretical analysis has been used.

Finite element method is used to find out the natural frequencies of the faulty cantilever beam. A fuzzy controller has been designed using trapezoidal, Gaussian as well as triangular membership function to find out the crack depth and crack location (as shown in Figure 3). Table 1 presents the linguistic terms used for membership functions and the range of the linguistic terms. For the fuzzy controller some fuzzy rules have been formulated, which is described in Table 2. The operation of fuzzy controller has been shown through an example as shown in Figure: 4. Table 3 represents the results of fuzzy logic controller as well as the deviation in result.

The above system is modeled and simulated in the Matlab Simulink environment for the valediction of the result which is shown above. It has been observed that the applied fuzzy controllers can predict the relative crack location, relative crack depth of the beam with a considerably less amount of computational time as compared to Finite element analysis.

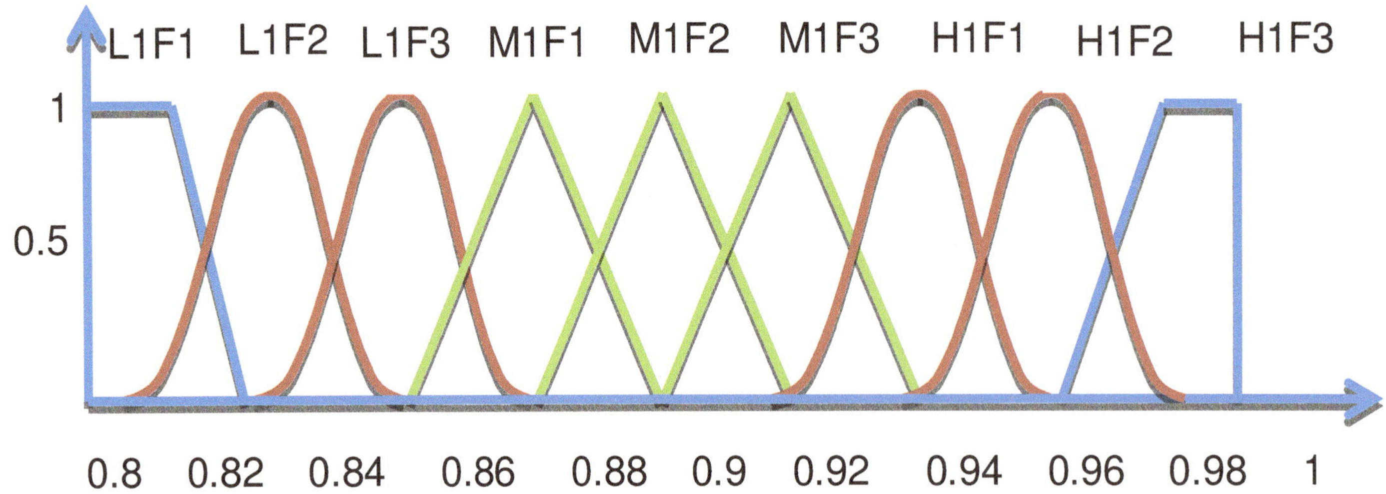

Figure 3(a). Hybrid Membership functions for relative natural frequency for 1st mode of vibration.

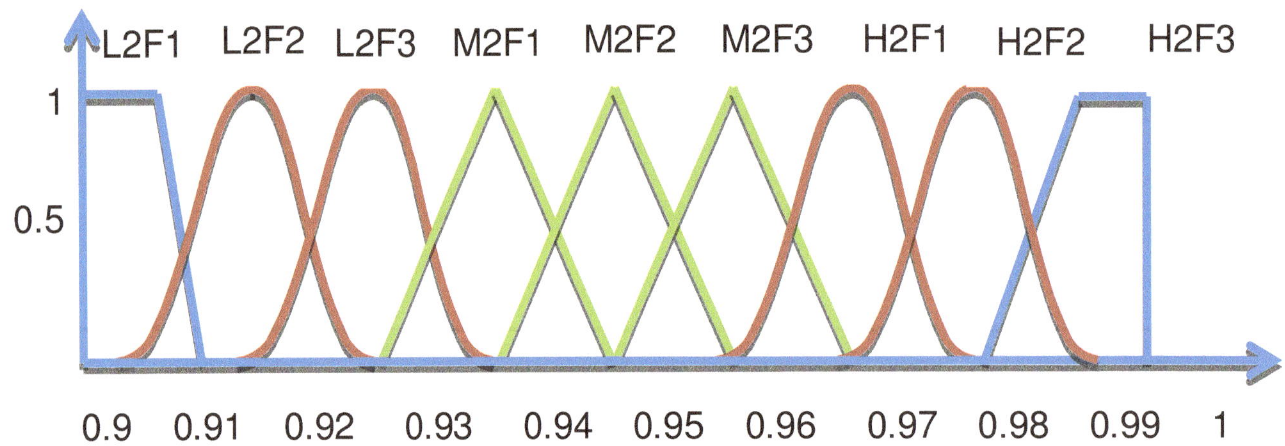

Figure 3(b). Hybrid Membership functions for relative natural frequency for 2nd mode of vibration.

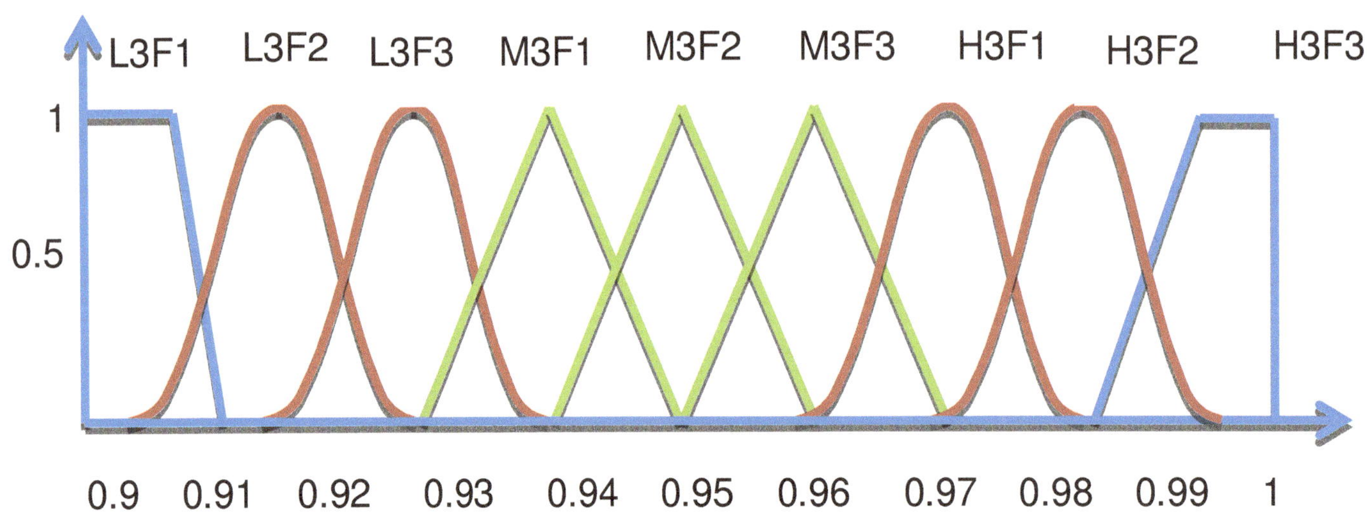

Figure 3(c). Hybrid Membership functions for relative natural frequency for 3rd mode of vibration.

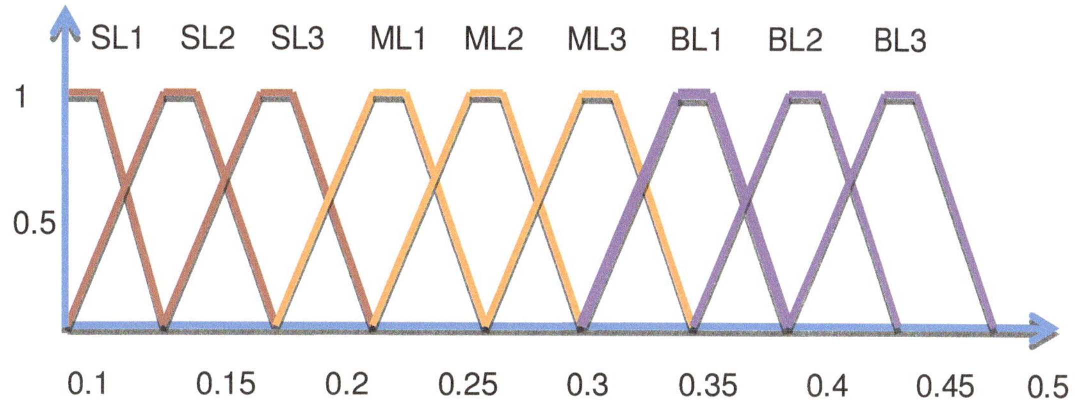

Figure 3(d). Trapezoidal Membership functions for relative crack location.

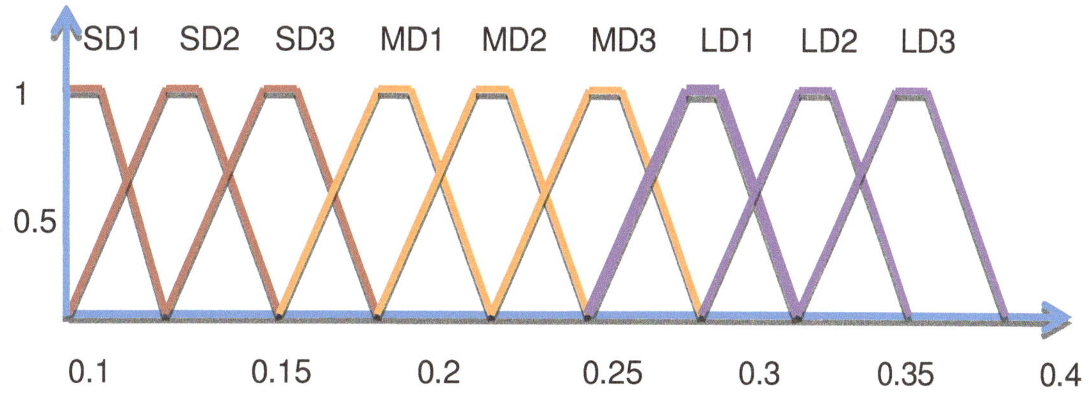

Figure 3(e). Trapezoidal Membership functions for relative crack depth.

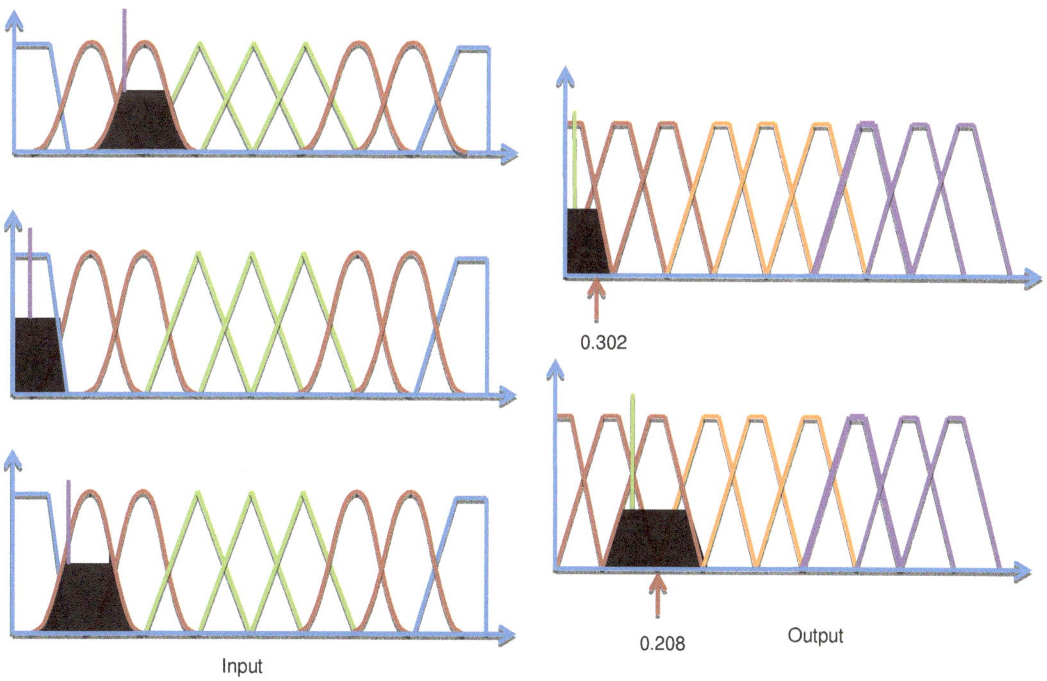

Figure 4. Rule No. 6 of Table 2 is activated.

Table 3. Example of Input data and results of fuzzy controller.

Sl.no	Relative first natural frequency	Relative second natural frequency	Relative third natural frequency	Relative crack depth	Relative crack location	Fuzzy controller relative crack depth	Fuzzy controller relative crack location	% Deviation in result (crack depth)	% Deviation in result (crack location)
1	0.870	0.914	0.985	0.225	0.135	0.216	0.139	4	2.96
2	0.878	0.945	0.992	0.312	0.257	0.307	0.265	1.602	3.12
3	0.885	0.969	0.995	0.282	0.220	0.273	0.230	3.19	4.54
4	0.904	0.973	0.974	0.265	0.235	0.257	0.229	3.01	2.55
5	0.915	0.981	0.979	0.230	0.270	0.226	0.267	1.73	1.11
6	0.936	0.991	0.955	0.216	0.3125	0.208	0.302	3.7	3.36
7	0.947	0.995	0.969	0.2	0.375	0.21	0.391	5	4.26
8	0.918	0.929	0.974	0.283	0.4	0.263	0.388	7.06	3
9	0.932	0.955	0.995	0.166	0.237	0.162	0.212	2.4	10.54
10	0.976	0.980	0.929	0.15	0.268	0.161	0.252	7.33	5.97

REFERENCES

Eiji M, Jyh-Shing RJ, Chuen-Tsai S (2007). 'Neuro-Fuzzy and Soft Computing', A Computational Approach to Learning and Machine Intelligence, Pearson Prentice Hall. p. 13-17.

Harish CD, Parhi DRK (2008). 'Online fuzzy logic crack detection of a cantilever beam', Int. J. Knowledge-based Intelligent Eng. Syst., 12(2): 157-171.

Harish CD, Parhi DRK (2009). 'Application of Neural network for fault diagnosis of cracked cantilever beam', World Congress on Nature and Biologically Inspired Computing. p. 1303-1308.

Mohammad T, Vakil B (2008). 'Crack detection in beam-like structures using genetic algorithms', Appl. Soft Comput., 8(2): 1150-1160.

Norhisham B, Hong H, Andrew JD (2007). 'Damage detection using artificial neural network with consideration of uncertainties', Eng. Struct., 29(11): 2806-2815.

Parhi DRK, Amiya KD (2009). 'Analysis of methodologies applied for diagnosis of fault in vibrating structures', Int. J. Vehicle Noise Vib., 5(4): 271-286.

Peng ZK, Lang ZQ, Billings SA (2007). 'Crack detection using nonlinear output frequency response functions', J. Sound Vib., 301: 777-788.

Ganguli R (2001). 'A Fuzzy Logic System for Ground Based Structural Health Monitoring of a Helicopter Rotor using Modal Data', J. Intelligent Mat. Syst. Struct.. 12(6): 397-407.

Saridakis KM (2008). 'Applying neural networks, genetic algorithms and fuzzy logic for the identification of cracks in shafts by using coupled response measurements', Comput. Struct., 86(11-12): 1318-1338.

Saridakis KM (2008) 'Applying neural networks, genetic algorithms and fuzzy logic for the identification of cracks in shafts by using coupled response measurements', Comput. Struct., 86(11-12): 1318-1338.

18

Stochastic redundancy allocation problem using simulation

Y. Santosh Kumar Reddy[1]*, C. Nadha Muni Reddy[2] and K. Hema Chandra Reddy[3]

[1]Department of Mechanical Engineering, Sri Venkateswara College of Engineering and Technology, Chittoor, Andhra Pradesh, India.
[2]Sri Venkatesa Perumal College of Engineering and Technology, Puttur, Chittoor (District) Andhra Pradesh, India.
[3]Academic Planning Unit, Jawaharlal Nehru Technological University, Anantapur Andhra Pradesh, India.

Reliability is one of the main parameters of the products considered to be important to survive in the competitive market. Reliability in its simple form means the probability that a failure may not occur in a given period of time that is, the component performs adequately without failure. The subsystems or components in a system are arranged in series or parallel depending on the space constraint. However, the reliability of the system with components arranged in parallel is more than that system for which the components are arranged in series. Redundancy is the method of arranging the components in a subsystem in parallel such that if one component fails then the other component automatically comes into operation. Through redundancy, any desired level of reliability can be obtained, but in doing so, we have to invest money or other material resources to achieve the desired reliability. A designer has to consider the economic views of the organization in designing a system with high reliability. In this paper, a complex system (series-parallel system) is considered with stochastic reliability for its components. For a particular configuration, reliabilities of the components are generated and evaluated by the system reliability, after which simulation is run and repeated for various runs, before finally consolidating the configuration reliability and resource utilization. The simulation is carried for various feasible configurations and each configuration evaluated. The configuration with best reliability within resource restrictions is selected.

Key words: Redundancy, allocation, reliability.

INTRODUCTION

A lot of work has been reported in series system to its simplicity in configuration. Federowicz and Muzumdar (1968) developed a geometric program to maximize reliability achieved by redundancy. Henley and Gandhi (1975) developed a procedure for generating the reliability function directly from the Boolean algebra transmission function. Tillman and Liittschwager (1967) presented a new method for the optimization of system reliability with linear constraints using parametric Approach.

In parallel configuration, Lijttschwager (1964) used dynamic programming for a solution of a multistage reliability problem. Banerjee (1976) developed a program to evaluate optimal redundancy allocation for non series – parallel networks. Anderson (1969) explained the theory of reliability of parallel systems with repair and switching. Masatoshi (1977) studied some reliability aspects of system design.

In this paper stochastic reliabilities of the components are considered and Monte Carlo simulation technique is employed to arrive at the optimal solution.

Redundancy is the provision of alternative means or parallel paths in a system for accomplishing a given task that all means must fail before causing the system failure. Application of redundancy in the system design is found in almost all types of systems due to numerous advantages over other methods of improving system reliability. The important ones are:

*Corresponding author. E-mail: y_skreddy@rediffmail.com.

(1) Any desired level of reliability can be achieved.
(2) Design through redundancy needs comparatively less skill on the part of the designer.
(3) It provides quick solution.
(4) This method can be employed in the event of failure of all other methods.

Various forms of redundancy – active (hot) redundancy, standby (cold) redundancy, warm redundancy, component redundancy, system redundancy, hierarchical redundancy etc. can be employed in a system, depending on the feasibility. One has to select a suitable form considering such factors as the type of components, type of systems, reliability requirements, resources Optimization is an act of obtaining the best results under given circumstances. In design, construction and maintenance of any engineering system, engineers have to take many technological and managerial decisions at several stages. The ultimate goal of all such decisions is to either minimize the effort required or to maximize the required benefit.

In most of the practical problems, decisions have to be made sequentially at different points of time, at different points in space and at different levels. The problems in which decisions are to be made sequentially are called sequential decision problems. Since these decisions are to be made at a number of stages, they are also referred to as multistage decision problems. Simulation programming is a logical extension of mathematical techniques, well studied for the optimization of multistage decision problems.

SIMULATION

Simulation is a quantitative technique developed for studying alternative courses of action by building a model of that system and then conducting a series of repeated trial and error experiments to predict the behaviour of the system over a period of time.

In situations where the mathematical formulation of the problem is not feasible, simulation technique is used by representing reality through a model that will respond in the same manner as in a real-life situation. In simulation, a certain type of mathematical model is formulated which describes the real system's operation. The system is divided into various segments and their inter-relationships with some predictable behavior in terms of probability distributions for each of the possible stages of the system are studied. The simulation experiment is then performed on the model of the system. The following are the reasons for adopting simulation in place of other known mathematical techniques:

(1) It can handle complex systems that require the modeling of interacting stochastic variants. These are the means for modeling empirical or theoretical distribution of real-world parameters.
(2) Simulation may be the only method available because it is difficult to observe the actual environment.
(3) Actual observations of a system may be too expensive.
(4) There may not be sufficient time to allow the system to operate extensively.
(5) Simulation enables one to study dynamically systems either in real time, compressed time or expanded time.
(6) Simulation can be used to experiment with new situations about which we have little or no information, so as to prepare for what may happen.
(7) Simulation analysis can be performed to verify analytical solutions.
(8) When new elements are introduced into a system, simulation can be used to anticipate bottlenecks and other problems that may arise in the behaviour of the system.

SOLUTION THROUGH SIMULATION APPROACH

This process starts with assigning 1000 random numbers (0.00 to 0.99) to each of these probability distributions. This is done in tabular column shown in Table 1. The cumulative probabilities and range of random numbers allotted are calculated from the input data. Then a random number between 0.00 and 0.99 is generated for each stage, it is compared with each range of random numbers specified for each stage and reliability, corresponding to the appropriate range of random numbers in which the random number falls, is assigned to the stage as its reliability. Generation of random numbers is done through a probabilistic mechanism. Most computer facilities include a generator of this kind in their software libraries. If there is 'n' number of stages in the system, 'n' random numbers are to be generated and stage reliability of stages R1, R2, R3, ... Rn is calculated. Then the system reliability is calculated for the given complex network as:

$$R = \{1-[1-(1-r_1^{n1})\,(1-r_2^{n2})\,(1-r_3^{n3})]\}r_4^{n4} \quad \text{......(1)}$$

In addition to reliability considerations, the design of the system may also be constrained by cost, and volume. Let c_i denote the cost of a component i, The total system cost is:

$$C = \sum_{i=1}^{m} c_i n_i$$

$n_i \geq 1$, integer i = 1, 2, 3,4

The cost of the components N1, N2, N3 and N4 are assumed as 1000, 1200, 1400 and 1600 respectively. The cost of the system is restricted to Rs13,000 and the volume is constrained to the maximum of four identical components.

The formulation of P_1 is designed to achieve maximum system reliability of the subject C_0, which is an upper limit on the total system cost.

Therefore, this process is formulated and repeated for different component combinations of the system. If there are 'n' stages and maximum of 'm' components in each stage, the total number of component combinations achieved is m^n. So totally nx (mxn) random numbers are generated.

The cost constraint is also taken into consideration. The component combination of the system with highest reliability and satisfying the cost constraint is considered as the optimal solution after first simulation run. Conducting simulation at once may not yield optimal solution, since each simulation run is like a single experiment conducted under a given set of conditions. So for effectiveness, the component combination that has occurred as the optimal solution for more number of times is taken as the final optimal solution after finite simulation runs.

PROBLEM DESCRIPTION

The network consists of components connected as shown in Figure 1 with reliabilities and probabilities as shown in Table 1.

CONCLUSIONS

It is cumbersome to apply optimization techniques or stoachastic NLP techniques to find the optimal solution for redundancy design problems in complex circuits. For the given complex network (series- parallel) , reliability equation can be written as Equation 1 which is difficult to optimize using dynamic programming approaches and hence Monte Carlo simulation technique is employed in this work and proved to be simplified approach/logic to tackle such problems. For the given network using simulation approach, optimal redundancy is evaluated to be n1=1; n2=1; n3=3; n4=4 by consuming cost resource of Rs 12,800/- and the reliability of the system is found to be 0.9946

```
%Mention the constraint value
CV=13000;
p = struc(1:4,1:4,1:4,1:4)
for i=1:size(p,1)
   n1=p(i,1);
   n2=p(i,2);
   n3=p(i,3);
   n4=p(i,4);
   for j=1:500
   r11(j)=rand;
   r22(j)=rand;
   r33(j)=rand;
   r44(j)=rand;
   if (r11(j)<=0.99)&&(r11(j)>=0.7)
      r1(j)=0.75;
      elseif (r11(j)<=0.69)&&(r11(j)>=0.3)
         r1(j)=0.7;
         else
            r1(j)=0.65;
   end
    if (r22(j)<=0.99)&&(r22(j)>=0.8)
      r2(j)=0.85;
      elseif (r22(j)<=0.79)&&(r22(j)>=0.4)
         r2(j)=0.8;
         else
            r2(j)=0.75;
   end
   if (r33(j)<=0.99)&&(r33(j)>=0.6)
      r3(j)=0.9;
      elseif (r33(j)<=0.59)&&(r33(j)>=0.4)
         r3(j)=0.85;
         else
            r3(j)=0.8;
   end

if (r44(j)<=0.99)&&(r44(j)>=0.7)
      r4(j)=0.8;
      elseif (r44(j)<=0.69)&&(r44(j)>=0.3)
         r4(j)=0.75;
         else
            r4(j)=0.7;
   end
C1=1000;
C2=1200;
C3=1400;
C4=1600;
TC(j)=n1*C1+n2*C2+n3*C3+n4*C4;
%create a loop to find the satisfying TC for a given n1,
n2, n3, n4
   if TC(j)<CV
      b1=(1-(1-r1(j)).^n1);
      b2=(1-(1-r2(j)).^n2);
      b3=(1-(1-r3(j)).^n3);
      b4=(1-(1-r4(j)).^n4);
      value(j)=(1-(1-(b1*b2))*(1-b3))*b4;
   else
      value(j)=0;
   end
  averagevalue=mean(value);
  j=j+1;
   end
avgvalue(i)=averagevalue;
maxavg=max(avgvalue);
end
%create a loop to find the combination.............
for i=1:256
   if(avgvalue(i)==maxavg)
      z=i;
```

Table 1. Cumulative probabilities and ranges of random numbers allotted.

Component	Reliability	Probability	Random number allocation
	0.65	0.3	00-29
1	0.7	0.4	30-69
	0.75	0.3	70-99
	0.75	0 .4	00-39
2	0 .8	0.4	40-79
	0.85	0.2	80-99
	0.8	0.4	00-39
3	0 .85	0 .2	40-59
	0.9	0.4	60-99
	0.7	0.3	00-29
4	0.75	0.4	30-69
	0.8	0.3	70-99

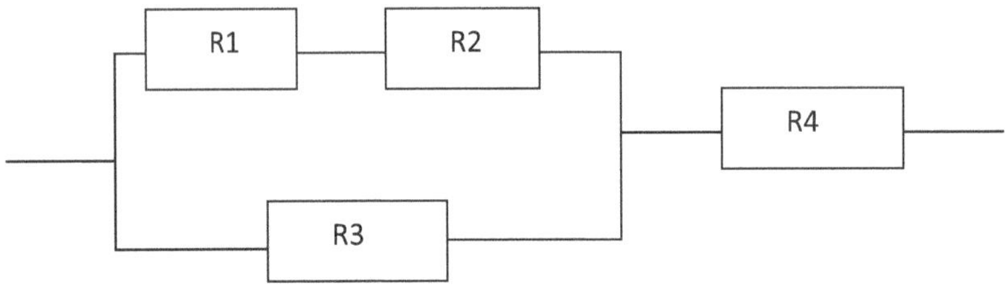

Figure 1. Network of components.

```
end
p(z,:)
%create a loop to find the reliabilities to the corresponding combination
a=p(z,:);
  n1=a(1);
  n2=a(2);
  n3=a(3);
  n4=a(4);
  for j=1:500
  r11(j)=rand;
  r22(j)=rand;
  r33(j)=rand;
  r44(j)=rand;
  if (r11(j)<0.99)&&(r11(j)>0.7)
    r1(j)=0.75;
    elseif (r11(j)<0.69)&&(r11(j)>0.3)
      r1(j)=0.7;
      else
        r1(j)=0.65;
  end
   if (r22(j)<0.99)&&(r22(j)>0.8)
    r2(j)=0.85;
 elseif (r22(j)<0.79)&&(r22(j)>0.4)
      r2(j)=0.8;
      else
        r2(j)=0.75;
  end
  if (r33(j)<0.99)&&(r33(j)>0.6)
    r3(j)=0.9;
    elseif (r33(j)<0.59)&&(r33(j)>0.4)
      r3(j)=0.85;
      else
        r3(j)=0.8;
  end
   if (r44(j)<0.99)&&(r44(j)>0.7)
    r4(j)=0.8;
    elseif (r44(j)<0.69)&&(r44(j)>0.3)
r4(j)=0.75;
      else

 r4(j)=0.7;
  end
C1=1000;
C2=1200;
C3=1400;
C4=1600;
TC(j)=n1*C1+n2*C2+n3*C3+n4*C4;
%create a loop to find the satisfying TC for a given n1, n2, n3, n4
```

```
  if TC(j)<=CV
    b1=(1-(1-r1(j)).^n1);
    b2=(1-(1-r2(j)).^n2);
    b3=(1-(1-r3(j)).^n3);
    b4=(1-(1-r4(j)).^n4);
    value(j)=(1-(1-(b1*b2))*(1-b3))*b4;
  else
    value(j)=0;
  end
 averagevalue=mean(value);
 j=j+1;
 maxt=max(averagevalue);
 end
 %create a loop to find the combination.............
p(z,:)
r11=r11';
r22=r22';
r33=r33';
r44=r44';
r1=r1';
r2=r2';
r3=r3';
r4=r4';
value=value';
R=[r11 r1 r22 r2 r33 r3 r44 r4 value];
disp('Cost of the system is.....');
disp(TC(j-1));
disp('Reliability of the system is.....');
disp(averagevalue);
% disp('number of components of r1 r2 r3 r4....');
disp('  RN1     R1     RN2    R2      RN3     R3
    RN4     R4    System reliability');
disp(R);
```

REFERENCES

Anderson DE (1969). Reliability of a Special Class of Redundant Systems. Reliab. IEEE Trans. R-18(1): 21-28.

Banerjee SK (1976). Optimal redundancy allocation for non series – parallel networks. Reliab. IEEE Trans. 25(2): 115-118.

Federowicz AJ, Majumdar M (1968). Use of geo-metric programming to maximize reliability achieved by redundancy. Oper. Res., 16(5): 948-954, Sept-Oct.

Masatoshi S (1977). Multiobjective reliability and redundancy optimization of a series parallel system by the surrogate worth trade off-method. Microelectron. Reliab., 17(4): 465-467.

Henley EJ, Gandhi SL (1975). Process reliability analysis. AIChE J., 21(4): 677-686.

Tillman FA, Liittschwager JM (1967). Integer programming formulation of constrained reliability problems. Manage. Sci.,13(4): 887-899, July.

Performance evaluation of bamboo with morter and concrete

H. M. A. Mahzuz*, Mushtaq Ahmed, Md. Ashrafuzzaman, Rejaul Karim and Raju Ahmed

Department of Civil and Environmental Engineering, Shahjalal University of Science and Technology, Sylhet, Bangladesh.

Bamboo has remarkable properties as a construction material, being both light weight and extremely strength and durable. Moreover it grows all over the Bangladesh; a South Asian lower-middle income country where rural people is the majority of the total population. Bamboo having considerable tensile and compressive strength as a construction material is used widely in Bangladesh. But the main problem that it has is less sustainability. But if it is mixed with some durable material like mortar and concrete than its durability as well as the strength taking ability will be much higher. This study concentrates on the strength test of composite members made accompanied by bamboo. It determines the compressive strength of bamboo, the strength of composite column made of bamboo and mortar in its hollow area, and the strength of composite column of bamboo and concrete. Different mixing ratios and cross-sectional areas are used for conducting the study. This study is performed mainly for the rural areas, where bamboo is of ample amount, steel is rare, expensive or transportation cost is high. In coastal area the economic condition of people is very poor. In such type of backward area, such study may be essential for their development as well as an assurance for low cost housing. After the study it is seen that samples constructed as aid of bamboo can offer respectable amount of strength that can be safely used for low-cost housing.

Key word: Compressive strength, column, cement, sand, sample, stone, stone chips.

INTRODUCTION

Bamboo is a natural perennial grass-like composite and contains ligno-cellulosic-based natural fibers. Generally it occurs in the natural vegetation of many parts of tropical, subtropical and mild temperature regions, with about 1250 species identified throughout the world. It reaches its full growth in just a few months and reaches its maximum mechanical resistance in just few years. Therefore Bamboo contains tremendous economical advantage as well (Ashrafuzzaman et al., 2010). Its height ranges from 10 cm to even 40 m in height. Bamboo plays an essential role in the daily life of millions of people in subtropical and tropical regions (Liese, 1987). It is seen that everyday about 2.5 billion people in Asia use bamboo for their everyday work (Scurlock et al., 2000). In Bangladesh, a Southeastern country, bamboo is a common material to build low cost houses especially in rural areas (Rashid, 2007; Shila, 2002). It is also used as a common fuel product in Bangladesh as well as different part of South Asia (Bhatt and Sachan, 2004; Leach, 1987; Miah et al., 2010). Construction industry is one of the most polluting industries in the world. Production of both concrete and steel causes considerable deterioration of the environment. For example cement requires over 1400°C by burning fossil fuel (CS Monitor, 2008; Aziz, 1995). Even the flame temperature may reach to 1650°C (Neville, 1995). Production of every ton cements results in emission of at least one ton of CO_2 (Baba, 2009). Similarly, production of per ton of steel releases over two tons of CO_2 in the atmosphere (Ghavami, 2007). The steel making process is very energy intensive. The majority of energy used in the production of steel comes from coke/coal, electricity, liquid oxygen, and the raw ingredients themselves (www.energytechpro.com, 2004). Even transportation of the construction material is also associated with the emission of CO_2 (Scientific American, 2008). Due to high

*Corresponding author. E-mail: mahzuz_211@yahoo.com.

Figure 1. Prepared bamboo sample for compression test.

cost of building materials for low cost housing search of low cost housing material is always a good area of research (Mahzuz et al., 2009). Wood based building materials such as bamboo are the oldest form of building materials. They are easy to work with, structurally strong construction material suitable for framing, flooring, roofing, lining etc (Joseph and Mc-Nally, 2010; Sun et al., 2003). Research and developments are also made for the effective utilization of natural fibres from coconut husk, sisal, sugarcane bagasse, bamboo, jute, wood, akwara, plantain and musamba for making concrete (Aziza, 1981). Due to its superior properties like high strength to weight ratio, high tensile strength and other factors like low cost, easy availability and harmless to the environment during service, bamboo has constantly attracted the attention of scientists and engineers for use as reinforcement in cement based composites (Ashrafuzzaman et al., 2010). Modern uses of bamboo as an alternative for construction products such as concrete, brick and (hard) wood in particular are in increasing demand. Researches are already made pointing how laminated bamboo-frame building could be a direct alternative for wood-frame building, bamboo having a great advantage in yearly yield per forest area compared to wood (Flander and Rovers, 2009). Even bamboo is used to make footbridge now a days (Laroque, 2007). Bamboo is a very kin interest for various researchers to use as reinforcement in concrete elements (Ghavami, 2005; Yao and Li, 2003; Ghavami, 1995; Brink and Rush, 1966). Different researchers have made their research about the mechanical properties of bamboo (Lakkad and Patel, 1981; Amada and Sun, 2001; Li et al., 2002; Chung and Yu, 2002). Diameter and age have significant effect on the compressive strength capacity of bamboo.

The sclerenchyma fibers were more closely packed at the top section of the bamboo, resulting in a higher strength capacity of bamboo. The fiber density of sclerenchyma tissue within the bamboo is a good indicator of the strength capacity of bamboo (Lo et al., 2004). Though the tensile strength of bamboo is much lower than that of steel yet it offers respectable amount of tensile strength (200.5 MN-m^{-2}) (Carrasco et al., 2002). On the other hand bamboo in terms of fiber composites and mat composites offers better strength as well (175.27 MN-m^{-2} and 110.5 MN-m^{-2}, respectively) (Jain et al., 1992). However, wide applications of bamboo in civil engineering were limited by its disadvantages in the past. The principal disadvantages in its natural form are its poor bond with concrete, low modulus of elasticity, high water absorption tendencies, low durability, and low resistance to fire. Nowadays, some of these shortcomings can be significantly improved by subjecting the bamboo to appropriate treatments (Ashrafuzzaman et al., 2010).

METHODOLOGY

In order to prepare the study, major steps like work procedure design, sample preparation, laboratory test and comparison of the conducted studies are followed. Four basic types of samples are prepared for study at this test; all are accompanied by bamboo at different form. The basic types of samples are:

a) Bamboo sample alone.
b) Composite sample of bamboo with mortar poured in its hole.
c) Composite sample of bamboo with the mixture of cement, sand and stone chips poured in its hole.
d) Bamboo reinforced column sample where bamboo sticks are used as an alternative of steel.

It is to be noted that all the bamboo samples used for this study were at least of 3 years of age. In the following section the preparation of those samples and the test procedures are discussed.

Preparation of bamboo sample

The bamboo samples selected for the study were air dried and worm free. Bamboo was sized into one feet length as shown in Figure 1.The internodal regions of the stem are hollow. The two ends of the bamboo were kept perpendicular to the surface/length of the bamboo. Three replicates were tested under compression machine and the average was taken as the result.

Preparation of composite sample of bamboo with mortar poured in its hole

Each of the bamboo samples was taken of one feet length as stated above. Then mortar is poured at the hole of the bamboo sample. After curing (28 days) it is prepared for the compression test. Water/cement ratio was fixed (0.45). The mortar was prepared in three (1:5, 1:4 and 1:3) mixing ratios of cement and sand. Three replicates were tested for each ratio. Total 9 samples were made. Curing of mortar was performed for 28 days as shown in Figure 2. Due to swelling and odorous problem of bamboo, at the 16th day of curing the samples were dried two days. Then the curing started

Figure 2. Curing of composite sample.

Figure 3. Prepared bamboo splints.

again. Before the final compression test the sample was air dried two days.

Composite sample of bamboo with the mixture of cement, sand and stone chips poured in its hole

Each of the bamboo samples was taken of one feet length as stated above. Then mixture of cement, sand and stone chips (of about 6.25 mm size) are poured at the hole of the bamboo sample. After drying and curing it is prepared for the compression test. Used water/cement ratio was 0.45 as well. The mixture was prepared in three (1: 1.5:3, 1:2:4 and 1:2.5:5) mixing ratios of cement, sand and stone chips. Three samples were tested for each ratio. That is in total 9 samples were made. After the preparation of the sample it was dried for one day in normal weather. Curing of concrete mixture was performed for 28 days as shown in Figure 3. Due to swelling and odorous problem of bamboo, at the 16th day of curing the samples were dried two days. Then the curing started again. Before the final compression test the sample was air dried two days.

Figure 4. Preparation of bamboo reinforcement case with tie bar.

Figure 5. Concrete and bamboo case placed mould.

Preparation of bamboo reinforced column sample

Here bamboo splints are made to test their effectiveness as an alternative of steel in concrete column. The moulds used to create the samples were circular in shape $(\text{Height}: 300\text{mm}, \text{Dia}: 150\text{mm})$. Each sample was provided with six number of bamboo splints. Three types of splints of cross sectional area of (12.5×12.5) mm^2, (8.75×8.75) mm^2 and (10.75×10.75) mm^2 were used. Two concrete mix ratios were used for study purpose (1:1.25:2.5, 1:1.5:3). Three samples were made for each mixing ratio and each splint size. That is in total 18 samples were made. For all samples water/cement ratio was taken as 0.45. GI weir was used to hold the bamboo splints in position while concrete is poured in mould (Figure 4). For all sample a clear cover of 12.5 mm was maintained. Concrete mix was prepared with sand, cement and stone$_{(18.75mm\ size)}$. After the preparation of sample it was dried for one day in normal weather. Curing was performed for 28 days. After 28 days excess portion of the bamboo splints were cut smoothly (Figures 5 and 6). Then the compression strengths of the samples were determined by the

Figure 6. Cutting of the excess portion of bamboo splints.

Table 1. Determination of the compressive stress of bamboo.

Sample no	Outer dia cm	Inner di cm	Cross sectional area cm^2	Load KN	Stress MPa
01	7.44	6.1	14.24	59.1	41.5
02	8.03	6.5	17.46	66.9	38.32
03	7.4	6.2	12.81	55.4	43.22
Average compressive strength of bamboo sample					41.02

Table 2. Determination of stress (ratio 1:5).

Sample no	Outer dia cm	Inner dia cm	Cross sectional area cm^2	Load KN	Stress MPa
01	7.12	5.9	39.8	36.3	9.2
02	6.82	5.65	36.53	28.6	7.83
03	8.1	6.67	51.53	50.9	9.88
Average compressive strength					8.97

compression test machine.

RESULTS

Relation between the ratio of morter and stress

Determination of the compressive stress of bamboo

Three samples of bamboo were tested, each of which was about one foot in height as shown in the Figure 1. Test results are shown in Table 1.

Determination of strength of composite sample of bamboo with mortar poured in its hole

Test results for the ratio 1:5: The results are tabulated in Table 2.

Test results for the ratio 1:4: The results are tabulated in Table 3.

Test results for the ratio 1:3: The results are tabulated in

Table 3. Determination of stress (ratio 1:4).

Sample no	Outer dia cm	Inner dia cm	Cross sectional area cm^2	Load KN	Stress MPa
01	8.01	6.51	50.39	80.1	15.9
02	7.91	6.42	49.14	68.6	13.96
03	7.17	5.69	40.38	29.6	7.33
Average compressive strength					12.39

Table 4. Determination of stress (ratio 1:3).

Sample no	Outer dia cm	Inner dia cm	Cross sectional area cm^2	Load KN	Stress MPa
01	5.606	4.43	24.68	41.2	14.91
02	5.94	5.09	27.71	44.5	16.1
03	8.28	6.49	53.846	57.9	10.753
Average compressive strength					13.92

Table 5. Determination of stress (ratio1:1.5:3).

Sample no	Outer dia cm	Inner dia cm	Cross sectional area cm^2	Load KN	Stress MPa
01	6.29	5.12	31.07	42.8766	13.80
02	7.25	5.93	41.28	58.4112	14.15
03	8.24	6.78	53.326	78.1226	14.65
Average compressive strength					14.20

Table 6. Determination of stress (ratio1:2:4).

Sample no	Outer dia cm	Inner dia cm	Cross sectional area cm^2	Load KN	Stress MPa
01	7.17	5.69	40.37	51.3	12.71
02	7.52	6.23	44.41	57.4	12.925
03	6.91	5.45	37.50	46.7	12.453
Average compressive strength					13.45

Table 4.

Relation between the ratio of (cement+sand+stone chip)

Determination of strength of composite sample of bamboo with the mixture of cement, sand and stone chips poured in its hole

Test results for the ratio 1:1.5:3. The results are tabulated in Table 5.

Test results for the ratio 1:2:4.The results are tabulated in Table 6.

Test results for the ratio 1:2.5:5.The results are tabulated in Table 7.

Relation between the ratio in the basis of their strength capacity

Test of bamboo reinforced sample

Test results for the ratio 1:1.25:2.5: (splint size of bamboo =(12.5×12.5) mm^2):

Total cross sectional area of bamboo = 6 x12.5x12.5 = 937.5 mm^2, the results are tabulated in Table 8.

Test results for the ratio 1:1.25:2.5 (splint size of bamboo = (10.75×10.75) mm^2):

Total cross sectional area of bamboo =6 x10.75 x 10.75 =693.5 mm^2

The results are tabulated in Table 9.

Table 7. Determination of stress (ratio1:2.5:5).

Sample no	Outer dia cm	Inner dia cm	Cross sectional area cm^2	Load KN	Stress MPa
01	8.12	6.42	51.78	63.21	12.21
02	7.4	5.93	43.01	53.31	12.4
03	7.06	5.43	39.21	49.9	12.73
Average compressive strength					12.44

Table 8. Determination of stress.

					For the 1:1.25:2.5		For the ratio 1:1.5:3	
Sample No.	Diameter (mm)	Cross sectional area (mm^2)	Bamboo reinforcement (cm^2)	Bamboo ratio (ρ_b) $\frac{Bamboo\ area}{Gross\ concrete\ area}$	Load KN	Stress MPa	Load KN	Stress MPa
01	150	17670	937.5		410.06	22.48	403.31	22.11
02	150	17670	937.5	0.053	360.81	19.78	366.28	20.08
03	150	17670	937.5		385.07	21.11	374.12	20.51
Average compressive strength						21.12		20.90

Table 9. Determination of stress.

					For the 1:1.25:2.5		For the ratio 1:1.5:3	
Sample No.	Diameter (mm)	Cross sectional area (mm^2)	Bamboo reinforcement (mm^2)	Bamboo ratio (ρ_b) $\frac{Bamboo\ area}{Gross\ concrete\ area}$	Load KN	Stress MPa	Load KN	Stress MPa
01	150	17670	693.5		347.13	19.03	391.82	21.48
02	150	17670	693.5		388.53	21.30	364.09	19.96
03	150	17670	693.5	0.039	409.69	22.46	365.91	20.06
Average compressive strength						20.93		20.50

Table 10. Determination of stress.

Sample No.	Diameter (mm)	Cross sectional area (mm^2)	Bamboo reinforcement (mm^2)	Bamboo Ratio (ρ_b) $\frac{Bamboo\ area}{Gross\ concrete\ area}$	For the 1:1.25:2.5 Load KN	For the 1:1.25:2.5 Stress MPa	For the ratio 1:1.5:3 Load KN	For the ratio 1:1.5:3 Stress MPa
01	150	17670	460	0.026	391.27	21.45	382.69	20.98
02	150	17670	460		374.85	20.55	336.55	18.45
03	150	17670	460		366.64	20.10	381.24	20.90
Average compressive strength						20.7		20.11

Test results for the ratio 1:1.25:2.5 (splint size of bamboo = (8.75×8.75) mm^2):

Total cross sectional area of bamboo = 6 × 8.75 × 8.75= 460 mm^2, the results are tabulated in Table 10.

DISCUSSION

The observed results mainly focus on the strength of the elements where bamboo is used at different forms. It is seen from the study that compression test of bamboo gives an average stress of 41.02 Mpa that is a good value to resist moderate loads especially of low cost buildings because load of low cost buildings are not so high. Such bamboo can be safely used as the feasible alternative of concrete column. If more than one bamboo is used than the resulting composite column can provide more resistance against more load. Table 1 shows the quantitative result of the compressive stress test of bamboo samples. The Composite sample of bamboo with mortar poured in its hole also shown respectable amount of stress before failure. At the mix ratio of 1:3, 1:4 and 1:5 the average stresses are 13.92 Mpa, 12.39 Mpa and 8.97 Mpa respectively. That is the stress got at the mix ratio of 1:3 is 12.35% higher than that of the mix ratio of 1:4, and the stress got at the mix ratio of 1:4 is 38.13% higher than that of the mix ratio of 1:5. It means that at lower stress ratio composite members provide better strength. Tables 2, 3 and 4 show the result of the tests of samples related to these types of sample. Finally the summary of the result is shown in Figure 7. Composite sample of bamboo with the mixture of cement, sand and stone chips poured in its hole have shown higher strength than the composite sample of bamboo with mortar poured in its hole. It is seen from the test results that at the mix ratio of 1:1.5:3, 1:2:4 and 1:2.5:5 the average stresses are 14.20, 13.45 and 12.44 Mpa respectively. That is the stress got at the mix ratio of 1:1.5:3 is 5.58% higher than that of the stress of the mix ratio 1:2:4, and the stress got at the mix ratio of 1:2:4 is 8.12% higher than that of the mix ratio of 1:2.5:5. It means that at lower stress ratio composite members provide better strength. Tables 5, 6 and 7 show the result of the tests of samples related to these types of sample. Finally the summery of the result is shown in Figure 8. At Bamboo reinforced concrete three types of bamboo area are used having the same gross area. Two types of mix ratios are used. The bamboo ratios are, 0.053, 0.039 and 0.026 respectively. It is to be noticed that at higher bamboo ratio there higher resistance against load. Tables 8, 9 and 10 show the result of the tests of samples related to these types of sample. Finally the summery of the result is shown in Figure 9.

Conclusion

Bamboo is one of the oldest materials used for the construction of houses and other structures. As an excellent building materials it is relatively cheap, easy to work with and readily available in most of the countries where bamboo grows. The importance of bamboo as a construction material particularly for housing has received a greater attention in recent years. In Asia, traditional bamboo houses of the low-income group use bamboo for supporting the structure. Even when other materials are used, bamboo forms a major part of the unit. Low cost bamboo houses are a cheap and safe alternative to the shelters of plastic, wood and stone that are currently being used by many homeless people. These houses are made from prefabricated and natural bamboo generally associated with other kinds of building materials like wood, mud, brick and concrete as per necessary so as to make house in different

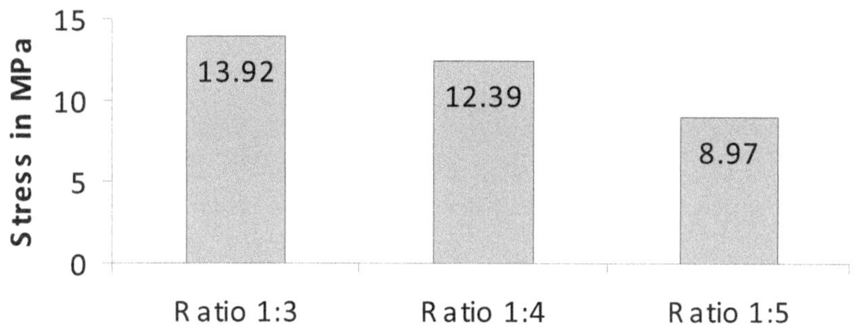

Figure 7. Strength of column at different ratio of mortar.

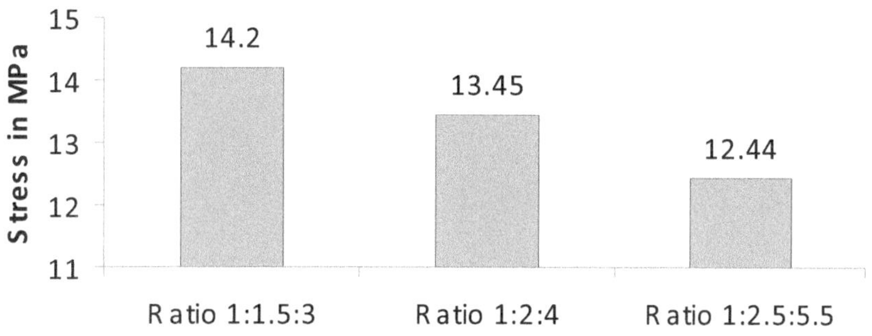

Figure 8. Strength of column at different ratio of concrete.

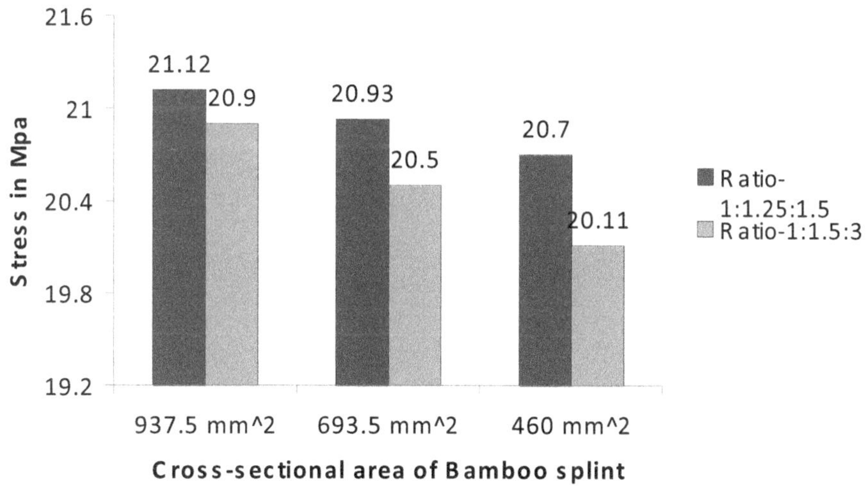

Figure 9. Comparison between the ratio in the basis of their strength capacity.

appearances and to increase durability. The global shortage of housing materials especially in the developing countries is such that it warrants serious consideration. Of all the advantages of bamboo housing technology, the most important is its low cost that does not sacrifice quality, durability or space. In stead it allows an option that is feasible for populations of scarce resourced areas like introducing bamboo in mortar and concrete based on the observed resting strength using different mixing techniques and ratios. Further studies are also necessary to find the more efficient use of bamboo for using it in low cost housing.

REFERENCES

Amada S, Sun US (2001), Fracture properties of bamboo, Composites Part B: Engineering, 32(5): 451-459, doi:10.1016/S1359-8368(01)00022-1.

Ashrafuzzaman M, Karim R, Ahmed R (2010). Effectiveness Evaluation of the Use of Bamboo with Mortar & Concrete, an Undergraduate Thesis, Department of Civil & Environmental engineering, Shahjalal University of Science and Technology, Sylhet, Bangladesh. p. 1.

Aziz MA (1995). Engineering Materials. Z and Z Computer and Printers, Dhaka, Bangladesh, p. 65.

Aziza MA, Paramasivamb P, Leec SL (1981). Prospects for natural fibre reinforced concretes in construction, Int. J. Cement Composites Lightweight Concrete, 3(2): 123-132, doi:10.1016/0262-5075(81)90006-3.

Baba M (2009). Mini Project Report on Shed Type Structures-Steel vs Bamboo, Dept. of Civil Engineering, Indian Institute of Technology, Delhi, p/5, http://www.bambootechnologies.org/pdf/Report_Moti.pdf.

Brink FE, Rush PJ (1966). Bamboo Reinforced Concrete Construction, U. S. Naval Civil Engineering Laboratory, Port Hueneme, California.

Bhatt BP, Sachan MS (2004). Firewood consumption pattern of different tribal communities in Northeast India, Energy Policy, 32(1): 1-6, doi:10.1016/S0301-4215(02)00237-9.

Carrasco A, Fronda J, MacRae B (2002). Mechanical Properties of Bamboo Undergraduate Civil Engineering Student Research, University of Southern California, Los Angeles, CA 90089-2531, Found in: www-classes.usc.edu/engr/ce /334/2002_10. ppt.

Chung KF, Yu WK (2002). Mechanical properties of structural bamboo for bamboo scaffoldings, Engineering Structures, 24(4): 429-442,doi:10.1016/S0141-0296(01)00110-9.

Monitor CS (2008). http://www.csmonitor.com/2008/0312/p14s01-stgn.html.

Flander KD, Rovers R (2009). One laminated bamboo-frame house per hectare per year, Cement & Concrete Composites, 23(1): 210-218.

Ghavami K (2007). Low cost and Energy saving construction materials, Proc. International on Modern Bamboo Structures, 28-30 Oct., Changsha, China, pp. 5-21.

Ghavami K (2005). Bamboo as Reinforcement in Structural Concrete elements, Cement & Concrete Composites, 27(6): 637-649.

Ghavami K (1995). Ultimate load behavior of Bamboo-reinforced light-weight concrete beams, Cement & Concrete Composites, 17(4): 281-288.

http://www.energytechpro.com/DemoIC/Business_Sector/Industrial/Steel_Production.htm (2004).

Jain S, Kumar R, Jindal UC (1992). Mechanical behaviour of bamboo and bamboo composite, J. Mater. Sci., ISSN 0022-2461 (Print) 1573-4803 (Online), 27(17): 4598-4604September, DOI 10.1007/BF01165993.

Joseph P, Mc-Nally ST (2010). Sustainable Non-metallic Building Materials, Sustainability, 2: 411, doi: 10.3390/su2020400, http://eprints.ulster.ac. uk/10111/1/_PJPub14.pdf.

Lakkad SC, Patel JM (1981). Technical note on Mechanical properties of bamboo, a natural composite, Fibre Sci. Technol., 14(4):319-322, doi:10.1016/0015-0568(81)90023-3.

Laroque P (2007). Design of A Low-Cost Bamboo Footbridge, MEngg Report, MIT.

Leach G (1987). Household energy in South Asia, Biomass, 12(3): 155-184, doi:10.1016/0144-4565(87)90034-5.

Li Z, Liu CP, Yu T (2002). Laminate of Reformed Bamboo and Extruded Fiber-Reinforced Cementations Plate, J. Mater. Civil Eng., ASCE research Library, 14(5): 359-365.

Liese W (1987). Research on bamboo, Wood Sci. Technol., 21(3):189-209, ISSN: 0043-7719 (Print) 1432-5225 (Online), DOI 10.1007/BF00351391.

Lo TY, Cui HZ, Leung HC (2004). The effect of fiber density on strength capacity of bamboo, Materials Letters, 58(21): 2595-2598, Aug,doi:10.1016/j.matlet.2004.03.029.

Mahzuz HMA, Alam R, Alam MN, Basak R, Islam MS (2009). "Use of arsenic contaminated sludge in making ornamental bricks" is published in the Int. J. Environ. Sci. Technol., 6(2): 291-298, ISSN: 1735-1472.

Miah MD, Kabir RRMS, Koike M, Akther S, Shin MY (2009). Rural household energy consumption pattern in the disregarded villages of Bangladesh, Energy Policy, 38(2): 997-1003, Feb. 2010, doi:10.1016/j.enpol.2009.10.051.

Neville AM (1995). Properties of Concrete, 4th edition, Pearson Education, ISBN 81-7758-587-8, p. 6.

Rashid R (2007). Traditional House of Bangladesh: Typology of houses according to materials and location, Virtual Conference on Sustainable Architectural Design and Urban Planning, AsiaSustainabilityNet.upc.edu, Sept., pp. 15-24.

Scurlock JMO, Dayton DC, Hame B (2000). Bamboo: an overlooked biomass resource? Biomass and Bio-energy, 19(4): 229-244.

Scientific American, http://www.sciam.com/artical.cfm?id=cement-from-carbondioxide, (2008).

Shila DV (2002). Bamboo Construction Technology for Housing in Bangladesh: Opportunities and Constraints of Applying Latin American Bamboo Construction Technologies for Housing, International Network for Bamboo and Rattan Eindhoven University of Technology ActionAid, page-108, 2002, found at: http://www.sheltercentre.org/library/Bamboo+Construction+Technology+Housing+Bangladesh+Opportunities+and+Constraints+Applying+Lat.

Sun M, Rydh CJ, Kaebernick H (2003). Material Grouping for Simplified product life cycle assessment, J. Sustainable product Des, 3: 45-48.

Yao W, Li Z (2003). Flexural behavior of bamboo-fiber-reinforced mortar laminates, Cement & Concrete Composites, 33: 15-19.

20

Effect of rapid quenching and aging on tensile behavior of commercially pure (CP) titanium implant material

B. K. C. Ganesh[1] and N. Ramanaih[2]

[1]Department of Mechanical Engineering, College of Engineering (Autonomous), Andhra University, Andhra Pradesh India.

[2]Department of Mechanical Engineering, College of Engineering (Autonomous), Andhra University, Vishakhapatnam, India.

Pure titanium and titanium alloys are widely used as implant materials in the medical and dental fields because of their superior bio-compatibility, corrosion resistance and specific strength compared with other metallic implant materials. Study of mechanical properties of these implant materials is very important during the course of their machining, fabrication and surface treatment operations. Titanium has good corrosion resistance due to the presence of an oxide layer on its surface which accounts for its superior biocompatibility. It has a poor wear resistance due to high coefficient of friction. An effective heat treatment which involves heating the material above the beta transus temperature (transformation temperature) will result in the formation of acicular martensite (α'). The formation of acicular martensite shows significant improvement in hardness and ultimate tensile strength of the implant material resulting in improved wear resistance. In this work, tensile behavior of commercially pure (CP) titanium implant material is studied when subjected to rapid quenching and aging procedure. The conditions under which the various mechanical properties change with respect to their crystallographic structures when subjected to the heat treatment process are discussed.

Key words: Biomaterial, phase transformation, mechanical properties, ultimate tensile strength.

INTRODUCTION

Biomedical materials with high technological and economical value are now attracting more and more attention and have extensive application prospect. Titanium and its alloys became an excellent choice for biomedical materials because of their low density and modulus, high strength, corrosion resistance and excellent biocompatibility (Cui, 2009).

The design and selection of biomaterials depend upon the intended medical application. The primary requirements of a biomaterial are to have good amount of wear resistance, fatigue strength, ultimate tensile strength and corrosive resistance. Development of new biomaterials is an interdisciplinary effort and is often a collaborative effort between material scientists and engineers, biomedical engineers, pathologists and clinicians.Titanium exists in various allotropic forms. At low temperatures, it has a closed packed hexagonal crystal structure known as α, whereas above 883°C, it has a body centered cubic structure known as β. The α to β transformation temperature of pure titanium either increases or decreases based on nature of the alloying elements. The alloying element such as aluminum, oxygen, nitrogen, etc that tend to stabilize are called alpha stabilizers and the addition of these elements increase beta transus temperature, while elements that stabilize β phase are known as β stabilizers (V, Mo, Nb, Fe and Cr etc) and addition of these elements depress the β transus temperature. Some of the elements that do not have marked effect on the stability of either of the phase but form solid solutions with titanium are termed as neutral elements (Zr and Sn). Geetha et al. (2009) have shown that the addition of Zr stabilizes the β phase in Ti-Zr-Nb system.

The α + β phases also form the basis for normally accepted classification of titanium alloys. Alloys having only 1 to 2% of β stabilizers and about 5 to 10% of β phase are termed as near α alloys containing higher amounts of β which results in 10 to 30% of β phase in the

*Corresponding author. E-mail: bkcganesh@yahoo.com.

Table 1. Tensile properties of C.P.Titanium.

Condition	UTS Mpa	Elongation (%)	Hardness (VHN)
As-received	303	15	114
Quenched	358	36	130
Aging	364	25	135

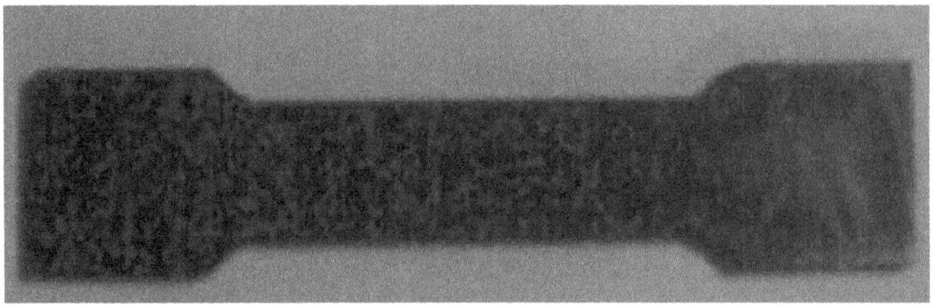

Figure 1. Image of tensile specimen.

microstructure are known as α + β alloys. Alloys with higher β stabilizers where β phase can be retained by fast cooling are known as metastable β alloys. These alloys decompose to α + β on aging. Most of the biomedical titanium alloys belong to α + β or metastable β class.

Mitsuo (2008) reported that the application of stress by rapid quenching results in the formation of martensitic transformation in steels, which contain residual austenite in their microstructures. This phenomenon is called stress or strain induced martensitic transformation, which enhances ductility or fracture toughness of steels. Deformation induced martensitic transformation also occurs in titanium where the unstable beta phase is retained at room temperature by rapid cooling such as water quenching from a high temperature near the beta transus temperature.

Venkatesh et al. (2009) reported an improvement in various mechanical properties when α + β alloys were heat treated above the beta transus temperature and cooled by water quenching, air cooling and furnace cooling. Sami et al. (2007) have studied the influences of heat treatment on Ti-6Al-7Nb implant alloy. In this work the effect of various types of heat treatment of CP Titanium above beta transus temperature with subsequent cooling in air, water and furnace is reported. The change in mechanical properties in respect of the above heat treatment and aging are also reported.

MATERIALS AND METHODS

Three tensile specimens of titanium Grade 2 Implant material were prepared according to ASTM E-8 specification by wire cut EDM as shown in Figure 1. The chemical composition of the material is as follows: Nitrogen, 0.13%, Oxygen, 0.15%, Iron, 0.32%, Titanium, 99.41%. The length of the specimen corresponds to 100 mm with a gauge length of 25 mm and with the length of reduced section of 45 mm.

These specimens were first heat treated in a muffle furnace at a temperature of 900°C which corresponds to the beta transus temperature of titanium. After heat treatment at this temperature for half-an hour, these samples quenched rapidly by water quenching. Then the samples were aged at 500°C for 3 h duration.

All the samples were tested by instron ultimate tensile testing machine under a constant strain rate of 0.01 s^{-1}. The results of the tests conducted on this material are given in Table 1. Hardness values of the specimens are calibrated by Vickers hardness testing machine.

RESULTS AND DISCUSSION

The microstructures that are obtained for the as-received base metal and quenched specimens are given in Figures 2 and 3. Figure 2 shows equi-axed grains whereas Figure 3 shows acicular martensite α' obtained as a result of rapid quenching in water. Stress strain curves corresponding to the tensile behavior under different conditions are given in Figure 4.

The material when subjected to tensile loading has failed by ductile overload fracture resembling cup and cone type of failure as shown in the Figure 5. The results have indicated that there is an increase in ultimate tensile strength and ductility when the specimen is subjected to rapid quenching from 900°C. This is primarily due to the existence of two different crystallographic structures and corresponding transformation of the metal from H.C.P to B.C.C. and vica- versa.

The increase in the case of ductility can be attributed to the ease of plastic deformation obtained during the transformation of the metal from HCP crystallographic structure to BCC crystallographic structure when the metal is heated to beta transus temperature at 900°C. This phenomenon also explains the limited plastic deformation of the hexagonally closely packed (hcp) α

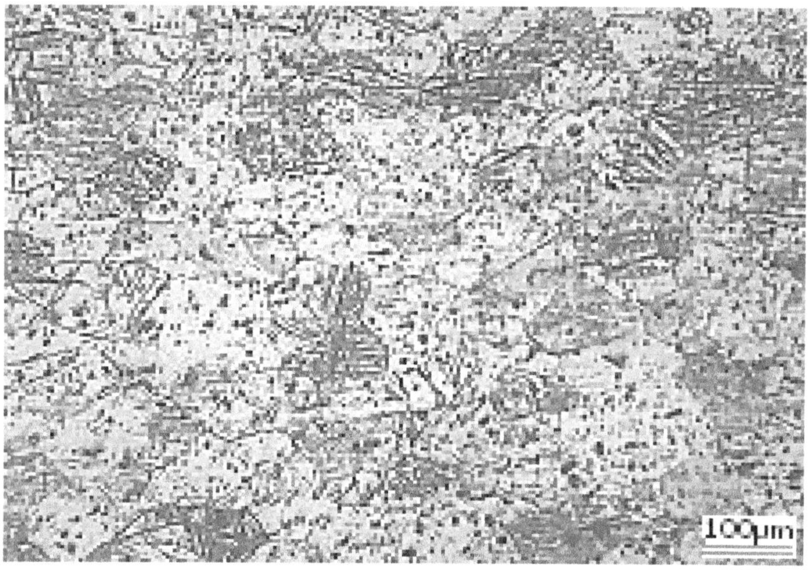

Figure 2. Microstructure of As-received material 100×.

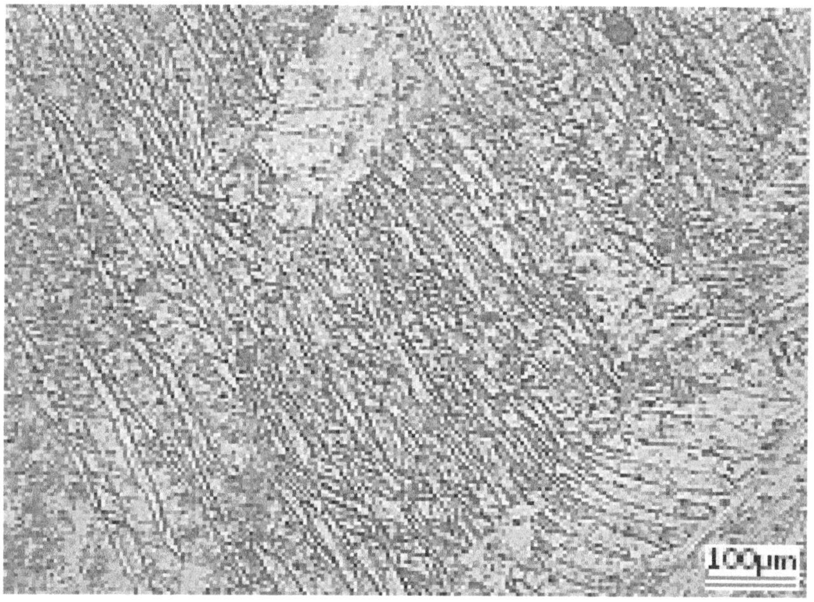

Figure 3. Microstructure of Quenched specimen100×.

titanium compared to body centered cubic crystallographic (bcc) β titanium. The number of slip systems which is equivalent to the number of dislocation glide opportunities is only three for the hcp structure while it is 12 for the bcc lattice. The number of slip systems is determined by the number of slip planes multiplied by the number of slip directions. Slip planes and its corresponding slip directions of both crystallographic structures are shown in Table 2.

The transformation of the bcc β to the hcp α modification of high purity titanium by cooling through the transformation temperature occurs by a nucleation and shear type process. The crystallographic orientation relationship for Zirconium has first been studied by burgers and is therefore named as burgers relationship:

$\{110\}_\beta // \{0001\}_\alpha$ and $<111>_\beta // <1120>_\alpha$

This relationship was confirmed later also for titanium (Newkirk and Giesler, 1953). According to this relationship, a bcc crystal can therefore transform to 12 hexagonal variants, having different orientations with regard to the parent crystal. This transformation can occur in principal either martensitically or by a nucleation and growth process, depending on alloy composition and cooling rate.

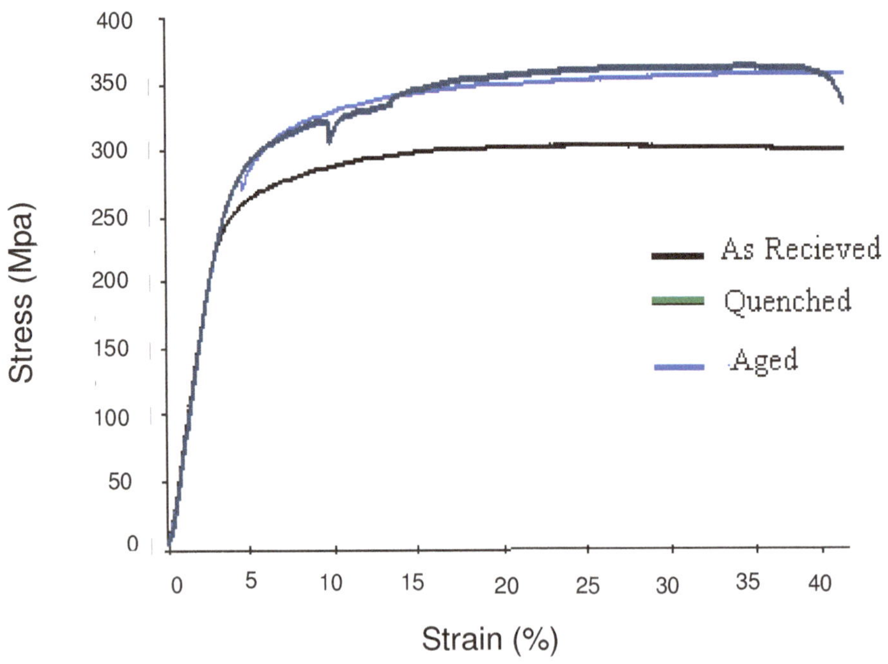

Figure 4. Stress strain curves under different conditions.

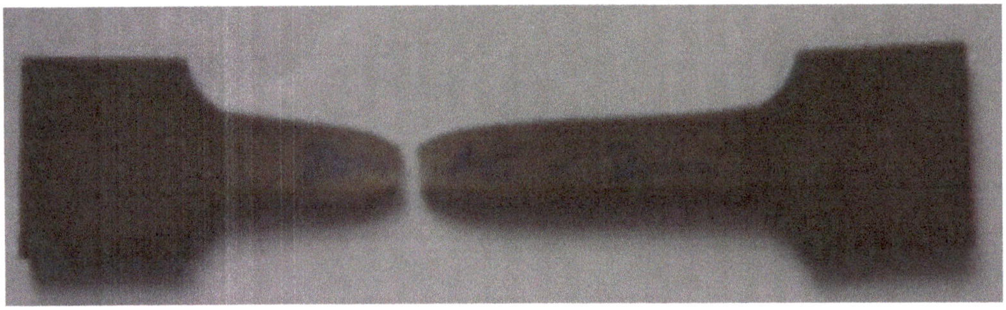

Figure 5. Material subjected to tensile loading.

The increase in the ultimate tensile strength of the material can be resulted due to cooling from the β phase where {110} plane of bcc titanium transform to the {0001} basal plane of hexagonal α phase (Christopher and Manfred, 2003). The distance between the basal planes in hexagonal closely packed α is slightly larger than the corresponding distance between the {110} plane in bcc β phase. Therefore, the transformation of beta to alpha during quenching and subsequent aging leads to a contraction of the c-axis relative to the a-axis in the hcp α and reduces the c/a ratio below the value of ideally closely packed hcp structure. This reduction in c-axis as well as c/a ratio well below the ideal c/a ratio would obstruct the movement of dislocations which results in an increase in yield and ultimate tensile stress of the material (Rolf, 2004). The hcp lattice structure which is less inclined to relaxation during heat treatment due to its lower symmetry as compared to cubic lattices is also playing an important role in improving the ultimate tensile strength (Orest and Roman, 1999). This along with fine grain structure obtained during the heat treatment process is determinant in improving the ultimate tensile strength. As expected, the strength of the aged specimen has slightly increased with a decrease in ductility.

The improvement in hardness is due to the presence of quenching process which results in the formation of acicular α'-martensite as shown in Figure 3. In general, the hardness increase as the rate of cooling increases. This is due to the formation of martensite arising from rapid cooling (Venkatesh et al., 2009). Subsequent aging has not resulted in any further improvement of the above mechanical properties which can be seen in α+ β alloys where transformation of retained martensite results in development of the wide variety of mechanical properties. Scanning electron microscope (SEM) images of all the three conditions viz As received, quenched and aged are shown in the Figures 6, 7 and 8.

Figure 6 shows fine conical dimples showing ductile

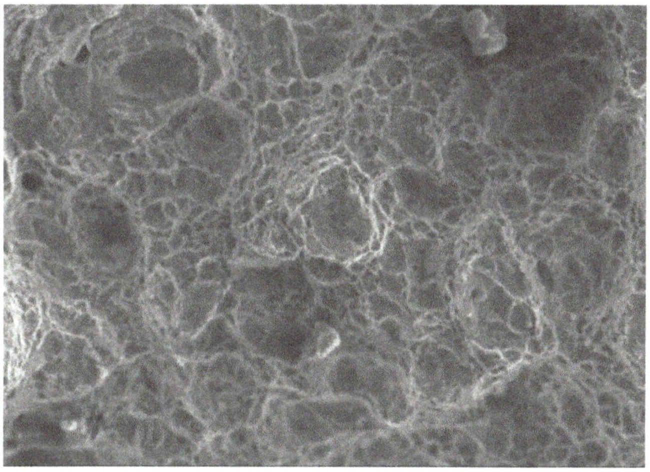

Figure 6. As received condition 500×.

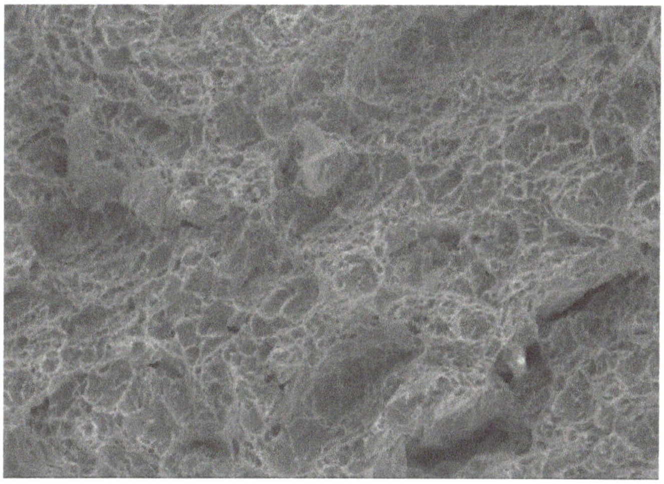

Figure 7. Quenched condition 500×.

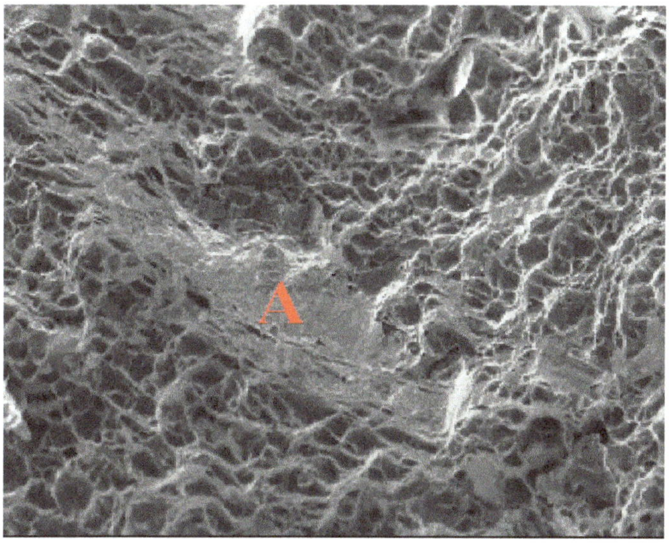

Figure 8. Aged condition 500×.

overload fracture. Figure 7 shows fine tear ridges during quenched condition which had resulted in drastic increase in ultimate tensile strength. Figure 8 shows aging SEM image showing coarse and thick tear ridges with cleavage facet marked as A. This indicates that the failure is induced by secondary cracking showing the influence of hydrogen embrittlement. The presence of coarse and thicker tear ridges indicates a decrease in ductility and a small increase in ultimate tensile strength when compared to the SEM image of quenched specimen.

Conclusion

Various mechanical properties of titanium implant can be tailored to achieve a desired combination of properties by changing its crystallographic structure and processing parameters. A change in the crystallographic structure and processing modifies the microstructure. This change is due to the phase transformation of various equilibrium and non-equilibrium phases present in the alloy system. The resultant mechanical properties depend on the nature of deformation, fracture of the micro structural constituents, and the interaction between constituents. The physical and electronic properties of the titanium atom, due to its position in the periodic table make it suitable for alloying with other elements to produce a wide range of alloys. All these factors result in obtaining a wide variety of microstructures and mechanical properties desired to manufacture an implant material suitable to work for longer duration without failure.

REFERENCES

Christopher L, Manfred P (2003)." Titanium and Titanium alloys. Fundamentals and Applications." WILEY-VCH Publications, Wiienham, Germany, 1: 5-6.

Cui W F, Jin Z, Guo AH, Zhou L (2009). High temperature deformation behavior of alpha-beta type biomedical titanium alloy Ti-6Al-7Nb. Mater. Sci. Eng. A., 499: 252-256.

Geetha M, Singh AK, Asokamani R, Gogia AK (2009). Ti based biomaterials, the ultimate choice for orthopedic implants - A review. Progress Mater. Sci., 54: 397-425.

Mitsuo N (2008). Mechanical biocompatibilities of titanium alloys for biomedical applications. Mech. Behav. Biomed. Mater., pp. 30-42.

Newkirk JB, Giesler AH (1953). The effect of phase transformations on the orientation of zirconium crystals. Acta Meterialia, 1: 370.

Orest MI, Roman T (1999). "Potential of Rapid Heat Treatment of Titanium alloys and Steels." Mater. Sci. Eng., 263: 142-154.

Rolf EH (2004). "Understanding Material Science." 2nd edition. Springer Publications, pp. 47-50.

Sami AA, Thair LA, Abdulsalam KS (2007). Influence of heat treatment conditions on microstructure of Ti-6al-7Nb alloy as used surgical implant materials. Eng. Technol., 25(3).

Venkatesh BD, Lchen D, Bhole SD (2009). Effect of heat treatment on mechanical properties of Ti-6Al-4Valloy. Mater. Sci. Eng., 506: 117-124.

Effect of immersion speed on the mechanical properties and micro-structure of oil quenched AISI 1020 steel

Mudasiru L. O.[1], Babatunde I. A.[1], Raheem W. A.[1] and Lasisi A. K.[2]

[1]Department of Mechanical Engineering, Ladoke Akintola University of Technology, Ogbomosho, Nigeria.
[2]Department of Mechanical Engineering, Kwara State Polytechnic, Ilorin, Kwara State, Nigeria.

Steel is one of the most important commodities in the engineering world due to its usefulness and availability. Its application is wider than any other metal, ranging from domestic use in the forms of utensils and many more to its use in construction and manufacturing industries. The wide range of application of steel in engineering practices has prompted the desire to increase certain mechanical properties to meet engineering requirements through heat treatment. In this work, samples of the specimen were purchased at a local market and the spectrometry analysis was carried out. The steel samples were machined into required shape, heat-treated and cooled under different immersion speeds. The effects of immersion speeds on the mechanical properties (hardness, % elongation, tensile strength, modulus of elasticity and yield strength) and microstructure of the quenched AISI 1020 were investigated. The result obtained revealed an improvement in the mechanical properties of the materials at different immersion speed.

Key words: Heat treatment, mechanical properties, microstructure, AISI 1020.

INTRODUCTION

Carbon steel is a type of steel where the main interstitial alloying constituent is Carbon in the range of 0.12 to 2.0% (Çalik, 2009; Sharma et al., 2012; Fadare, 2011). The American Iron and Steel Institute (AISI) defines Carbon steel as a steel in which no minimum content is specified or required for Chromium, Cobalt, Molybdenum, Nickel, Niobium, Titanium, Tungsten, Vanadium, Zirconium, or any other element added to obtain a desired alloying effect. According to Odusote et al. (2012), steels with Carbon content varying from 0.25 to 0.65% are classified as medium Carbon, while those with Carbon content less than 0.25% are termed low Carbon. The Carbon content of high Carbon steels usually ranges within 0.65 to 1.5%.

Generally, the mechanical properties of plain Carbon

Table 1. Composition of the low-Carbon steel.

C (%)	Si (%)	Mn (%)	S (%)	P (%)	Cr (%)	Ni (%)	Cu (%)
0.189	0.207	0.497	0.021	0.022	0.101	0.073	0.174

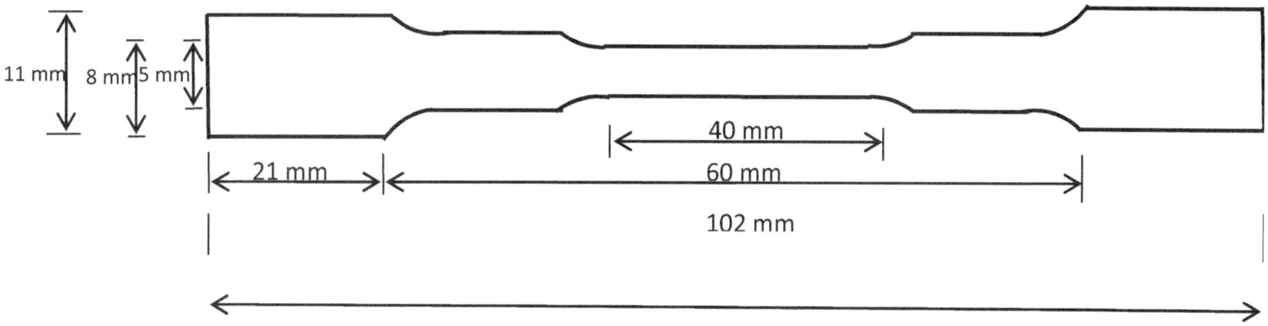

Figure 1. Machined specimen for tensile test.

steels increases with the increased Carbon concentration dissolved in austenite prior to quenching during hardening heat treatment, which may be due to transformation of austenite into martensite (Odusote et al., 2012). Therefore, the mechanical strength of medium Carbon steels can be improved by quenching in appropriate medium. The major influencing factors in the choice of the quenching medium are the kind of heat treatment to be employed, composition of the steel, the sizes and shapes of the parts (Odusote et al., 2012).

Quenching is a common manufacturing process aiming to produce components with desirable properties such as low residual stresses and distortions, avoidance of cracks, specific hardness, and achievement of improved properties (Mackerle, 2003). Low Carbon steel which was employed in the course of this study has high malleability and ductility, good machinability, good surface finish, good formability and general availability.

MATERIALS AND METHODS

Samples of steel bar purchased from a local market in Oyo State, South-Western Nigeria, were cut into 25 mm diameter and 100 mm long. Also, the determination of the chemical composition of the specimen was carried out using Minipal 4 Spectrometer (Table 1).

Specimen preparation

Each specimen was prepared for hardness and tensile tests. The samples were machined into standard specifications for tensile testing using Monsanto Tensiometer, following standard test procedures in accordance with the ASTM E8M-91 standards (1992) (Figure 1). An optical metallurgical microscope XJL-17 model with magnification 10×10 was used for taken the micrograph of the samples.

Heating and quenching process

Heating of the specimen was carried out at the metrology laboratory of the Mechanical Engineering Department, Ladoke Akintola University of Technology, Nigeria. This process involved heating the tensile specimen to an austenising temperature of 850°C. The material was maintained at this temperature for 1 h and then quenched at various immersion speeds of 0.3102, 0.3241, 0.3384, and 0.4167 m/s respectively, using oil as the quenchant.

Mechanical tests

After quenching operation, the materials were returned to the Engineering Materials Development Institute (EMDI) Akure, where the effects of the quenching at different immersion speeds on the physical and microstructural properties of the steel samples were determined.

Microstructure analysis

The microstructural analysis was obtained using an optical metallurgical microscope. 20 mm cylindrical rods were cut from each sample, grounded, polished and etched in Nital prior to microstructural analysis. The microstructure was taken with magnification of 500.

Hardness test

Prior to testing, specimens were mounted using phenolic powder, grinded and polished to obtain a smooth surface. A direct load of

Table 2. Mechanical properties of quenched low Carbon steel at different immersion speeds.

Immersion speed (m/s)	Tensile strength (N/mm^2)	% elongation	Modulus of elasticity	Yield strength (N/mm^2)	Hardness (Case)	Hardness (Core)
0.3102	82.18	55.8	5674.3	312.1	166.25	162.55
0.3241	71.05	53.1	8836.9	298.7	158.35	132
0.3384	81.68	46.5	4642.9	215.4	136.3	110.95
0.4167	70.39	42.6	10576.8	241.2	157.85	115.3

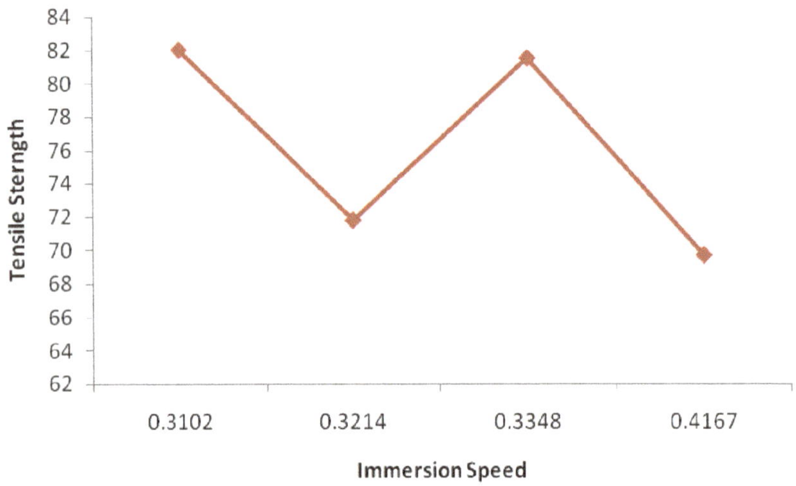

Figure 2. Tensile strength (N/mm^2) against Immersion speed (m/s).

490 N was applied for a dwell time of 10 s and hardness readings were evaluated using Leitz 8299 micro-hardness tester. The indentation was square shaped. Diagonal length of square was measured by scale attached on the microscope. Hardness was obtained directly from the chart given in the manual for corresponding load and diagonal length. Multiple hardness tests were performed on each sample and the average of the values was taken as a measure of the hardness value for each specimen.

Tensile test

P2000 electronic tensiometer was used to conduct the test following standard test procedures in accordance with the ASTM E8M-91 standards (1992). The samples were tested at a nominal strain rate of 10^{-3}/s until fracture. The tensile properties evaluated from the engineering stress-strain curves developed from the tension test are – the ultimate tensile strength (σ_u), the yield strength (σ_y), and the strain to fracture (ε_f).

RESULTS AND DISCUSSION

Effect of heat treatment on mechanical properties

The effect of immersion speed on the mechanical properties (ultimate tensile strength, hardness, toughness, percentage elongation, and percentage reduction) of the quenched steel is shown in the Table 2.

Effect on tensile strength

As the immersion speed of the quenched steel increases, the tensile strength of the samples fluctuates, resulting in variation between the tensile strength and speed of immersion. In Figure 2, it was observed that at low immersion speed of 0.3102 m/s, maximum tensile strength of 82.18 N/mm^2 was obtained. At the highest immersion speed of 0.4167 m/s, the tensile strength reduces to 70.39 N/mm^2 which shows no stable relationship between the immersion speed and the tensile strength.

Effect on percentage elongation

Figure 3 shows the relationship between the percentage elongation and immersion speed. As immersion speed

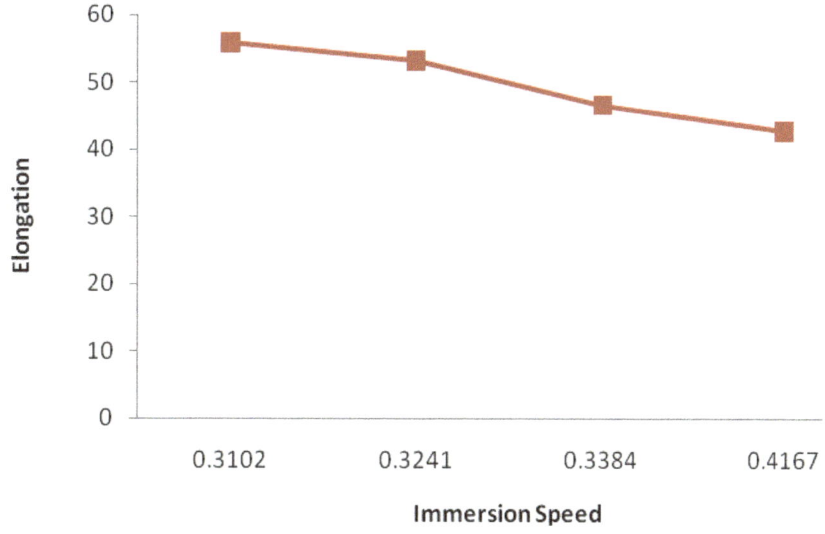

Figure 3. Percentage elongation against immersion speed (m/s).

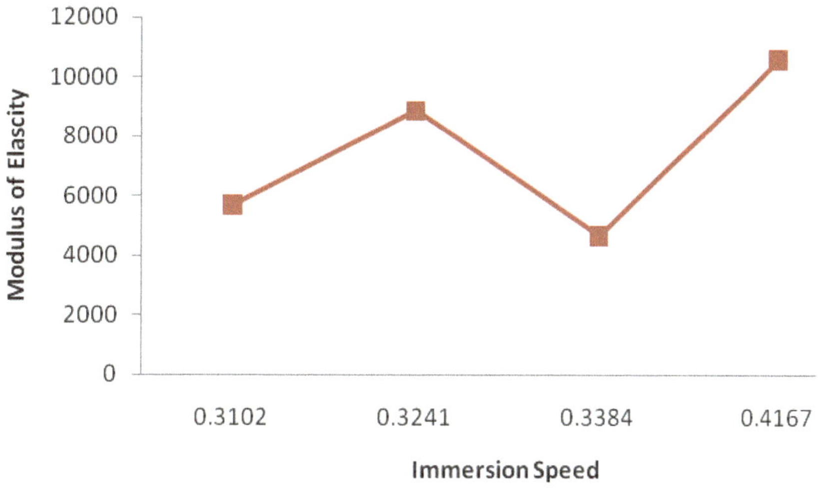

Figure 4. Modulus of elasticity (N/mm^2) against Immersion speed (m/s).

increases, the percentage elongation of the specimens decreases. This shows an inverse relationship between the percentage elongation and immersion speed. At a low immersion speed of 0.3102 m/s, high percentage elongation of 55.15% was obtained above which the percentage elongation of the samples decreases.

Effect on elasticity

The modulus elasticity of the quenched mild steel samples also fluctuates in relation with increased immersion speed. From Figure 4, the modulus of elasticity increases from 5674.3 to 8836.9 as the immersion speed increases from 0.3102 to 0.3214 m/s. For the next immersion speed of 0.3384 m/s, the modulus of elasticity drops drastically to 4642.9 and increases to 10576.8 for the increased immersion speed of 0.4167 m/s.

Effect on yield strength

Figure 5 shows the relationship between the yield

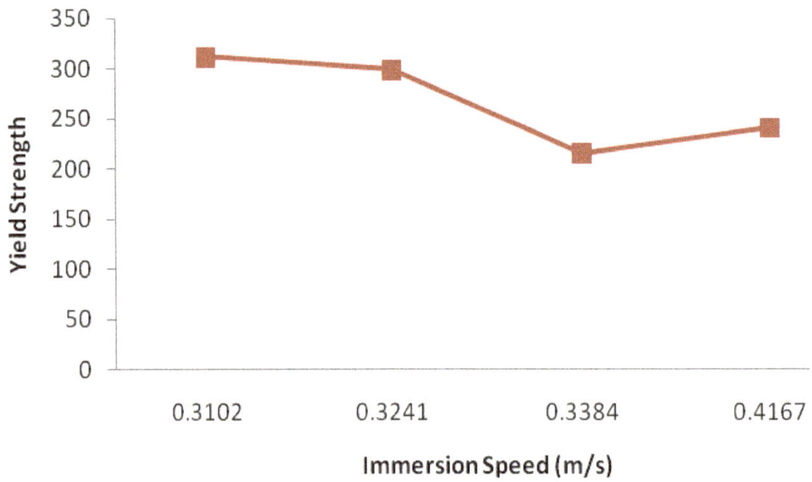

Figure 5. Yield Strength (N/mm^2) against Immersion speed (m/s).

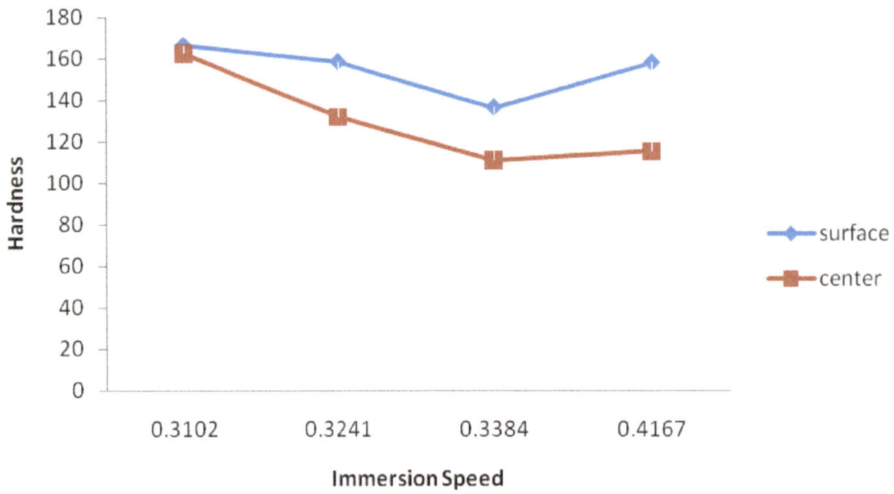

Figure 6. Hardness value (BHN) against immersion speed (m/s) (surface and center).

strength and immersion speed. The yield strength of the steel samples reduces with increase in immersion rate from 0.3102 to 0.3384 m/s. Optimum yield strength of 310 was obtained at low immersion speed of 0.3102 m/s.

Effect on hardness

There is a notable difference between the hardness of the steel samples at the surface and at the center (Figure 6). Although, the hardness of the specimens had an inverse relation with the immersion speed, i.e. the higher the speed of immersion, the lower the hardness value at both the surface and the center. It was evidence that the hardness values at the surface of the samples are higher than that at the center. This can be attributed to the fact that quenching occurs faster and more rapidly at the surface than at the center.

Effect of immersion speed on the microstructure of the steel

The micrographs of the steel samples quenched at different immersion speed were shown in Figures 7 to 10. The massive martensite structure of hardened sample,

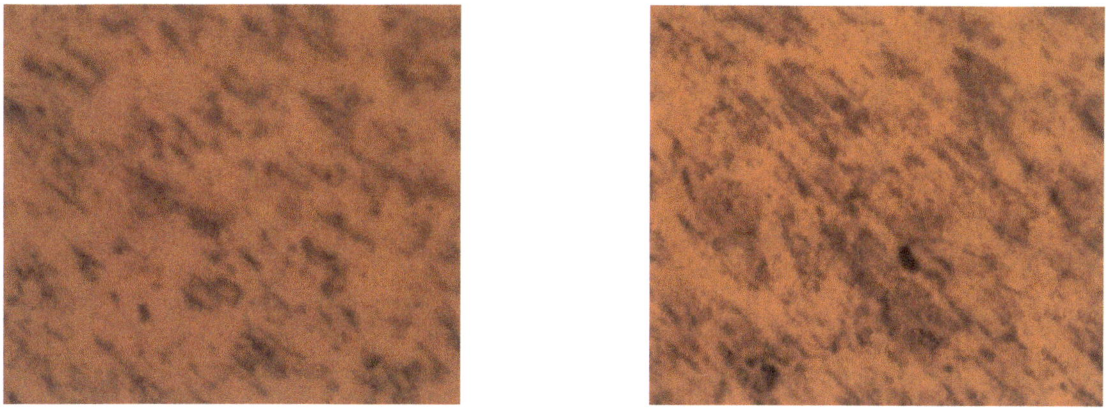

Surface Core

Figure 7. Microstructure of oil quenched steel at immersion rate of 0.3102 m/s.

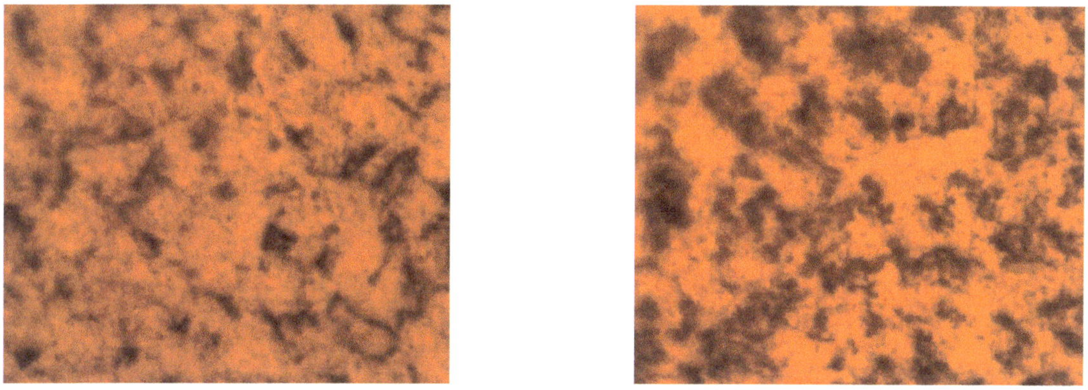

Surface Core

Figure 8. Microstructure of oil quenched steel at immersion rate of 0.3241 m/s.

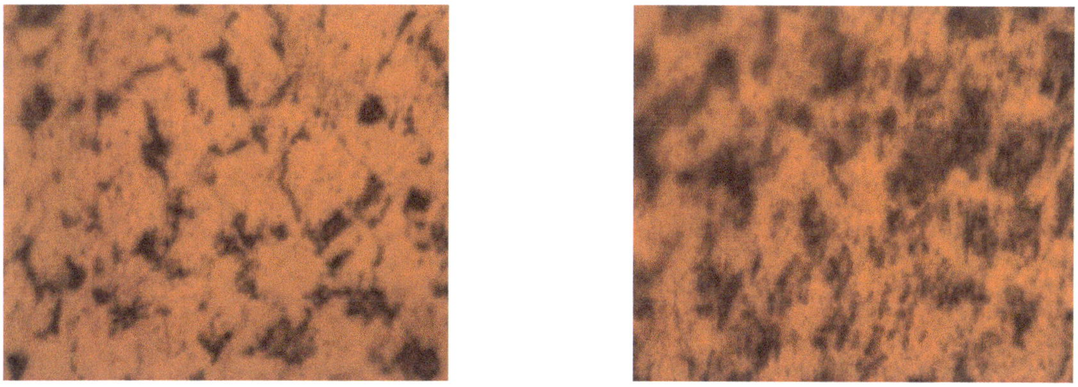

Surface Core

Figure 9. Microstructure of oil quenched steel at immersion rate of 0.3384 m/s.

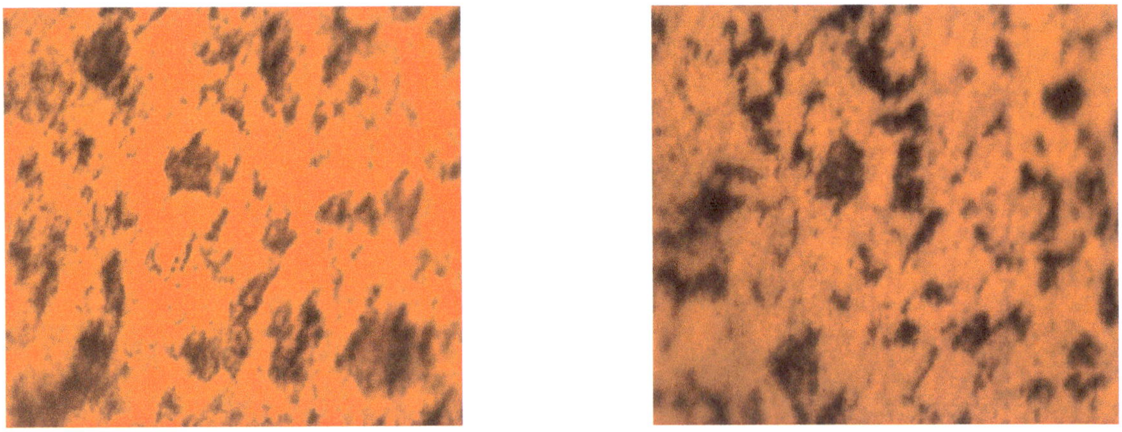

Surface Core

Figure 10. Microstructure of oil quenched steel at immersion rate of 0.4167 m/s.

when low Carbon steels are rapidly quenched from its austenite temperature to room temperature, the austenite will decompose into a mixture of some medium Carbon martensite and fewer pearlite as a result increases the tensile strength, hardness and little reduction in ductility of the material. In this study, it was observed that the surface (outer layer) of the steel samples cooled more rapidly resulting in large martensitic formation. Little pearlite were noticed at the surface compared to the center (inner layer). This varied effect may be due to the fact that the surface has contact with the oil quenchant; hence, cools faster than the center. The high martensitic composition may make the material very hard and brittle. As the speed of immersion decreases, the martensitic formation also decreases, giving room for more pearlite formation, hence decrease in the hardness as the speed of immersion increases. Obviously, the rates of immersion bring about a variation in the hardness of the material.

Conclusions

The speeds of immersion have varying effects on individual mechanical properties of AISI 1020. Mechanical properties such as yield strength and percentage elongation decreases while young modulus increases as the material is quenched. The tensile strength of the mild steel samples varied as the immersion speeds. The effect of immersion speeds on the hardness of the steel differs at the center and at the core. As the immersion speed increases, the hardness at the surface increases as compared to the hardness value at the center of the mild steel samples.

Conflict of Interest

The authors have not declared any conflict of interest.

REFERENCES

Çalik A (2009). Effect of cooling rate on hardness and microstructure of AISI 1020, AISI 1040 and AISI 1060 Steels. Int. J. Phys. Sci. 4(9):514-518.

Fadare DA, Fadara TG, Akanbi OY (2011). Effect of Heat Treatment on Mechanical Properties and Microstructure of NST 37-2 Steel." J. Min. Mater. Character. Eng. 10(3):299-308.

Odusote JK, Ajiboye TK, Rabiu AB (2012). Evaluation of Mechanical Properties of Medium Carbon Steel Quenched in water and Oil. J. Technol. 15(4):218-224.

Mackerle J (2003). Finite element analysis and simulation of quenching and other heat treatment processes. A bibliography (1976-2001): Comput. Mater. Sci. 27:313-332.

Sharma JP, Chaturvedi S, Arora P (2012). Studies on effect of percentage of Carbon on the tensile and compressive strength of structural steel. Int. J. Eng. Sci. Technol. 4:5.

Effect of battery effluent on plasticity and swelling characteristics of expansive soil

S. Bali Reddy[1] and A. V. Narashima Rao[2]

[1]Indian Institute of Technology Guwahati, India.
[2]Department of Civil Engineering, S. V. University, Tirupathi, India.

Ground pollution is perpetuated by humans due to many reasons. Industrial activity is necessary for the socio-economic progress of a country, but at the same time, it generates large amount of solid and liquid wastes. Among various means available, disposal through land is simple and widely used. All types of pollution have either direct or indirect effect on soil properties. Behaviour of any contaminant in soil depends upon the physical and chemical properties of the contaminant as well as its interactivity with that of soil. The effect of battery effluent on plasticity, swelling characteristics of black cotton soil has been presented in this paper. The soil used in this investigation falls under "SC" group as per I.S. classification and its differential free swell index is 254.54% indicating very high degree of expansiveness. The battery effluent used in this investigation is a colourless liquid and soluble in water.

Key words: Plasticity, expansive soil, battery effluent.

INTRODUCTION

The index and engineering properties of the ground gets modified in the vicinity of the industrial plants mainly as a result of contamination by the industrial wastes disposed. The major sources of surface and subsurface contamination are the disposal of industrial wastes and accidental spillage of chemicals during the course of industrial operations. The leakages of industrial effluent into subsoil directly affect the use and stability of the supported structure.

Extensive damage to the floors, pavements and foundations of a light industrial building in Kerala State was reported by Sridharan et al. (1981; 2002). Joshi et al. (1994) reported that severe damage occurred to the interconnecting pipe of a phosphoric acid storage tank in particular and also to the adjacent buildings due to differential movements between pump and acid tank foundations of fertilizer plant in Calgary, Canada. A similar case of accidental spillage of highly concentrated caustic soda solution as a result of spillage from cracked drains in an industrial establishment in Tema, Ghana caused considerable structural damage to a light industrial building in the factory, in addition to localized

Table 1. Properties of soil.

S/No	Property		Value
		Atterberg limits	
1	(a)	Liquid Limits	77%
	(b)	Plastic limit	30%
	(c)	Plasticity index	47%
	(d)		
		Compaction characteristics	
2	(a)	Maximum dry Unit Weight	18.48 kN/m^3
	(b)	Optimum Moisture Content	12.8%
3	Specific Gravity		2.76

Table 2. Chemical composition of battery effluent.

S/No.	Parameter	Value
1.	Color	White
2.	pH	8.45
3.	Sulphates	250 mg/L
4.	Chlorides	30 mg/L
5.	Lead monoxide	27.77%
6.	Lead sulfate	63.08%
7.	Free lead	7.44%
8.	Total lead	75.42%
9.	BOD	110 mg/L
10.	COD	320 mg/L

subsidence of the affected area (Kumaplay and Ishola, 1985). Therefore, it is better to start ground monitoring from the beginning of a project instead of waiting for complete failure of the ground to support human activities and then start the remedial actions.

Expansive soils have high shrinkage and swelling characteristics. In general, these soils are very much sensitive to changes in environment. The environment includes the stress system, the chemistry of pore water in the system, the seasonal variations in ground water table and temperature variations. Hence, an attempt is made in this investigation to study the effect of battery effluent on the plasticity and swelling characteristics of an expansive soil.

MATERIALS AND METHODS

The soil used for this investigation is obtained from Tirupati (India). The soil is classified as 'SC' as per I.S. Classification indicating that it is clayey sand. It is highly expansive as the free swell index is 254.5%. The properties of the soil are given in Table 1. Battery effluent is a colorless liquid and soluble in water in proportions. The chemical properties of the effluent are shown in Table 2.

Procedure for contamination

The soil from the site is dried and the pebbles and vegetative matter present, if any, are removed by hand. It is further dried and pulverized and sieved through a sieve of 4.75 mm to eliminate gravel fraction, if any. The soil mixed with different percentages of battery effluent, from 0 to 100%, in increments of 20%. The contaminated soil prepared thus is stored for a day in air tight containers for uniform distribution of battery effluent. The soil effluent mixtures are mixed thoroughly before testing.

Tests conducted

The following tests are conducted in the presented investigation:

Liquid limit tests

Liquid limit tests are conducted at various percentages of battery effluent. About 120 g of an air-dried sample passing through 425 µ I.S. sieve is taken in a dish and mixed with certain amount of water to form a uniform paste. A portion of this paste is placed in the cup of the liquid limit device and surface is smoothened and leveled with a spatula to a maximum depth of 10 mm. A groove is cut through the sample along the symmetrical axis of the cup, preferably in one stoke, using a standard grooving tool. After the soil pat has been cut by a proper grooving tool, the handle is turned

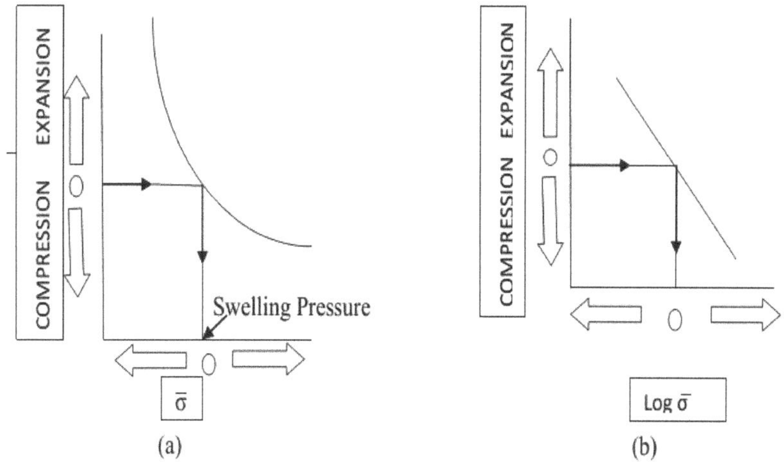

Figure 1. Pressure-volume change plots.

at a rate of 2 revolutions per second until the two parts of the soil sample come into contact at the bottom of the groove along a distance of 12 mm. About 15 g of soil near the closed groove is taken for water content determination. The liquid limit is the water content at which the soil is sufficient fluid to flow when the device is given 25 blows. As it is difficult to get exactly 25 blows for the sample to flow, the test is conducted at different water contents so as to get blows in the range of 10 to 40. The soil in the cup is transferred to the dish containing the soil paste and mixed thoroughly after adding more water. The soil sample is again taken in the cup of the liquid limit device and the test is repeated.

Plastic limit tests

About 30 g of soil, passing through 425 μ I.S. Sieve, is taken in evaporating dish. It is mixed thoroughly with water till it becomes plastic, and can be easily moulded with fingers. About 10 g of the plastic soil mass is taken in one hand and a ball is formed. The ball is rolled with fingers on a glass plate to form a soil thread of uniform diameter. The rate of rolling is kept about 80 to 90 strokes per minute. If the diameter of the thread becomes approximately 3 mm and if it starts just crumbling that water content is known as the plastic limit.

SWELLING CHARACTERISTICS

Differential free swell index

This test is conducted on the local soil contaminated with battery effluent in varying percentages from 0 to 100% in increments of 20%. Two samples of the dried soil weighing 10 g each passing through 425 μ I.S. sieve are taken. One sample is put slowly in a 100 ml graduated glass cylinder having kerosene (a non-polar liquid). The other sample is similarly put in another 100 ml glass cylinder having distilled water. Both the samples are left for 24 h and then their volumes are noted. Differential free swell index is calculated by the formula given below:

$$DFSI = \frac{(V_1 - V_2)}{V_2} \times 100$$

V_1 = Soil volume in distilled water, V_2 =Soil volume in kerosene.

Swelling pressure

This test is conducted on the local soil contaminated with battery effluent in varying percentages from 0 to 100% in increment of 20%, such as 0, 20, 40, 60, 80 and 100%. Two methods are in common use for measuring the swelling pressure of soils in the laboratory. In the first method, the swelling pressure of an undisturbed or a remolded soil is measured for 'no volume change' condition. The method requires continuous adjustment of pressure on the soil specimen taken in a consolidation cell, so that the soil volume at any time is equal to its initial volume. Details of the test procedure are given in IS: 2720 (Pert XLI)- 1977. Remolded specimens are taken at the density and moisture content of the field soil, as for example, in an earth embankment where the moisture content of compaction and the required compacted density are known. Undisturbed specimens, taken carefully from the field soil, are tested for estimating the swelling behavior of an existing deposit.

The second method consists of taking a few (more than three) initially identical soil specimens in consolidation cells of fixed ring type, subjecting them to different magnitudes of pressures and then allowing the soil to saturate. Under the different load intensities, some of the soil specimens would compress after saturation while some others would swell. In fact, the load intensities ought to be properly selected to produce this kind of differential behavior. After the volume change (compression or swelling) is complete and has been recorded, load intensity versus volume change plot is obtained (Figure 1a). From this plot, the pressure corresponding to zero volume change is read and is denoted as the swelling pressure for the soil. It is much more convenient to a plot the load intensity to a logarithmic scale, as this would produce a straight line (Figure 1b).

RESULTS AND DISCUSSION ON PLASTICITY CHARACTERISTICS

Consistency represents the relative ease with which the soil can be deformed. This term is mostly used for

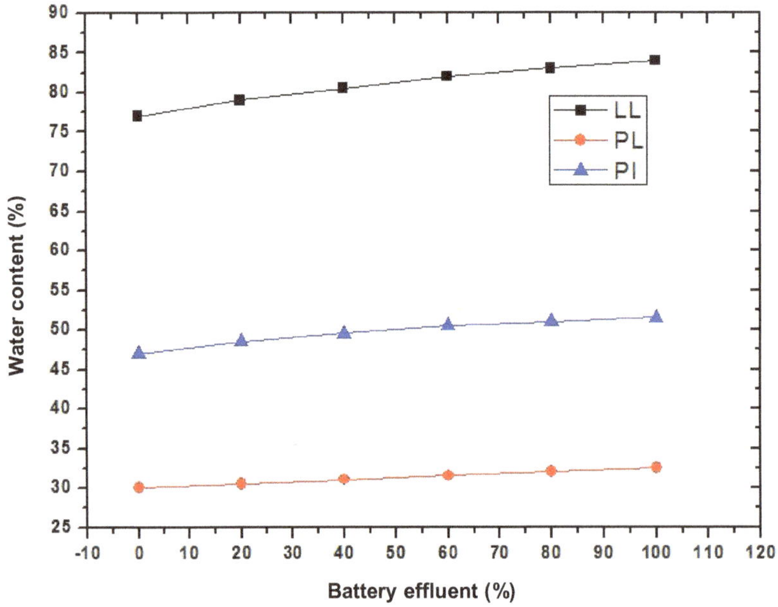

Figure 2. Variation of LL, PL and PI with battery effluent.

fine-grained soils for which the consistency is related to a large extent to water content. Atterberg (1911) formally distinguished the following states of consistency – liquid, plastic, semi-solid and solid. The water contents at which the soil passes from one of these states to the next have been arbitrarily designated as consistency limits-liquid limit, plastic limit, and shrinkage limit, in that order. Liquid limit is the water content corresponding to the arbitrary limit between liquid and plastic states of consistency of a soil. The minimum water content at which the soil made a standard grove by a Casagrande tool will flow together for a distance of 12.5 mm under the impact of 25 numbers of blows in a carasgrande apparatus. It is also defined as the maximum water content at which the soil is still in the liquid state, but has a small shearing strength against flowing, which can be measured by standard available means. Plastic limit is the water content at which a soil will just begin to crumble when rolled into a thread of approximately 3 mm in diameter. Plasticity index is defined as the numerical difference between the liquid limit and plastic limit of a soil. When soil is contaminated with battery effluent at various percentages such as 0, 20, 40, 60, 80 and 100%, the liquid limit, plastic limit and plastic index may vary.

The results of liquid limit and plastic limit tests conducted at different percentage of battery effluent are presented in Figure 2. From this figure, it is found that the liquid limit value of the contaminated soil increases slightly with the increase in percentage of battery effluent. The plastic limit value of the uncontaminated soil is 30%. From this figure, it is found that plastic limit value of contaminated soil increases with increase in percentage of battery effluent. The plasticity index of uncontaminated soil is 47%. The plasticity index values of the contaminated soil increase with percentage increase in battery effluent. Consistency limits (liquid limit, plastic limit, and plastic index) and swelling pressures increases due to absorption of sulphates that is known to be present in battery effluent on to the clay surface. Absorption of sulphates causes expansion of double layer leading to increase in consistency limits and swelling pressure.

Swelling characteristics

Expansive soils or swelling soils are those soils, which have the tendency to increase in volume when water is available and to decrease in volume if water is removed. These volume changes in swelling soils are the cause of many problems in structures that come into their contact or constructed out of them. From the engineering point of view, it will be sufficient to get an indication of the possible swelling behavior by performing simple tests. Almost all over the world, the surface/surficial deposits consist of expansive (block cotton) soils, which are found to be problematic for engineering construction. These expansive (black cotton) soils on coming into contact with water heave considerably and lose strength. If swelling is not allowed, swelling pressure of varying order occurs. Swelling pressure of the order of 150 to 300 kPa is commonly encountered. Though problematic for engineering construction, nevertheless, a large number of major and minor irrigation projects must necessarily be

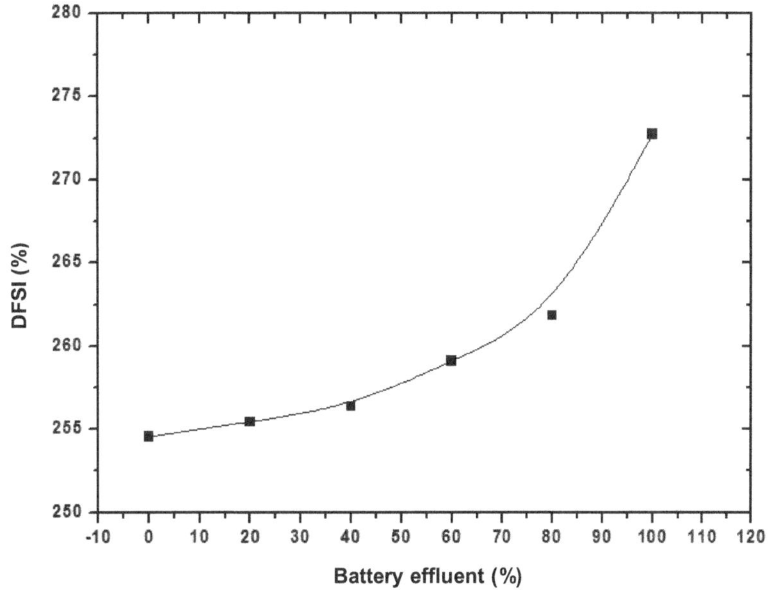

Figure 3. Variation of DFSI with battery effluent.

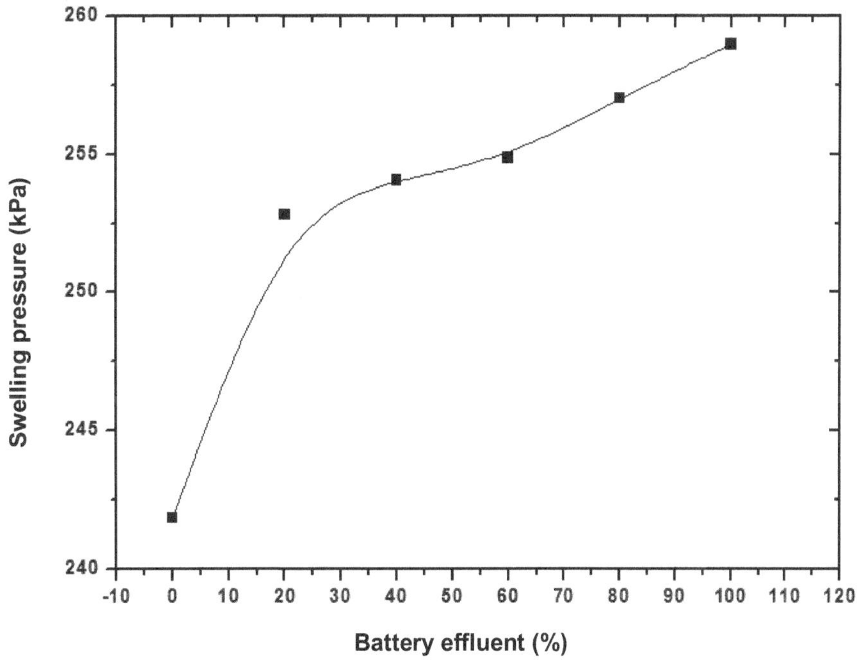

Figure 4. Variation of swell pressure with battery effluent.

executed in such soils. They are almost mainly encountered in all the regions of the world. The variation of differential free swell index with percent battery effluent is shown in Figure 3. From the figure, it is observed that the differential free swell index increases slightly with percent increase in battery effluent. The percent increase in the differential free swell index is about 7 at 100% of battery effluent.

Swelling pressure of the soil admixed with the various percentages of battery effluent are determined and presented in Figure 4. The value of swelling pressure of natural soil is 246 kPa. From the figure, it is observed that there is slightly increasing trend from 0 to 100% battery effluent.

Conclusions

Based on experimental results, the following conclusions are drawn. If increasing battery effluent, liquid limit and plastic limits are increased. Swelling characteristic like differential free swell index and swelling pressure are also increased with increasing battery effluent. Further more study is required in micro level to under stand the behaviour of soil when mixing with battery effluent.

Conflict of Interests

The author(s) have not declared any conflict of interests.

REFERENCES

Atterberg A (1911). Uber die physikalische Bodenunter suchung, and under die plastizitat de to no" International mitteilugegs for Boden kundle verlag fur Fachiliterabur Vol-1 Berlin.

Joshi RC, Pan X, Lohinta R (1994). Volume change in calcareous soils due to phosphoric acid contamination. Proc. XIII CSMFE, New Delhi 4:1569-1574.

Kumapley NK, Ishola A (1985). The effect of chemical contamination on soil strength. Proc. XI ICSMFE, San Fransico, A. A. Balkema, Rotterdam 3:1199-1201.

Sridharan A, Nagaraj TS, Sivapullaiah PV (1981). Heaving of soil due to acid contamination. ICFMFE, Stockholm, 6:383-386.

Sridharan, A, El-Shafei A, Miura N (2002). Mechanisms controlling the undrained strength behavior of remolded Ariake marine clays. In: Mar. Georesour. Geotechnol. 20(1):21-50.

I.S: 2720 (Part XLI)-1977: Swelling Pressure Test.

23

Effect of Argon plasma and ion beam on the morphology and wettability of polyethylene terephthalate (PET)

A. Atta[1], A. M. Abdel Reheem[2] and M. M. Abdel Rahman[2]

[1]Physics Department, National Center for Radiation Research and Technology (NCRRT), Atomic Energy Authority (AEA), Cairo Egypt.

[2]Accelerators and Ion Sources Department, Nuclear Research Center, Atomic Energy Authority, P. O. Box 13759 Inchas, Cairo, Egypt.

Plasma and ion beams are considered tools that can improve existing materials or create new materials. Their technology is widely used to alter the surface properties of polymers without affecting their bulk properties. The treated polymers have found various applications in microelectronics, biomedical and chemical industries. In this work, the use of Argon plasma and Argon ions produced from cold cathode ion source as a method for the morphology and wettability of Semi crystalline polyethylene terephtalate (PET) foils are investigated. A cold cathode ion source was used as a preparation tool of the surface of polyethylene terephthalate PET polymer substrate to be ready for various applications as coating or thin film deposition. Argon ions of 1.5 keV are produced from the cold cathode ion source with an operating gas pressure of 2 × 10^{-4} mbar. The induced effects in the structure, surface morphology, and surface wettability were investigated by means of the Fourier transform infrared spectroscopy (FTIR), scanning electron microscope (SEM) and by the contact angle methods, respectively. The change in the contact angle on PET as a function of Argon plasma time and ion beam fluence was also studied. Changes in surface free energy of polymer surface after Argon plasma irradiation were observed.

Key words: Cold cathode ion source, Argon plasma and ion beams, polyethylene terephthalate (PET) foils.

INTRODUCTION

Polymer materials with improved mechanical, optical or electrical properties have been widely used in a variety of industrial applications. However, their use is sometimes limited by undesired properties of the surface, contrary to very useful characteristics of the bulk, such as low weight, chemical inertness and high impact resistance. Therefore, it is interesting to modify the surface in a controlled manner to enhance wettability, printability, adhesion with other materials or with biological components, compatibility (as in the production of material blends using two immiscible polymers), etc. (Cho et al., 2003; Ektessabi and Sano, 2000; Spohr, 1990; Ektessabi and Yamaguchi, 2000). Plasma treatment does not produce toxic waste in contrast to the treatments using chemicals. The plasma is capable to exert four major effects (Chen, 2001; Reznickova et al., 2011) surface cleaning, surface ablation or etching, surface cross-linking, and modification of the surface chemical structure, occurring both *in situ* or after subsequent exposure to the atmosphere. These effects depend on a presence of the active species in plasma (electrons, ions, radicals, photons) which interact with polymer surfaces and modify their chemical and physical properties. Ion irradiation is an established tool for modifying the chemical structure and physical properties of the polymers Pelagad et al, 1980; Chico et al., 2012). The ion beam irradiation technique has proven more effective in modifying polymer than other particle beam irradiation techniques because of its higher cross section for ionization and larger linear energy transfer (Drabik et al., 2007). Our aim of this work was intended to use plasma and ion beam currents produced from cold cathode ion source as a method for the morphology and wettability of semi crystalline poly ethylene terephtalate (PET) foils are treated and investigated.

EXPERIMENTAL OUTLINES

A schematic diagram of the cold cathode ion source and its associated electrical circuit is shown in Figure 1. It consists essentially of stainless steel anode cylinder with 30 mm length and 22 mm diameter connected with positive voltage power supply of +10 kV, cathode disc of diameter 22 mm and 2 mm thickness, and a semitransparent 50% as exit aperture. The extractor electrode connected to negative voltage power supply, Faraday cup placed at 30 mm distance from an exit aperture. PET sample put on faraday cup at distance 3 cm from the cylindrical cathode. A PET samples are immersed in acetone liquid and put it in ultrasonic apparatus to clean it. Cold cathode ion source was used to obtain the Argon plasma and ion beam and consequently irradiation PET surface using Argon plasma and ions. PET is polyester having a high melting point due to the presence of aromatic ring and has a very good mechanical strength. It is semi-crystalline in nature and is resistant to heat and moisture and virtually unattached by many chemicals. It has extensive use in textile fibers.

The working gas is admitted to the ion source through a hose fixed in a Perspex flange at the upper side of the anode. A complete vacuum system is used to evacuate the ion source chamber. It consists of stainless steel mercury diffusion pump of speed 270 l/s provided with electrical heater and backed by 450 l/min rotary fore-line vacuum pump. The rotary pump is used to evacuate the system to pressure of 10^{-2} to 10^{-3} mbar, while the mercury oil diffusion pump is used to evacuate the ion source vacuum chamber to the order of 10^{-5} mbar. A liquid nitrogen trap is fixed between the ion source chamber and the mercury oil diffusion pump in order to prevent the mercury vapor from entering the ion source chamber. The working gas is transmitted to the ion source from a gas cylinder through a needle valve to regulate the rate of gas flow. The plasma generation for this ion source is based on the ionization mechanism produced by primary electrons colliding with gas molecules due to a potential difference between the anode and the cathode. Therefore, high output ion beam current can be extracted axially in a direction normal to the discharge region. The anode is made of stainless steel material which is featured by high ionization coefficient and it is cylinder to improve the stability of the discharge. PET samples were put on Faraday cup at distance 3 cm from the anticathode. A PET samples are immersed in acetone liquid and put it in ultrasonic apparatus to clean it. Cold cathode ion source was used to obtain the plasma and Argon ion beams and consequently irradiation PET surface using plasma and Argon ions. PET is polyester having a high melting point due to the presence of aromatic ring and has a very good mechanical strength (Soliman et al., 2013). It is semi-crystalline in nature and is resistant to heat and moisture and virtually unattached by many chemicals. It has extensive use in textile fibers.

RESULTS AND DISCUSSION

In this study, the system is evacuated to about 3×10^{-5} mbar to remove the residual gases before the Argon gas injection in the ion source. The ion source apparatus was cleaned before introducing inside the vacuum system. It was polished, and washed by acetone. The polishing of the electrode parts should remove the irregular parts from their surfaces and the contamination due to the erode materials of the discharge. The plasma and the extracted Argon ion beam from the cold cathode ion source were used to irradiate Polyethylene terephthalate (PET). The sample considered in our investigations was PET (polyethylene terephthalate), provided by Goodfellow Company. The polymer samples, 50 μm thickness foils, were cut into 10×10 mm pieces, ultrasonically cleaned in alcohol to remove organic material and dried with hot air before treatment. The PET films are exposed to Argon plasma exposure process. We use plasma technique "reactive ion etching (RIE) system" with the Argon plasma parameters are the working pressure is 2×10^{-4} mbar and discharge voltage 900 volte, g = 180 gauss varied time exposure from 0 to 4 min and Argon ion beam for 0 to 4 h.

In the present work, discharge and output characteristics of this ion source using Argon plasma and Argon ion beams were done and this type of ion sources was used as a preparation tool of the surface of PET polymer substrate to be ready for coating or thin film deposition. Modifications in optical and surface properties of PET polymer induced by Ar plasma and Ar ion beam irradiation have been investigated in Atomic Energy Authority – Egypt), by:

1) FTIR spectra of the pristine and the irradiated samples were investigated using (FTIR-Beckman-4250) spectrophotometer in the range 400 cm^{-1} to 4000 cm^{-1} at National Center for Radiation Research and Technology (NCRRT), AEA, Cairo, Egypt. The irradiated PET films were exposed to atmosphere for a few days before the FTIR experiments were conducted.
2) SEM (Model, JEOL, Japan) at NCRRT, AEA, Cairo, Egypt, was used to investigate the surface morphology of the pristine and PET surfaces after bombarded by Argon plasma and Argon ion beam.

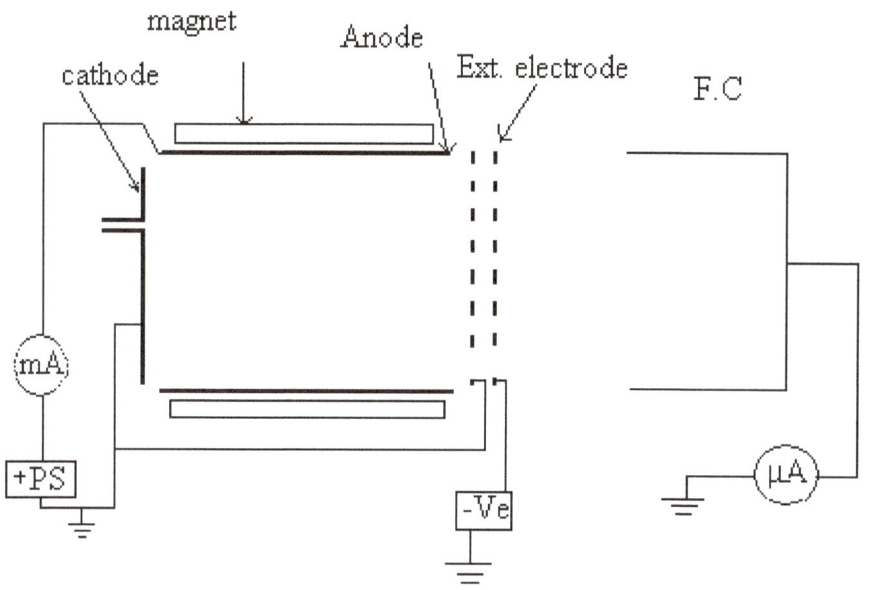

Figure 1. Experimental arrangement for irradiation of polymer samples using Ar plasma and Ar ion beam.

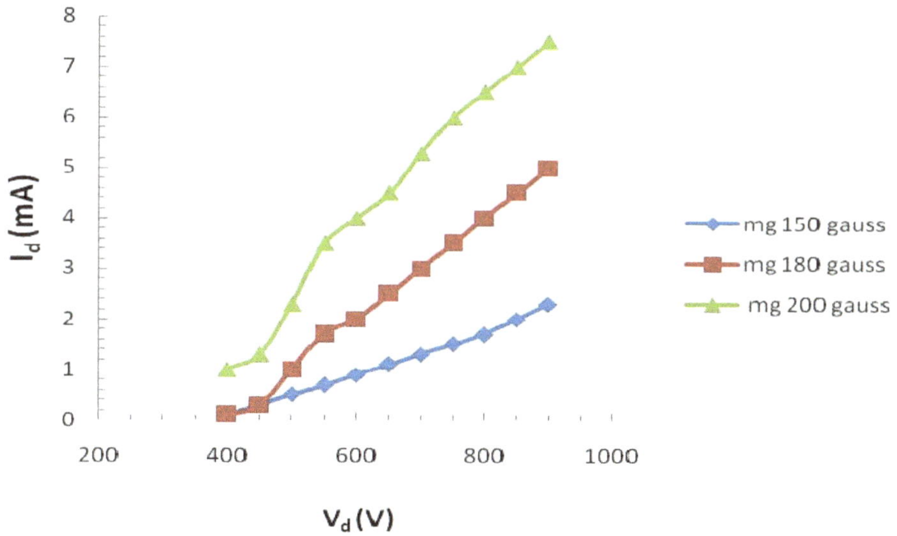

Figure 2. The relation between the electrical discharge current and the electrical discharge voltage with different magnetic field intensities.

3) The contact angle in a sessile drop method was measured by a contact angle meter (Cam Micro, Tantec). Each value of the contact angles was taken as an average from the five different samples fabricated under the same modification conditions. Surface free energy, that is, the sum of the polar force and the dispersion force, was calculated by measuring the contact angles of three different polar liquids (water and diiodomethane). From the measured contact angles, the polar forces and the dispersion forces were calculated using the Owen method (Soliman et al., 2013).

Ion source characteristics

Figure 2 shows the relation between the electrical discharge current, I_d, and the electrical discharge voltage, V_d, at pressures P = 2 × 10^{-4} mbar using Argon gas. It is clear that by increasing the discharge voltage, the discharge current increases at different magnetic field

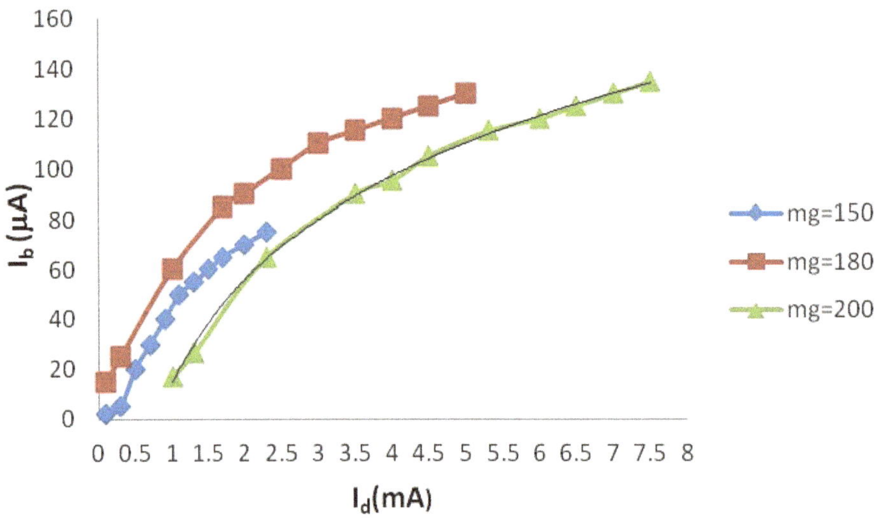

Figure 3. The relation between the electrical discharge current and the output ion beam current, with different magnetic field intensities.

intensities where the characteristic of such discharge is characterized by abnormal glow (Abdelrahman et al., 2008). This magnetic field is created an electromagnet coil wrapped around the cylindrical anode. Whereas, this magnetic field provides an increase of electron trajectories, hence, increasing the electron path which enhances the probability of ionization and then increase the discharge current. It was also found that an increase in magnetic field intensity was accompanied by an increase in the discharge current. This increase may be due to an increase in magnetic field vector parallel anode surface, and this leads to minimization in electron current in the anode region (Kotov 2004).

Figure 3 shows the relation between the electrical discharge current, I_d and the output ion beam current, I_b at 150, 180 and 200 gauss magnetic field intensities without extraction voltage. It could be seen that from this figure, maximum output ion beam current was obtained at magnetic field intensity of 180 gauss.

Figure 4 shows the relation between the electrical discharge current, I_d and the output ion beam current, I_b at 180 gauss magnetic field intensity with extraction voltage V_{ext} = - 600 V. It could be found from this figure that, an increase of the discharge current was accompanied by an increase of the output ion beam current at magnetic field intensity of 180 gauss and Argon gas pressure, $P_{Ar} = 2\times10^{-4}$ mbar.

Plasma and ion beam irradiation of polymers

The irradiation of all the polymer samples used in the present study were carried out by locally designed, a modified saddle field ion source at Egyptian Atomic Energy Authority, Egypt. The Ar ion beam irradiation was carried out in a vacuum chamber of 10^{-5} mbar, using a modified saddle field ion source. The ion fluence was estimated by time of irradiation and beam current as (Soliman et al., 2013):

$$I = Q/t = Dqe/t = \phi Aqe/t$$

Where I ion current (A), Q total charge, D: dose (ion fluence in ions / cm^2 × area (A) of irradiation in cm^2. and q is the charge state, e electron charge (1.6×10^{-19} C), finally t is the irradiation time in seconds.

FTIR spectral studies

FTIR spectroscopy has been found to be an important technique to understand the changes in the molecular bonds after irradiation. The changes have been observed from the relative increase or decrease in the intensity of the typical bands associated to the functional group present in the polymers. FTIR spectrum of pristine, Argon plasma and Argon ion beam irradiated PET are shown in Figure 5. By bombardment, the PET polymers by Argon plasma and ion beam were observed to decrease in the band intensities. It might be ascribed to the possible transient recrystallization due to Argon plasma and ion beam effect. The reduction in the intensity of the typical bands indicates some degradation of PET polymer by plasma and ion beam (Zhiyong et al., 2009). The infrared absorption spectra of the PET substrates in a wave number range of 2000 to 4000 cm^{-1} reveal the presence of an absorbance band at about case of Argon plasma than Argon ion beam.

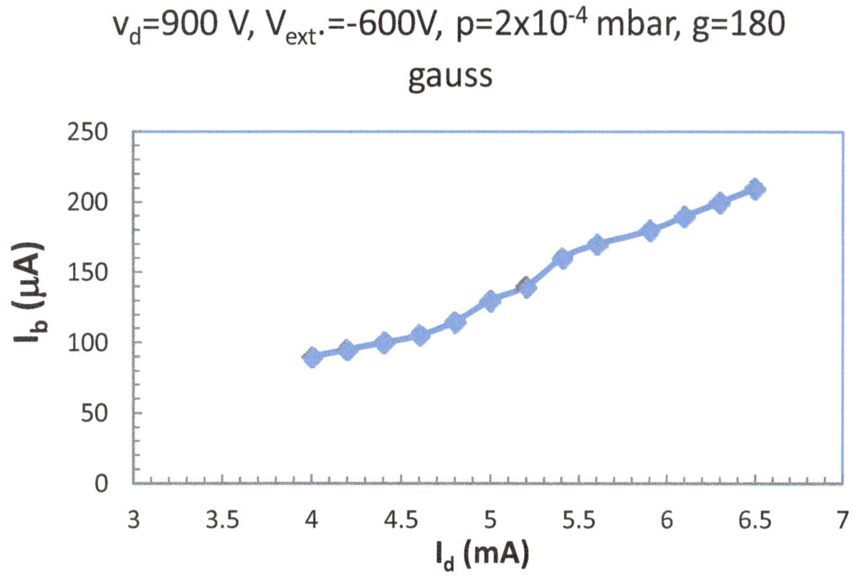

Figure 4. Output ion beam current versus discharge current using Argon gas at P = 2×10^{-4} mbar.

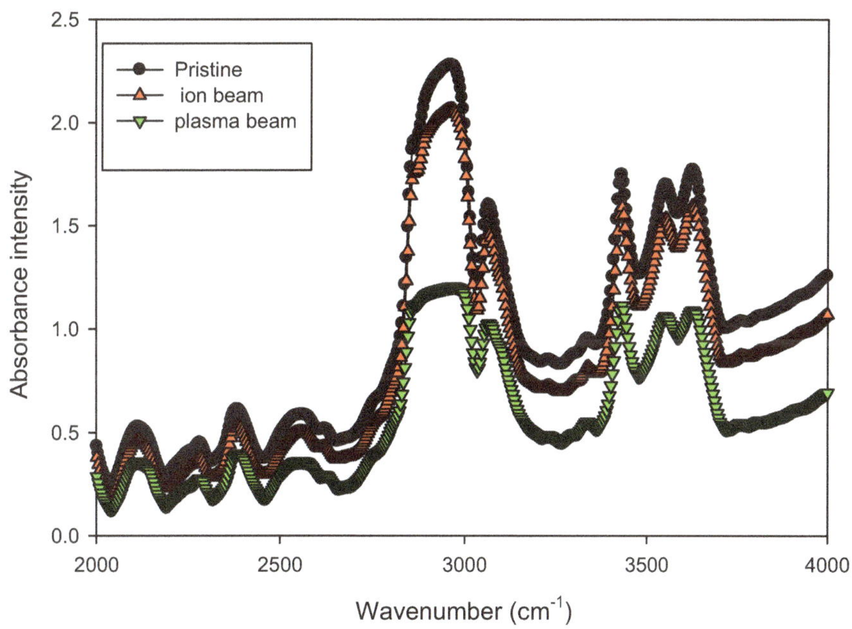

Figure 5. FTIR spectra of the pristine and irradiated PET films using Argon plasma and ion beams.

Scanning electron microscopy (SEM)

Polyethylene terephthalate (PET) film is transparent and the surface is smooth (Costa et al., 1997). The color of films gradually changes with irradiation from. The change in the color of polymer could be attributed to the formation of hydrogenous carbon clusters. The pristine sample showed rather relatively smooth surface as shown in Figure 6a. On the other hand, the SEM micrographs of irradiated samples by Argon ion beam display numerous small voids on the surface of the PET shown in Figure 6b. By using irradiated plasma, there was clear roughness on the surface of the irradiated samples and the resulting voids become large sized as

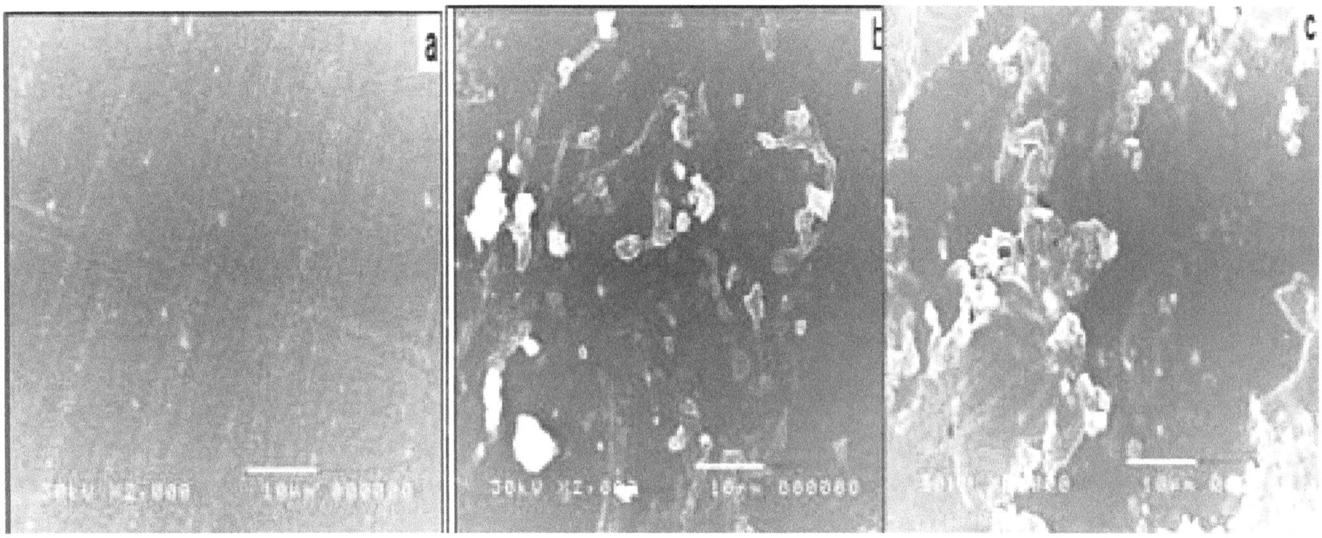

Figure 6. SEM micrographs for the pristine and irradiated PET by using Argon plasma and ion beams.

shown in Figure 6c. Bombardment of PET by Argon plasma is more efficient than Argon ion beam.

Surface wettability

Wettability and adhesion of polymers / films can be controlled by using surface modification techniques such as exposure to plasma, ion beams, flame, chemicals, enzymes etc. Wettability can also be regulated by changing chemical composition of the surface. In addition, surface roughness can be an important factor for enhancing the adhesion and wettability. Wettability is governed by molecular interaction of the outermost surface layer of a few angstrom units.

Contact angle is one from several independent methods that has been used to estimate solid surface tension (Liqing et al., 2009). Contact angle measurement is easily performed by establishing the tangent (angle) of a liquid drop with a solid surface.

$$\frac{y_l(1+\cos\theta)}{2\sqrt{y_l^d}} = \sqrt{y_s^d} + \sqrt{y_s^p}\cdot\sqrt{\frac{y_l^p}{y_l^d}} \quad (1)$$

Equation 1 contains two unknowns' dispersive (γ_s^d) and polar (γ_s^p) surface energies of the solid (Soliman et al., 2013). To obtain these unknown parameters, one has to use contact angle measurements of at least two different liquids from Table 1.

The change in the contact angle on PET as a function of plasma expoure time and ion beam fluence are shown in Figure 7a and b, respectively. It can be noted that as the exposure time increases, the contact angle decreases. The water contact angle decreases from 75 to 2.9° in case of plasma and decreased to 44° by using ion beam as shown in Table 2. The change of contact angles for the irradiated polymers is due to the formation of hydrophilic groups rather than the change of surface roughness. Hydrophilic group formation is composed of a two-step process. The first one is the creation of free radicals on a polymer surface by plasma irradiation. In the second step, interaction between free radicals in polymer chains and oxygen atoms results in the formation of polar groups such as carboxyl, carbonyl, hydroxyl and ester groups (Fernando et al., 2007). Thus, the decrease of the contact angle as a function of exposure time is due to the formation of new hydrophilic groups and the oxidized layer on the surface of polymer (Costa et al., 1998).

Surface free energy

The change in the surface free energy of PET as function of Argon plasma exposure time (min) and ion beam fluency were observed in Figure 8a, b, respectively. The surface free energy increased from 35 to 75 and 62.6 mJ/m^2 by using Argon plasma and Argon ion beam, respectively. Also, one notes that the irradiations effectively increase the polar component and the dispersive component as shown in Table 3. Increasing the polar component in polymer is mainly due to the formation of polar and/or hydrophilic groups. From these results, the large decrease in contact angle can be attributed to the dominant increase of the polar component in the surface free energy of the modified polymers.

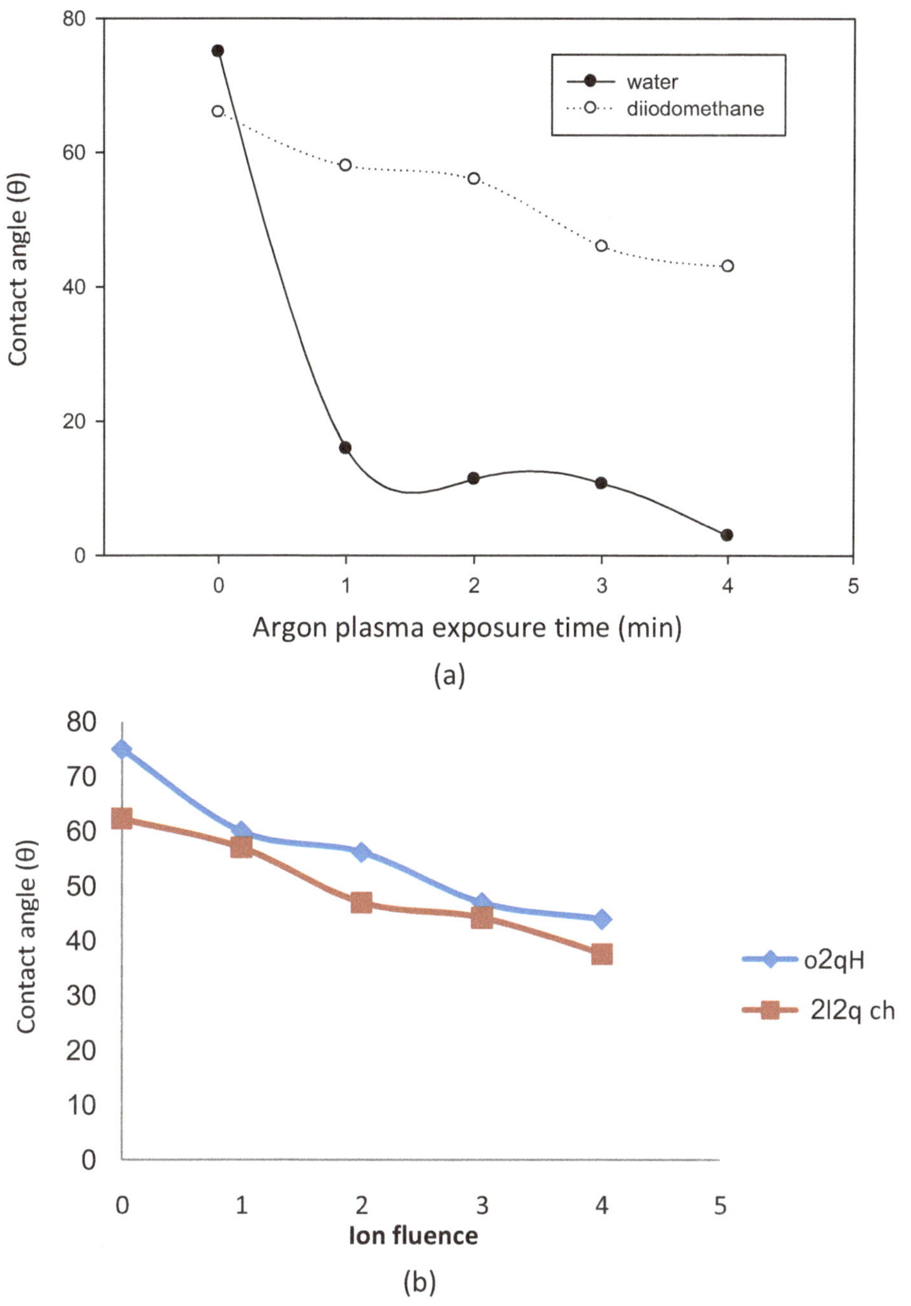

Figure 7. a) Change in the contact angle of PETas a function of Argon plasma exposure time; b) Change in the contact angle of PET as a function of Argon ion beam fluence.

Conclusions

In this work, it was concluded that, the characteristics of cold cathode ion source were studied with different operating parameters to get the optimum conditions for use in plasma and ion beam irradiation of some polymer materials. It could be found from this figure that, an increase of the discharge current was accompanied by an increase of the output ion beam current at magnetic field intensity of 180 gauss and Argon gas pressure, P_{Ar} = 2×10^{-4} mbar. Such results showed that this cold cathode ion source is quite suitable for impact processes, as it gives high discharge and ion beam currents with a wide beam diameter up to 2.2 cm. The effect of the Argon plasma and ion beam treatment on the surface properties of polyethylene terephtalate (PET) polymers was studied. The infrared absorption spectra of the PET substrates in a wave number range of 2000 to 4000 cm^{-1} reveal the presence of an absorbance band at about 2930 cm^{-1}. The intensity of bands is more reduction in case of

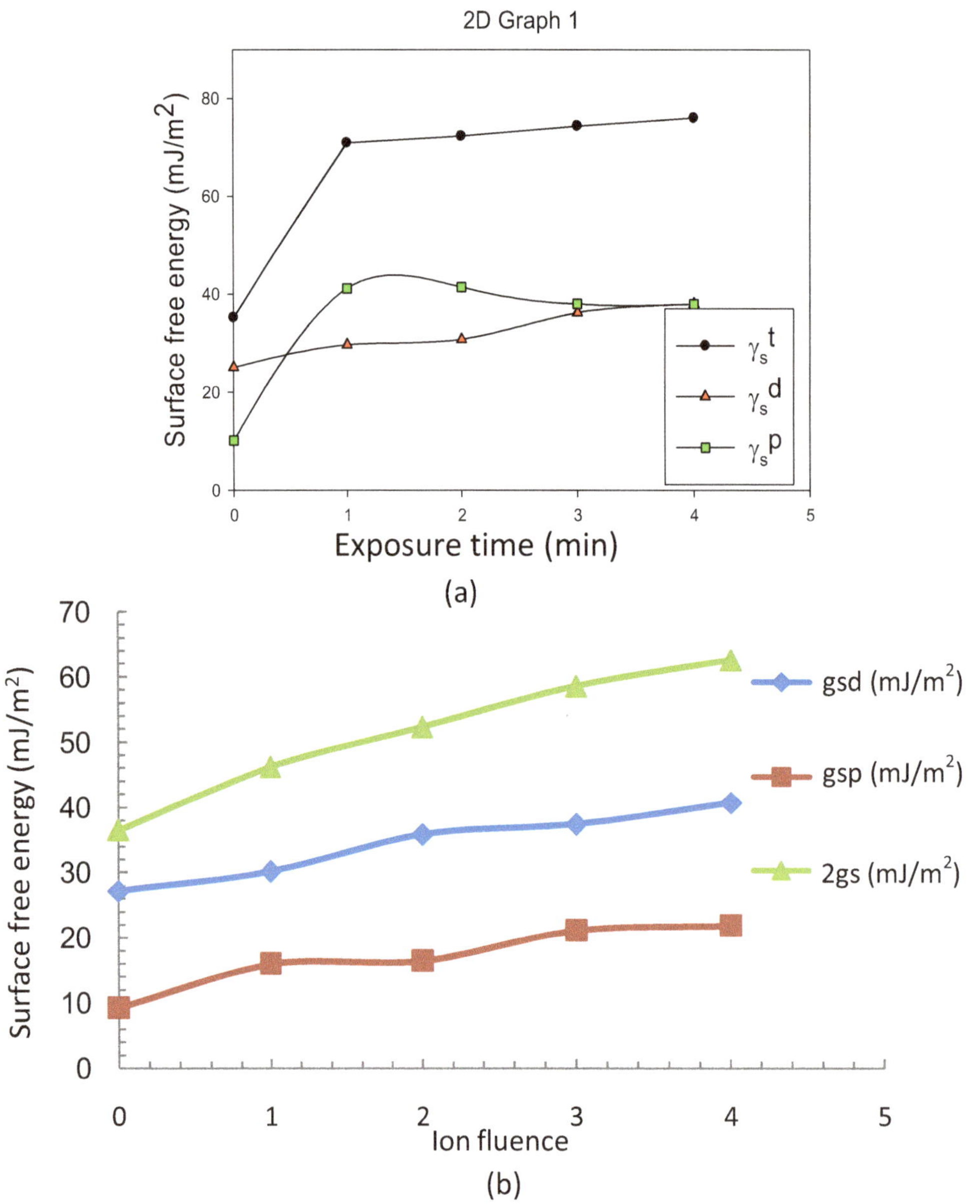

Figure 8. (a) Change in the surface free energy for PET as a function of Argon plasma exposure time; (b) Change in the surface free energy for PET as a function of Argon ion beam fluence.

Table 1. Surface free energy of the proposed liquids (Liqing et al., 2009).

Liquid	Surface free energyγ_l (Nm)	Polar componentγ_l^p (Nm)	Dispersion component γ_l^d (Nm)
Water	72.1	52.2	19.9
Ch_2I_2	50.8	0	50.8

Argon plasma than Argon ion beam.

The SEM micrographs of irradiated samples by Argon ion beam display numerous small voids on the surface of the PET. By irradiated using plasma, the roughnesses clearly on the surface of the irradiated samples and the voids have large size. Bombardment of PET by Argon plasma is more efficient than Argon ion beam; the surface free energy increased from 35 to 75 and 62.6 mJ/m^2 by using Argon plasma and Argon ion beam, respectively. The large decrease in contact angle can be

Table 2. The contact angles (θ) for PET surface as a function of plasma exposure time and ion beam fluence.

Exposure time (min.)	θ_{H2o}	θ_{ch2I2}	Ion fluence	θ_{H2o}	θ_{ch2I2}
Pristine	75.00	66.00	0	75.00	66.00
1	15.90	58.00	$1x10^{-18}$	60	57
2	11.40	56.00	$2x10^{-18}$	56.5	47
3	10.70	46.00	$3x10^{-18}$	47	44.25
4	2.90	43.00	$4x10^{-18}$	44	37.625

Table 3. The dispersive surface free energy (γ_σ^δ) and the polar surface free energy (γ_σ^π) for PET surface as a function of plasma exposure time and ion beam fluency.

Exposure time (min)	γ_σ^δ (mJ/m^2)	γ_σ^π (mJ/m^2)	γ_σ (mJ/m^2)	Ion fluence	γ_σ^δ (mJ/m^2)	γ_σ^π (mJ/m^2)	γ_σ (mJ/m^2)
Pristine	25.10	10.15	35.25	0	25.10	10.15	35.25
1	29.70	41.22	70.92	1×10^{-18}	30.25	16	46.25
2	30.80	41.47	72.28	2×10^{-18}	35.88	16.48	52.36
3	36.26	38.07	74.33	3×10^{-18}	37.45	21.16	58.61
4	38.01	37.97	75.98	4×10^{-18}	40.70	21.90	62.61

attributed to the dominant increase of the polar component in the surface free energy of the modified polymers in Argon plasma than Argon ion beam.

Conflict of Interests

The author(s) have not declared any conflict of interests.

REFERENCES

AbdelRahman MM, Helal A, Moustafa OA and Abdel Salam FW (2008). High efficiency glow discharge ion source. J. Nucl. Radiat. Phys. 3(1):1-9.

Chen JR (2001). Low-temperature Plasma Chemistry and Application, Science Press, Beijing p. 1.

Cho S, Han S, Kim KH, Beag YW, Koh SK (2003).Surface modification of polymers by ion-assisted reaction. Thin Solid Films. 445:332-341. http://dx.doi.org/10.1016/S0040-6090(03)01176-3

Chico B, Martinez L, Perez FJ (2005). Nitrogen ion implantation on stainless steel: AFM study of surface modification. Appl. Surf. Sci. 243(1-4):409-414. http://dx.doi.org/10.1016/j.apsusc.2004.09.097

Costa L, Luda MP, Trossarelli L, Brach del Prever EM, Crova M, Gallinaro P (1998). In vivo UHMWPE biodegradation of retrieved prosthesis. Biomater. 19(15):1371-85. http://dx.doi.org/10.1016/S0142-9612(98)00013-1

Drabika M, Kousala J, Pihosha Y, Choukourova A, Biedermana H, Slavinskaa D, Mackovac A, Boldyrevac A, Pesickad J (2007). Composite SiOx/hydrocarbon plasma polymer films prepared by RF magnetron sputtering of SiO2 and polyimide. Vacuum. 81(7):920–927. http://dx.doi.org/10.1016/j.vacuum.2006.10.013

Ektessabi AM, Yamaguchi K (2000). XPS study of ion beam modified polyimide films. Thin Solid Films 377:621-625. http://dx.doi.org/10.1016/S0040-6090(00)01444-9

Fernando SS, Christensen PA, Egerton TA, White JR (2007). Carbon dioxide evolution and carbonyl group development during photodegradation of polyethylene and polypropylene. Polym. Degrad. Stabil. 92(12):2163-2172. http://dx.doi.org/10.1016/j.polymdegradstab.2007.01.032

Kotov DA (2004). Broad beam low-energy ion source for ion-beam assisted deposition and material processing. Rev. Sci. Instr. 75(5):1934-1936. http://dx.doi.org/10.1063/1.1702109

Liqing Ya, Jierong C, Yafei G, Zheng Z (2009). Surface modification of a biomedical polyethylene terephthalate (PET) by air plasma. Appl. Surf. Sci. 255(8):4446-4451. http://dx.doi.org/10.1016/j.apsusc.2008.11.048

Reznickova A, Kolska Z, Hnatowicz V, Stopka P, Svorcik V (2011). Comparison of glow Argon plasma-induced surface changes of thermoplastic polymers. Nucl. Instrum. Methods B. 269(2):83-88. http://dx.doi.org/10.1016/j.nimb.2010.11.018

Pelagad SME, Singh NL, Qureshi A, Rane RS, Mukherjee S, Deshpande UP, Ganesan V, Shripathi T (2012). Investigation of surface properties of Ar-plasma treated polyethylene terephthalate (PET) films. Nucl. Instrum. Methods B. 289:34-38. http://dx.doi.org/10.1016/j.nimb.2012.08.010

Soliman BA, Abdel Rahman MM and Abdelsalam FW (2013). Irradiation effect on PET surface using low energy Argon ion beam. J. Nucl. Mater. 432(1-3):444-449. http://dx.doi.org/10.1016/j.jnucmat.2012.08.006

Zhiyong Z, Changlong L, Youmei S, Jie L, Yuhua T, Yucfan J, Junli D (2002). Modification of polyethylene terephthalate under high-energy heavy ion irradiation. Nucl. Instrum. Methods Phys. Res. B. 191(1-4):723-727. http://dx.doi.org/10.1016/S0168-583X(02)00641-9

Multi-criteria decision making using fuzzy logic approach for evaluating the manufacturing flexibility

S. R. Ansari[1], P. K. Mittal[2] and Ritu Chandna[2*]

[1]Department of Mathematics, H. N. B. Garhwal University, Srinagar (Garhwal), India.
[2]Department of Mathematics and R&D, Graphic Era University, Dehradun, Uttarakhand, India.

Manufacturing flexibility is the ability of a manufacturing system to cope with environmental changes effectively and efficiently. Most operation managers cannot provide exact numerical values to express opinions based on human perception due to ill defined and ambiguity of flexibility assessment. However, fuzzy logic provides a useful tool to deal with problems in which the phenomena are imprecise and vague. The purpose of this study is to measure the flexibility of manufacturing system based on multi-criteria decision making using fuzzy logic approach. In this approach, the performance ratings and importance weights of different flexibility capabilities assessed by experts are expressed in linguistic terms. Fuzzy performance-importance index of each flexibility capability is also devised to help managers identify the main adverse factors and calls for managers to institute an appropriate action plan to improve the flexibility level. An example is also used to illustrate the approach developed.

Key words: Manufacturing flexibility, fuzzy logic, multi-criteria decision, performance rating, importance weights.

INTRODUCTION

Manufacturing flexibility (MF) is an effective way to face up to the uncertainties of the rapidly changing environment and it is defined as the ability to absorb various disturbances which occur in production systems, as well as the ability to incorporate and exploit new technological advances and work practices. Although there have been tremendous efforts to define the meaning of manufacturing flexibility (Sethi and Sethi, 1990; Sarker et al., 1994; Beach et al., 2000; Golden and Powell, 2000; Vokurka and O'Leary-Kelly, 2000), the flexibility concept still remains incomplete or too abstract for operational applications.

Measurement of manufacturing flexibility imparts a great deal of insight at both the strategic and operational levels of a firm that equips managers to deal with current problems such as shrinking product life cycles, fierce market competition, and the ever-increasing demand for product variety (Gerwin, 1993). Nevertheless, several frameworks have been suggested for its measurement such as entropy (Chang et al., 2001; Shuiabi et al., 2005), graph theory (Kochikar and Narendran, 1992), Petri nets (Barad and Sipper, 1988) together with other mathe-matical programming approaches that are often difficult for operations managers to interpret (Gupta and Goyal, 1989; Parker and Wirth, 1999, Gupta, 1993; Bernardo and Mohamed, 1992). However, these techniques do not record and utilize human knowledge and perceptions about flexibility in its measurement. Moreover these methods fail in putting together the various dimensions of flexibility and do not identify the adverse factors for improving flexibility levels. Most operation managers cannot provide exact numerical values to express opinions based on human perception: more realistic measurement uses linguistic assessments instead of numerical values (Beach et al., 2000; Gerwin, 1993; Herrera et al., 2000; Vokurka and O'Leary-Kelly, 2000). After Zadeh (1965) introduced fuzzy set theory to deal with vague problems, linguistic labels have been used in approximate reasoning within the framework of fuzzy set

*Corresponding author. E-mail: rituchandna15@gmail.com.

Abbreviations: MF, Manufacturing flexibility; **FFI,** Fuzzy-flexibility-index; **FPII,** fuzzy performance-importance index.

theory (Zadeh, 1975) to handle the ambiguity in evaluating data and the vagueness of linguistic expression. Using fuzzy concepts, evaluators can use linguistic terms to assess the indicators in a natural language expression and each linguistic term can be associated with a membership function. Tsourveloudis and Phillis (1998), Van Hop and Ruengsak (2005), Wang and Chuu (2004), Beskese et al. (2004); Das and Caprihan (2007) are some of the attempts that have revealed several advantages of using fuzzy models for measuring flexibility elements in terms of expressing imprecise data pervading real-world problems.

However, the above models for fuzzy flexibility measurement simply focus on fuzzifying existing flexibility elements instead of incorporating other possible underlying elements and also do not identify the principal adverse factors of flexibility to institute appropriate amending measures early on to enhance flexibility more effectively.

From this review, to assist managers in better achieving a flexible enterprise, a model on the basis of fuzzy logic is purposed to provide a means of both measuring how flexible an enterprise is and identifying the principal obstacles to improve the flexibility level. By referring to the factors proposed in previous studies together with the approach used by Lin et al. (2005), an alternate framework for manufacturing flexibility measurement have been exploited in this paper. In this approach, the performance ratings and importance weights of different flexibility capabilities assessed by experts are expressed in linguistic terms.

Then appropriate fuzzy numbers are used to present the linguistic values, and a simple fuzzy arithmetical operation is employed to synthesize these fuzzy numbers into one fuzzy number, which is called the fuzzy-flexibility-index (FFI). Also, the FFI is matched with appropriate linguistics, thereby enabling the flexibility level to be expressed in linguistic terms. After that the fuzzy performance-importance index (FPII) of each flexibility capability is devised to help managers identify the main adverse factors and calls for managers to institute an appropriate action plan to improve the flexibility level. This model is developed from the concept of multi-criteria decision making.

FLEXIBILITY TYPES AND CAPABILITIES

Manufacturing flexibility is a vague notion, exhibiting a polymorphism that makes quantification a difficult exercise. For the sake of analysis, flexibility has been categorized into several distinct types. Many reviews have considered definitions of MF, requests for MF, classificatory dimensions of MF, measurement of MF, choices for MF, and interpretations of MF (Beach et al., 2000; Gupta and Goyal, 1989; Gupta, 1993; Sarker et al., 1994; Sethi and Sethi, 1990; Vokurka and O'Leary-Kelly, 2000; Koste et al., 2004; Upton, 1994).

Product flexibility

Product flexibility is the ability to change over to produce a new set of products economically.

Operational flexibility

It refers to the capability of producing a part in different ways by changing the sequence of the operations which were originally scheduled.

Routing flexibility

Routing flexibility is the ability of a production system to manufacture a part using alternative routes in the system.

Process flexibility

Process flexibility describes the ability to change over in order to produce a given set of part patterns with different batch sizes.

Machine flexibility

It deals with the ease of making changes among operations required to produce a number of products.

Volume flexibility

Volume flexibility describes the ability to operate profitably at different production volumes.

Expansion flexibility

Expansion flexibility describes the capability to expand a system's capacity with minimal effort.

Labor flexibility

Labor flexibility is the ease of moving personnel around various departments within an organization.

BASIC CONCEPT OF FUZZY SET THEORY

For the purpose of application, the basic properties of fuzzy set theory needed in this study are introduced.

Additional discussion can be found in Klir and Yuan (1995).

Euclidean distance method

The Euclidean distance method consists of calculating the Euclidean distance from the given fuzzy number to each of the fuzzy numbers representing the natural-language expressions set.

Suppose the natural-language expression set is flexibility level (FL). Then the distance between the fuzzy number fuzzy-flexibility-index (FFI) and each fuzzy number member $FL_i \in FL$ can be calculated as:

$$d(FFI, FL_i) = \left\{\sum_{x \in p} \left(f_{FFI}(x) - f_{FL_i}(x)\right)^2\right\}^{1/2} \quad (1)$$

Where p = $\{x_0, x_1, \ldots x_m) \subset [0, 10]$ so that $0 = x_0 < x_1 < \ldots\ldots < x_m = 10$. To simplify, let p = {0, 0.5, 1, 1.5, 2, 2.5, 3, 3.5, 4, 4.5, 5, 5.5, 6, 6.5, 7, 7.5, 8, 8.5, 9, 9.5, 10}. Then, the distance from the FFI to each of the members in the set FL can be calculated.

Fuzzy weighted average

Let $R_1, R_2, \ldots, R_n$ and $W_1, W_2, \ldots W_n$ denote, respectively, the fuzzy ratings and the fuzzy importance weights of the criteria. The fuzzy weighted average of R_i and W_i is defined as:

$$Y = \sum_{i=1}^{n} W_i R_i \,/ \sum_{i=1}^{n} W_i \quad \ldots\ldots\ldots\ldots(2)$$

Rank fuzzy numbers

Since fuzzy numbers do not always yield a totally ordered set as real numbers do, it is very difficult to rank fuzzy numbers for prioritizing. To resolve this problem, many fuzzy numbers ranking methods that can be used to compare fuzzy numbers have been developed (Chen and Hwang 1992). In terms of the effectiveness for calculating, the fuzzy numbers ranking method using fuzzy mean and spread is adopted here. According to the fuzzy mean and spread method a triangular fuzzy number M = (l, m and u), and let DV be the defuzzification value of the M; then DV can be computed as:

$$DV = (l + 2m + u)/4 \quad \ldots\ldots\ldots\ldots\ldots\ldots\ldots\ldots (3)$$

FUZZY FLEXIBILITY EVALUATION (FFE) APPROACH

The fuzzy flexibility evaluation (FFE) framework is composed of two major parts. The first part is the business operation environments' evaluation and flexibility capabilities' identification. The purpose of the business environment survey is to collect and analyze the flexibility drivers which are the changes in the business environment that drive a company to reconsider the company's position, strategy and process and in sequence maybe used to reset new strategies when running their business and building flexibility capabilities. The company's flexibility capabilities are the vital abilities that would provide the required strength to make appropriate responses to changes taking place in its business, so that flexibility capabilities will provide for flexibility measuring of a company. The second part of the framework is to evaluate flexibility capabilities and synthesize the ratings and the weights to obtain an FFI of a flexible enterprise and to match the FFI with an appropriate flexibility level and to make an improvement analysis. The main step description of the model developed from the concept of fuzzy multi-criteria decision making is given as follows:

Aggregate ratings and weighting to gain fuzzy- flexibility index and fuzzy merit-importance index of enable-factor

Suppose a committee of m analysts conducts the flexibility assessment and n flexible-enable attributes for flexibility assessment, then fuzzy-flexibility-index (FFI) represents the integrated merit of the flexibility-enable attributes of the enterprise is given by Equation 2.

Consequently, for each flexibility element capability *ijk*, the fuzzy performance-importance index $FPII_{ijk}$, is defined as:

$$FPII_{ijk} = [(1, 1, 1) \ominus W'_{ijk}] \otimes AC_{ijk}, \quad \ldots\ldots\ldots(4)$$

Where W_{ijk} is the fuzzy importance weight of the flexibility element capability *ijk*.

Translate FFI into linguistic flexibility term

Several methods for matching the membership function with linguistics terms have been proposed (Eshragh and Mandani, 1979; Schmucker, 1985).

There are basically three techniques: (1) Euclidean distance method, (2) successive approximation, and (3) piecewise decomposition. It is recommended that the Euclidean distance method be utilized because it is the most intuitive form of human perception of proximity (Guesgen and Albrecht, 2000). In this case the natural-language expression set FL = {Extremely Flexible [EF], Very Flexible [VF], Flexible [F], Fairly Flexible [FF], Slowly [S]} is selected for labeling, and the linguistics and corresponding membership functions are shown in Figure 1.

Then, by using the Euclidean distance method, the Euclidean distance D from the FFI to each member in set FL is calculated by Equation 1.

Rank fuzzy merit- importance index of enable-factors

Since fuzzy numbers do not always yield a totally ordered set as real numbers do, all the FPIIs must be ranked. All FPII can be defuzzified by using Equation 3 and can be ranked. This will help to analyze and identify the principal obstacles to improvement and enhance flexibility level.

AN ILLUSTRATIVE EXAMPLE

The FFE approach is demonstrated with a hypothetical an example: Manufacturing flexibility is the ability of a manufacturing system to

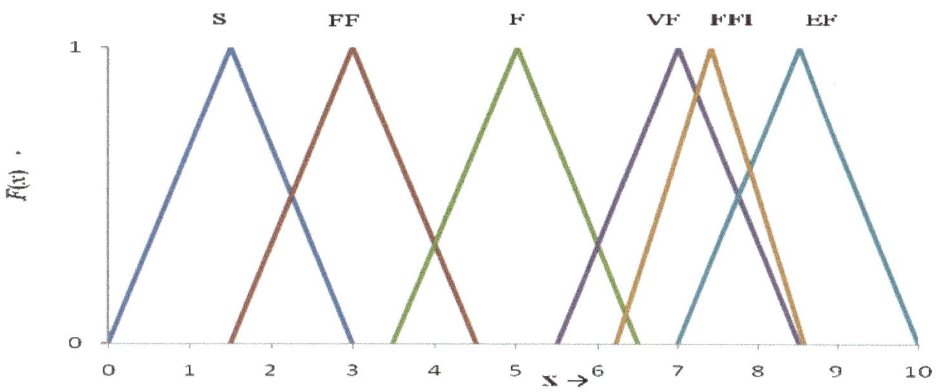

Figure 1. Linguistic levels to match fuzzy-flexibility-index [EF (7, 8.5, 10); VF (5.5, 7, 8.5); F (3.5, 5, 6.5); FF (1.5, 3, 4.5); S (0, 1.5, 3)].

Table 1. Flexibility capabilities for flexibility index evaluating in manufacturing system.

1- Grade index	2- Grade index	3- Grade index
	Operational flexibility (MF_{11})	No. of production sequence with minimal switching costs (MF_{111}), No. of Production Sequence with minimal switching times (MF_{112})
Product flexibility (MF_1)	Material handling flexibility (MF_{12})	Rerouting factor (MF_{121}) Load variety (MF_{122}) Transfer speed (MF_{123}) No. of connected elements (MF_{124})
	Sequence flexibility (MF_{13})	Operation Commonality (MF_{131}) Substitutability (MF_{132}) Average no. of machines to process different part type operations (MF_{133})
	Machine flexibility (MF_{21})	No. of diff operations performed (MF_{211}), Time taken for each Operation (MF_{212}), Set up time (MF_{213}), Versatility (MF_{214}), Adjustability (MF_{215}), Output quality and reliability (MF_{216}), Throughput from machine (MF_{217})
Process flexibility (MF_2)	Volume flexibility (MF_{22})	Range of volume (MF_{221}) Time required to increase or decrease the output (MF_{222}), Cost required to increase or decrease the volume of output (MF_{223})
	Expansion flexibility (MF_{23})	Modularity index (MF_{231}) Expansion ability (MF_{232})
	Labor flexibility (MF_{24})	Training level (MF_{241}) Job rotation (MF_{242})

cope with environmental changes effectively and efficiently. Suppose a committee comprised of few experts is formed to conduct the flexibility evaluation. After a series of activities consisting assessment of marketplace nature, competition circumstance, technology changing situation, customer requirements, social/cultural changes, product/process complexity, critically of relations with

Table 2. Aggregated performance rating and aggregated importance weight of flexibility capabilities.

MF_i	MF_{ij}	MF_{ijk}	W_i	W_{ij}	W_{ijk}	R_{ijk}
MF_1	MF_{11}	MF_{111}	VH	VH	VH	VG
		MF_{112}			H	G
	MF_{12}	MF_{121}		FH	FH	VG
		MF_{122}			H	VG
		MF_{123}			FH	G
		MF_{124}			FH	F
	MF_{13}	MF_{131}		VH	H	VG
		MF_{132}			VH	VG
		MF_{133}			VH	E
MF_2	MF_{21}	MF_{211}	H	H	H	VG
		MF_{212}			VH	VG
		MF_{213}			FH	F
		MF_{214}			H	G
		MF_{215}			FH	P
		MF_{216}			H	G
		MF_{217}			VH	G
	MF_{22}	MF_{221}		VH	VH	VG
		MF_{222}			H	VG
		MF_{223}			VH	VG
	MF_{23}	MF_{231}		VH	VH	G
		MF_{232}			VH	VG
	MF_{24}	MF_{241}		H	VH	G
		MF_{242}			H	G

criticality of relations with suppliers and flexibility discussion between analysts, the committee selects the criteria shown in Table 1 for evaluation.

The next step is to determine the appropriate linguistic scale to assess the performance ratings and importance weights of the flexibility capabilities. Furthermore, on the basis of linguistic level bank as shown in Table 3, the linguistic assessments of performance ratings and importance weights of the flexibility capabilities as shown in Table 2 are approximated by fuzzy numbers.

Then, by using the formulas in Equation 2, the fuzzy index of the flexibility 2-grade-capability MF_{ij} is obtained. For example, the fuzzy index of the flexibility 2-grade capability, information management flexibility MF_{11}, is calculated as:

$$MF_{11} = [(0.85, 0.95, 1.0) \otimes (7, 8, 9) \oplus (0.7, 0.8, 0.9) \otimes (5, 6.5, 8)] / [(0.85, 0.95, 1.0) \oplus (0.7, 0.8, 0.9)] = (6.10, 7.31, 8.53)$$

Applying the same Equation, other fuzzy indexes of flexibility 2-grade-capabilities MF_{ij} and the flexibility 1-grade-capabilities MF_i are obtained as listed in Table 4.

Furthermore, aggregate fuzzy ratings with fuzzy weights to obtain a FFI of the manufacturing system. FFI is an information fusion, which consolidates the fuzzy ratings and fuzzy weights of all of the factors that influence flexibility. FFI represents overall enterprise flexibility. Enterprise flexibility increases with increasing FFI. Thus, the membership function of FFI is used to determine the flexibility level.

Finally, applying Equation 2 again, the FFI of the manufacturing system is calculated as:

$$FFI = [(6.55, 7.67, 8.75) \otimes (0.85, 0.95, 1.0) \oplus (5.85, 7.10, 8.35) \otimes (0.7, 0.8, 0.9)] / [(0.85, 0.95, 1.0) \oplus (0.7, 0.8, 0.9)] = (6.23, 7.41, 8.56)$$

Further match the FFI with an appropriate flexibility level. Once the FFI has been obtained, to identify the level of flexibility, the FFI can be further matched with the linguistic label whose membership function is the same as (or closest to) the membership function of the FFI from the natural-language expression set of flexibility label (FL). In this case the natural-language expression set FL = {Extremely Flexible [EF], Very Flexible [VF], Flexible [F], Fairly Flexible [FF], Slowly [S]} is selected for labeling, and the linguistics and corresponding membership functions are shown in Figure 1. Then, by using the Euclidean distance method, the Euclidean distance D from the FFI to each member in set FL is calculated by Equation 1:

D(MF, EF) = 2.3017, D(MF, VF) = 0.7366, D(MF, F) = 1.9336, D(MF, FF) = 1.9336, D(MF, S) = 1.6549.

Table 3. Fuzzy numbers for approximating linguistic variable values.

Performance-rating		Importance-weighting	
Linguistic variable	**Fuzzy number**	**Linguistic variable**	**Fuzzy number**
Worst (W)	(0, 0.5, 1.5)	Very Low (VL)	(0, 0.05, 0.15)
Very Poor (VP)	(1, 2, 3)	Low (L)	(0.1, 0.2, 0.3)
Poor (P)	(2, 3.5, 5)	Fairly Low (FL)	(0.2, 0.35, 0.5)
Fair (F)	(3, 5, 7)	Medium (M)	(0.3, 0.5, 0.7)
Good (G)	(5, 6.5, 8)	Fairly High (FH)	(0.5, 0.65, 0.8)
Very Good (VG)	(7, 8, 9)	High (H)	(0.7, 0.8, 0.9)
Excellent (E)	(8.5, 9.5, 10)	Very High (VH)	(0.85, 0.95, 1)

Table 4. Fuzzy index of each grade of flexibility dimensions.

MF_i	MF_{ij}	R_i	R_{ijk}
	MF_{11}		(6.10, 7.31, 8.53)
MF_1	MF_{12}	(6.55, 7.67, 8.75)	(5.64, 6.94, 8.27)
	MF_{13}		(7.53, 8.53, 9.34)
	MF_{21}		(5.13, 6.45, 7.79)
MF_2	MF_{22}	(5.85, 7.10, 8.35)	(7, 8, 9)
	MF_{23}		(6, 7.25, 8.5)
	MF_{24}		(5, 6.5, 8)

Thus, by matching a linguistic label with the minimum D, the flexibility index level of the manufacturing system can be identified as "very flexible", as shown in Figure 1.

Finally, by applying Equation 4 the fuzzy merit- importance indices of enable-factors are obtained as listed in Table 5. Although the flexibility index level of manufacturing system is "very flexible" (according to the evaluation), there were obstacles within the organization which could have impacted the flexibility of the manufacturing system. In order to identify the principal obstacles for improving flexibility level, a fuzzy performance importance index (FPII) of flexibility element capability, which combines the performance rating and importance weight of each flexibility element capability, represents an effect which will contribute to the flexibility level of an organization. The lower the FPII of a factor is, the lower the degree of contribution for this factor. Thus, the score of the FPII of a factor is used for identifying the principal obstacles.

Then, by using the formulas in Equation 4, the FPIIs of each flexibility element capability are obtained and defuzzified by applying Equation 3, the results are listed in Table 5.

For example, the FPII of the perfect degree enterprise information system MF_{111}, is calculated as:

$$FPII_{111} = [(1, 1, 1) \ominus (1, 0.95, 0.85)] \otimes (7, 8, 9) = (0, 0.4, 1.35)$$

On the basis of experts' knowledge/ experience, the assessment committee set a scale (suppose 0.5) as the management threshold. As shown in Table 5, three capabilities have a lower performance than the others, namely: (1) Throughput from machine, (2) Modularity index and (3) Training level.

Furthermore, according to this identification, managers can select appropriate flexibility providers from Table 1 to improve the adverse factors and to implement better flexibility level.

RESULTS

1. The fuzzy flexibility index FFI of the manufacturing system was (6.23, 7.41, 8.56).
2. By matching a linguistic label with the minimum Euclidean distance, the flexibility index level of the manufacturing system was identified as "very flexible'.
3. Three capabilities had a lower performance than the others, namely: (1) Throughput from machine, (2) Modularity index, and (3) Training level.

DISCUSSION AND CONCLUSION

Flexibility has recently emerged as a key competitive priority in the present – day manufacturing environment. Manufacturing flexibility is an effective way to face up to the uncertainties of the rapidly changing environment and it is the ability to absorb various disturbances which occur in production systems, as well as the ability to incorporate and exploit new technological advances and work practices. However, in embracing MF there are many important questions to be asked concerning flexibility, such as: how to measure the flexibility of a company? How to identify the principal adverse factors for improving the flexibility level? How to assist in enhancing flexibility

Table 5. Fuzzy performance- importance indexes of 23 flexibility capabilities.

Flexibility capability	Aggregated fuzzy performance rating	(1, 1, 1)-W'_{ijk}	Fuzzy performance-importance index FPPI	Ranking score
MF_{111}	(7, 8, 9)	(0, 0.05, 0.15)	(0, 0.4, 1.35)	0.5375
MF_{112}	(5, 6.5, 8)	(0.1, 0.2, 0.3)	(0.5, 1.3, 2.4)	1.375
MF_{121}	(7, 8, 9)	(0.2, 0.35, 0.5)	(1.4, 2.8, 4.5)	2.875
MF_{122}	(7, 8, 9)	(0.1, 0.2, 0.3)	(0.7, 1.6, 2.7)	1.65
MF_{123}	(5, 6.5, 8)	(0.2, 0.35, 0.5)	(1.0, 2.275, 4.0)	2.3875
MF_{124}	(3, 5, 7)	(0.2, 0.35, 0.5)	(0.6, 1.75, 3.5)	1.9
MF_{131}	(7, 8, 9)	(0.1, 0.2, 0.3)	(0.7, 1.6, 2.7)	1.65
MF_{132}	(7, 8, 9)	(0, 0.05, 0.15)	(0, 0.4, 1.35)	0.5375
MF_{133}	(8.5, 9.5, 10)	(0, 0.05, 0.15)	(0, 0.475, 1.5)	0.6125
MF_{211}	(7, 8, 9)	(0.1, 0.2, 0.3)	(0.7, 1.6, 2.7)	1.65
MF_{212}	(7, 8, 9)	(0, 0.05, 0.15)	(0, 0.4, 1.35)	0.5375
MF_{213}	(3, 5, 7)	(0.2, 0.35, 0.5)	(0.6, 1.75, 3.5)	1.9
MF_{214}	(5, 6.5, 8)	(0.1, 0.2, 0.3)	(0.5, 1.3, 2.4)	1.375
MF_{215}	(2, 3.5, 5)	(0.2, 0.35, 0.5)	(0.4, 1.225, 2.5)	1.3375
MF_{216}	(5, 6.5, 8)	(0.1, 0.2, 0.3)	(0.5, 1.3, 2.4)	1.375
MF_{217}	(5, 6.5, 8)	(0, 0.05, 0.15)	(0, 0.325, 1.2)	0.4625
MF_{221}	(7, 8, 9)	(0, 0.05, 0.15)	(0, 0.4, 1.35)	0.5375
MF_{222}	(7, 8, 9)	(0.1, 0.2, 0.3)	(0.7, 1.6, 2.7)	1.65
MF_{223}	(7, 8, 9)	(0, 0.05, 0.15)	(0, 0.4, 1.35)	0.5375
MF_{231}	(5, 6.5, 8)	(0, 0.05, 0.15)	(0, 0.325, 1.2)	0.4625
MF_{232}	(7, 8, 9)	(0, 0.05, 0.15)	(0, 0.4, 1.35)	0.5375
MF_{241}	(5, 6.5, 8)	(0, 0.05, 0.15)	(0, 0.325, 1.2)	0.4625
MF_{242}	(5, 6.5, 8)	(0.1, 0.2, 0.3)	(0.5, 1.3, 2.4)	1.375

more effectively? Measurement of flexibility is difficult due to the multidimensionality and vagueness of the concept of flexibility and therefore the conventional assessment approaches cannot suitably or effectively be applied. Thus, in this paper a knowledge-based framework based on concept of multi criteria decision making and fuzzy logic for the measurement of manufacturing flexibility has been proposed. The evaluation procedure include: identifying flexibility capabilities, selecting linguistic variables for assessing and interpreting the values of the linguistic variables, fuzzy rating and fuzzy weights integrating, fuzzy index labeling, and defuzzifying FPII in order to identify the main adverse factors which can influence flexibility achievement. In addition, an example is given to illustrate the use of this method, which demonstrates the method can provide the analyst more convincing results.

The measurement framework proposed in this paper appears to have the following advantages. Firstly, it is adjustable by the user and enables analysts' linguistic assessment which may involve uncertainty. Managers can establish their own unique membership function by fitting in with their specific environment and consideration. Secondly this method can give the analyst relatively realistic and informative information. The FFI is expressed in a range of values. This provides an overall picture about the possible flexibility of an organization and ensures that the decision made in selection will not be biased. As an example of this study, the flexibility index has a fuzzy value (6.23, 7.41, 8.56). Finally this method can systematically identify the weak factors within an organization and provide the means for a manager to formulate a comprehensive plan for improvement. Therefore, the method can be further used in self-assessment.

Moreover, despite the above benefits for using fuzzy-logic for the measurement of flexibility, this approach has some limitations and it does not focus on finding an optimal design but only addresses flexibility level measurement. An objective of future research can be to investigate the influence of more rules on the value of flexibility and to develop a programming model subject to company constraints.

REFERENCES

Barad M, Sipper D (1988). Flexibility in manufacturing systems: definitions and Petri net modeling. Int. J. Prod Res. 26(2):237–248.

Beach R, Muhlemann AP, Price DHR, Paterson A, Sharp JA (2000). A review of manufacturing flexibility. Eur. J. Oper. Res., 122:41–57.

Bernardo JJ, Mohamed Z (1992). The measurement and use of operational flexibility in the loading of flexible manufacturing systems. Eur. J. Oper. Res., 60:144–155.

Beskese A, Kahraman C, Irani Z (2004). Quantification of flexibility in

advanced manufacturing systems using fuzzy concept. Int. J. Prod. Econ. 89:45–56.
Chang A, Whitehouse DJ, Chang S, Hsieh Y (2001). An approach to the measurement of single-machine flexibility. Int. J. Prod. Res., 39(8):1589–1601.
Chen SJ, Hwang CL (1992). Fuzzy Multiple Attribute Decision Making Methods and Application. Springer, Berlin, Heidelberg.
Das A, Caprihan R (2007). A rule-based fuzzy-logic approach for the measurement of manufacturing flexibility. Int. J. Adv. Manuf. Technol., 38:1098–1113.
Gerwin D (1993). Manufacturing flexibility: A strategic perspective. Manage. Sci., 39 (4):395–410.
Golden W, Powell P (2000). Towards a definition of flexibility: in search of the holy grail? Omega – Int. J. Manage. Sci., 28:373–384.
Guesgen HW, Albrecht J (2000). Imprecise reasoning in geographic information systems. Fuzzy Sets Syst., 113:121–131.
Gupta D (1993). On measurement and valuation of manufacturing flexibility. Int. J. Prod. Res. 31(12):2947–2958.
Gupta YP, Goyal S (1989). Flexibility of manufacturing systems: concepts and measurements. Eur. J. Oper. Res., 43:119–135.
Herrera F, Herrera-Viedma E (2000). Linguistic decision analysis: Steps for solving decision problems under linguistic information. Fuzzy Sets Syst., 115: 67– 82.
Klir GJ, Yuan B (1995). Fuzzy Sets and Fuzzy Logic: Theory and Applications. Prentice-Hall, Englewood Cliffs, NJ.
Kochikar VP, Narendran TT (1992). A framework for assessing the flexibility of manufacturing systems. Int. J. Prod. Res., 30 (12):2873–2895.
Koste LL, Malhotra MK, Sharma S (2004). Measuring dimensions of manufacturing flexibility. J. Oper. Manag. 22: 171–196.
Lin CT, Chen CT (2004). New product Go/NoGo evaluation at the front end—a fuzzy linguistic approach. IEEE Trans. Eng. Manag., 51 (2):197–207.
Lin CT, Chiu H, Tseng YH (2005). Flexibility evaluation using fuzzy logic. Int. J Prod. Econ., 101:353–368.
Parker RP, Wirth (1999). A manufacturing flexibility: measures and relationships. Eur. J. Oper. Res., 118:429–449.
Sarker BR, Krishnamurthy S, Kuthethur SG (1994). A survey and critical review of flexibility measures in manufacturing systems. Prod. Plan. Control, 5:512–523.
Schmucker KJ (1985). Fuzzy Sets, Natural Language Computations and Risk Analysis. Computer Science Press, USA.
Sethi AK and Sethi SP (1990). Flexibility in manufacturing: a survey. Int. J. Flex. Manuf. Syst., 2: 289–328.
Shuiabi E, Thomson V, Bhuiyan N (2005). Entropy as a measure of operational flexibility. Eur. J. Oper. Res., 165:696–707.
Tsourveloudis NC, Phillis YA (1998). Manufacturing flexibility measurement: a fuzzy-logic framework. IEEE Trans. Robot Autom., 14. (4):513–524.
Upton DM (1994). The management of manufacturing flexibility. California Management Review 72–89.
Van Hop N, Ruengsak K (2005). Fuzzy estimation for manufacturing flexibility. Int. J. Prod. Res., 43(17):3605–3617.
Vokurka RJ, O'Leary-Kelly SW (2000). A review of empirical research on manufacturing flexibility. J. Oper. Manag., 18:485–501.
Wang R-C, Chuu S-J (2004). Group decision-making using a fuzzy linguistic approach for evaluating the flexibility in a manufacturing system. Eur. J. Oper. Res., 154:563–572.
Zadeh LA (1965). Fuzzy set, Information and Control. 8:338–353.
Zadeh LA (1975). The concept of a linguistic variable and its application to approximate reasoning. Inform. Sci., 8:199–249, 301–357; 9 (1976) 43–58.

Parameter optimization of electrical discharge machining process by using Taguchi approach

Sameh S. Habib

Mechanical Engineering Department, Shoubra Faculty of Engineering, Benha University, Cairo, Egypt.

Electrical discharge machining (EDM) is one of the most extensively used non-traditional material removal processes for difficult-to-cut materials. The full potential of the EDM process has not yet been exploited due to its complicated discharge mechanism. Though a lot of research has been done to improve the process performance, optimal selection of process parameters for the best performance measures still remains a challenge. Parameter optimization is one of the techniques used in manufacturing processes to achieve best manufacturing conditions, which is an essential need for industries towards manufacturing of higher quality products at lower cost. In this paper, the cutting of hot work tool steel 2714 using electro discharge machining process with copper and graphite electrodes has been investigated. In this work L_{27} (3*4) orthogonal array based on the Taguchi experimental design is utilized to plan the experiments. Raw data is assessed by the Analysis of Variance (ANOVA) to find optimal conditions for response parameters. The main machining parameters such as pulse-on time, pulse-off time, discharge current, average machining voltage are chosen to determine the EDM response parameters such as material removal rate, surface roughness and gap size. Response tables and graphs are used to find the optimal parameter levels in the EDM process.

Key words: Electrical discharge machining (EDM), tool steel 2714, Taguchi approach, material removal rate, surface roughness and gap size.

INTRODUCTION

Electrical discharge machining (EDM) is an extremely prominent machining process among newly developed non-traditional machining techniques for difficult-to-cut materials. EDM is a thermo-electric process in which material is removed from work piece by the erosion effect of a series of electric discharges (sparks) between two electrodes (tool and workpiece) immersed in a dielectric liquid. The location of the discharge is determined by the narrowest gap between the two electrodes (McGeough, 1998). The workpiece and the tool should be made of electrically conductive material. EDM is especially well-suited for cutting intricate contours or delicate cavities that would be difficult to produce with a grinder, an end mill or other cutting tools. Physical and metallurgical properties do not create any limitation for the materials to be machined on EDM as there is no physical contact between tool and work piece (Yan et al., 2000). The EDM process has a very strong stochastic nature due to the complicated discharge mechanisms (Pandit and Mueller, 1987) making it difficult to optimize the sparking process. The optimization of the process often involves relating the various process variables with the performance measures maximizing material removal rate, while minimizing gap size and yielding desired surface roughness. Though a lot of researches have been done to improve the EDM process performance, optimal selection of the process parameters for the best performance measures still remains a challenge. Although studying of these parameters has been performed by many researchers, most of the studies do not much consider the optimization of the EDM process using both engineering philosophy design of experiments (DOE) and S/N ratio together with the analysis of variance (ANOVA) techniques which are used to measure the amount of deviation from the desired performance measures and identify the crucial process variables affecting the process responses (Ranjit, 1996, 2001).

Traditionally, the selection of the most favorable process parameters was based on experience or handbook values, which produced inconsistent machining performance. However, the optimization of parameters now relies on process analysis to identify the effect of operating variables on achieving the desired machining characteristics. Other works have applied the Taguchi approach to analyze and design the ideal EDM process. Taguchi design is a powerful tool for parametric design of performance characteristics; it recognizes that not all factors that cause variability can be controlled in practice. Using Taguchi's parameter design, significant machining parameters affecting the performance measures are identified. The Taguchi design has been employed to obtain the optimum factor/level combination of process parameters (Ranjit, 1996, 2001; Peace, 1993).

Taguchi modeling is helpful in the study of interaction of the different processing parameters, which would give a more precise image of the experimental results (Lorelei et al., 2012). Renjie et al. (2010, 2012) adopted the L_{25} orthogonal array based on Taguchi method, and evaluated the experimental data statistically by analysis of variance and stepwise regression. Renjie et al. (2010) optimized the machining parameters of silicon carbide ceramics with ED milling and mechanical grinding combined process. Mahapatra and Patnaik (2006) studied the optimization of wire electrical discharge machining (WEDM) process parameters using Taguchi method. They demonstrated that the WEDM process parameters can be adjusted to achieve better metal removal rate, surface finish and cutting width simultaneously. Marafona and Wykes (2000) used the Taguchi method to improve the TWR by introducing high carbon content to the electrode prior to the normal sparking process. Optimization of the operating parameters for EDM process of AISI D3 steel material has been studied using Taguchi method (Nipanikar, 2012). Lin et al. (2000) employed Taguchi approach with a set of fuzzy logic to optimize the process parameters taking the various performance measures into consideration. The implementation of Taguchi method on the EDM process of Tungsten Carbide has been reported (Mohd et al., 2009). Also, the effects of the EDM process variables on the material removal rate have been investigated using Taguchi parameter design approach (Amit and Pradeep, 2012). Tzeng and Chen (2003) optimized the high-speed EDM process by making use of dynamic signal-to-noise (S/N) ratio to classify the process variables into input signal, control and noise factors generating a dynamic range of output responses. Yan-Cherng et al. (2009) investigated EDM machining performance and optimizing machining parameters of Al_2O_3–TiC ceramics using Taguchi method. Utilizing L18 orthogonal array based on the Taguchi experimental design is used to plan the experiments and raw data assessed by the Analysis of Variance (ANOVA) to find optimal conditions for material removal rate (Singh et al., 2011).

In this work, the Taguchi method is used to determine the optimal machining parameters for maximum material removal rate, minimum surface roughness and minimum gap size in EDM operations. The level of influence of input parameters (on performance measures) has also been identified after experimental investigation with the help of ANOVA. Experimental verification of results achieved demonstrates that process performance may be improved significantly by this technique.

EXPERIMENTAL WORK

The experimental studies were performed on a NC die sinking EDM (Sodick AQ35LR) as shown in Figure 1. A kerosene type working fluid (Sodick hightech VITOL2) was used. Short cylindrical bars of both copper and graphite (Ibiden ED-3) materials of 8 mm in diameter were used as electrodes. Different settings of four controllable factors such as pulse-on time, pulse-off time, peak current and average machining voltage were used in the experiments. The electrode polarity was positive when using graphite electrodes and copper electrodes. The working EDM conditions are as listed in Table 1. A series of EDM experiments with varying discharge conditions were carried out.

Hot work tool steel 2714 is chosen as the workpiece material. Tool steel 2714 is a chromium-nickel-molybdenum alloy steel (56 Ni Cr Mo V 7) with the chemical composition listed in Table 2. The hardness of this steel as delivered ranges from 1250 to 1400 Mpa. This material can be used in forge and press dies, shear blades, extrusion rams die holders and hot deburring plates. The work piece is weighed before and after each experiment using an electric balance with a resolution of 0.01 mg to determine the value of material removal rate. For each set of values, three experiments are performed in randomized sequence in order to eliminate the influence of systematic errors, as recommended by Taguchi. Material removal rate is calculated as,

$$MRR = \frac{(W_i - W_f) \times 1000}{(D_w \times t)} \tag{1}$$

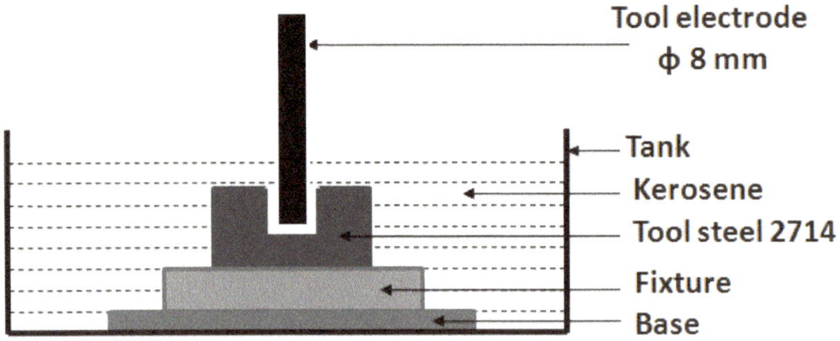

Figure 1. Principle of an EDM process.

Table 1. EDM conditions.

Condition name	Electrode type	
	Copper	**Graphite**
Pulse-on time (μs)	50, 100 and 150	20, 60 and 100
Pulse-off time (μs)	40, 50 and 60	10, 40 and 60
Discharge current (Amp.)	1, 4 and 6	1, 5 and 10
Average machining voltage (V)	40, 50 and 60	40, 45 and 50
Working fluid	Kerosene	
Workpiece material	Tool steel 2714	

Table 2. Chemical composition of 2714 tool steel.

C	Cr	Ni	Mo	V	St
0.56	1.10	1.70	0.50	0.10	Rem.

(Unit: mass %).

Where MRR is the material removal rate (mm^3/min), W_i is the initial average weight of the workpiece (g), W_f is the final average weight of the workpiece (g), D_w is the density of the workpiece (g/cm^3), t is the time of machining (min).

The surface roughness (SF) value Rz (μm) was measured using a 'Mitutoyo Talysurf (SJ-201) portable surface measuring unit with stylus radius of 5 μm. The cutoff length for each measurement was taken as 0.8 mm. The surface roughness values were measured for each specimen three times and the average was calculated. The resulted EDM cavities diameters were measured using an optical microscope (Keyence VH-7000 optical digital, USA). The gap size (GS) value is determined based upon the average of three results from Equation 2.

$$GS = \frac{D_H - D_{TE}}{2} \quad (2)$$

Where D_H is the hole cavity diameter in mm and D_{TE} is the tool-electrode diameter in mm.

Design of experiments (DOE)

In the present study, four process parameters namely, pulse-on time, pulse-off time, discharge current and average machining voltage are considered, although a large number of factors could be considered for controlling the EDM process. Tables 3 and 4 show the design factors along with their levels for both copper and graphite electrodes respectively. Three levels, having equal spacing, within the operating range of the parameters are selected for each of the factors.

In this study, the Taguchi method, an effective and powerful tool for experimental design of the performance characteristics was used to determine the optimal machining parameters for maximization of MRR and minimization of SF and (GS) in EDM process. Taguchi designs use orthogonal arrays, which estimate the effects of factors on the response mean and variation. Orthogonal array is a statistical method of defining parameters that converts test areas into factors and levels. Test design using orthogonal array creates an efficient and concise test suite with fewer test cases without compromising test coverage. The control factors are used to select the best conditions for stability in design of manufacturing process, whereas the noise factors denote all factors that cause variation. In this work, it is planned to study the

Table 3. Design factors and their levels for tool steel 2714 workpiece – Cu electrode.

Control parameter	Coding	Level		
		1	2	3
		Minimum	Intermediate	Maximum
Pulse-on time (T_{ON}) in µs	A	50	100	150
Pulse-off time (T_{OFF}) in µs	B	40	50	60
Discharge current (I) in Amp	C	1	4	6
Average machining voltage (V) in V	D	40	50	60

Table 4. Design factors and their levels for tool steel 2714 workpiece – Gr electrode.

Control parameters	Coding	Level		
		1	2	3
		Minimum	Intermediate	Maximum
Pulse-on time (T_{ON}) in µs	E	20	60	100
Pulse-off time (T_{OFF}) in µs	F	10	40	60
Discharge current (I) in Amp	G	1	5	10
Average machining voltage (V) in V	H	40	45	50

behavior of four control factors, (A, B, C, and D) for copper electrode and (E, F, G and H for graphite electrodes, and two interactions such as A×C and A×D for copper electrode and E×G and E×H for graphite electrode. The experimental observations are further transformed into a signal-to-noise (S/N) ratio. There are several S/N ratios available depending on the type of characteristics. The characteristic of higher value represents better machining performance, such as MRR, and is termed 'higher is better, HB'. Inversely, the characteristic of lower value represents better machining performance, such as surface roughness, and is termed 'lower is better, LB. Therefore,"HB" for the MRR 'LB" for the SF and "LB" for the (GS) were selected for obtaining optimum machining performance characteristics. The loss function (L) for objective of HB and LB is defined as follows:

$$L_{HB} = \frac{1}{n}\sum_{i=1}^{n}\frac{1}{y_{MRR}^2} \tag{3}$$

$$L_{LB} = \frac{1}{n}\sum_{i=1}^{n} y_{SF}^2 \tag{4}$$

$$L_{LB} = \frac{1}{n}\sum_{i=1}^{n} y_{GS}^2 \tag{5}$$

Where the terms y_{MRR}, y_{SF} and y_{GS} denote the response for metal removal rate, surface finish and gap size, respectively, and n denotes the number of performed experiments.

The S/N ratio response parameter can be calculated as a logarithmic transformation of the loss function as shown below.

$$\frac{S}{N}\,ratio\ for\ MRR = -10\ log_{10}(L_{HB}) \tag{6}$$

$$\frac{S}{N}\,ratio\ for\ SF = -10\ log_{10}(L_{LB}) \tag{7}$$

$$\frac{S}{N}\,ratio\ for\ GS = -10\ log_{10}(L_{LB}) \tag{8}$$

The orthogonal array chosen to set the control parameters and evaluate the process performance is the L_{27} (3^{13}), which has 27 rows corresponding to the number of experiments with 13 columns at three levels. It considers four process parameters (without interaction), (A, B, C, and D) for copper electrode and (E, F, G and H for graphite electrodes, to be varied in three discrete levels.

The plan of experiments when using copper electrodes is as follows: the first column was assigned to pulse-on time (A), the second column to pulse-off time (B), the third column to peak current (C), the sixth column to average machining voltage (D), the fourth column and fifth columns are assigned to (A×C) and (B×C), respectively, to estimate interaction between pulse-on time (A), pulse-off time (B) and discharge current (C), the seventh column and the eighth column are assigned to (A×D) and (B×D), respectively, to estimate interaction between the pulse-on time (A), pulse-off time (B) and average machining voltage (D). When using graphite electrodes, the plan of experiments look like it for copper electrodes. According to the Taguchi design concept, a L_{27} orthogonal array table was chosen for the experiments as shown in Table 5. The Taguchi analysis was made using the popular software specifically used for design of experiment applications known as MINITAB 15.

Table 5. Taguchi's L_{27} OA design.

Ex. No.	A or E	B or F	C or G	D or H
1	1	1	1	3
2	1	1	2	2
3	1	1	3	1
4	1	2	1	3
5	1	2	2	2
6	1	2	3	1
7	1	3	1	3
8	1	3	2	2
9	1	3	3	1
10	2	1	1	3
11	2	1	2	2
12	2	1	3	1
13	2	2	1	3
14	2	2	2	2
15	2	2	3	1
16	2	3	1	3
17	2	3	2	2
18	2	3	3	1
19	3	1	1	3
20	3	1	2	2
21	3	1	3	1
22	3	2	1	3
23	3	2	2	2
24	3	2	3	1
25	3	3	1	3
26	3	3	2	2
27	3	3	3	1

RESULTS AND DISCUSSION

Here, the use of an orthogonal array to reduce the number of cutting experiments for determination of optimal cutting parameters is presented. Results of the cutting experiments are studied by using the S/N and ANOVA analyses. Based on the results of these analyses, optimal cutting parameters for maximum material removal rate and minimum surface roughness and gap size are obtained and verified.

After DOE process, a set of 27 experiments are carried out in electrical discharge machine for each type of electrode. After each experiment material removal rate was calculated and also, surface roughness and gap size were measured. A quality characteristic for MRR is higher is better (HB) and lower is better (LB) is the quality characteristic for both surface roughness and gap size. Signal-to-noise (S/N) ratios were calculated for each experiment with the help of software Minitab 15 as shown in Tables 6 and 7. The output characteristic, MRR, SR and GS are analyzed by software Minitab 15 and ANOVA is formed, which shows the percentage contribution of each influencing factor on MRR, SR and GS.

Effect of input factors on MRR

Main effect plots for means and main effect plots for signal to noise ratio of material removal rate for copper and graphite electrodes are plotted with the help of Minitab 15 software as shown in Figures 2 and 3, respectively. The interaction plots for MRR between pulse-on time and both discharge current and average machining voltage for copper and graphite electrodes are also shown in Figures 4 and 5, respectively. In the main effects plot, if the line for a particular parameter is near horizontal, then the parameter has no significant effect. On the other hand, a parameter for which the line has the highest inclination will have the most significant effect. It is very much clear from the main effects plot that parameter pulse-on time (A) is the most significant parameter, while average machining voltage (D) has some contribution. As far as the interaction plots are concerned, estimating an interaction means determining the non-parallelism of the parameter effects. Thus, if the lines on the interaction plots are non-parallel, interaction occurs and if the lines cross, strong interactions occur between parameters.

From Figure 3, it can be seen that there are moderate interaction between the parameters: pulse on time (A) and discharge current (C) and strong interaction between the parameters pulse-on time (A) and average machining voltage (D). Thus from the present analysis it is clear that the pulse-on time (A) is the most influencing parameter for the multiple parameters of EDM process. According to the higher-the-better quality characteristic for MRR of copper electrode, based from the maximum point on the graph, the optimum condition for each factor indicated is A3 (150 μs), B3 (60 μs), C3 (6 A), D1 (40 V). The optimal process parameter combination for maximum metal removing rate is found to be pulse on time (A) at highest level, pulse off time (B) at highest level, discharge current (C) at highest level and average machining voltage (D) at lowest level.

The S/N ratio corresponds to the larger variance of the output characteristics around the desired value. The mean S/N ratio for each level of the cutting parameters is summarized and called the mean S/N response table for material removal rate (Table 8). When using copper electrodes, the effect of cutting parameters can be ranked as follows (pulse-on time, discharge current, pulse-off time and average machining voltage). However, the effect of cutting parameters can be ranked as follows (pulse-on time, pulse-off time, discharge current and average machining voltage) when using graphite electrodes.

ANOVA is a statistical technique that can infer some important conclusions on the basis of analysis of the experimental data. The method is very useful for revealing the level of significance of influence of factor(s) or interaction of factors on a particular response. It separates the total variability of the response into

Table 6. Experimental design using L_{27} orthogonal array for copper electrodes.

Exp. no.	A	B	C	D	MRR	S/N ratio	SR	S/N ratio	GS	S/N ratio
1	50	40	1	60	0.005	-46.021	8.564	-18.654	37	-31.364
2	50	40	4	50	0.018	-34.895	10.216	-20.186	47	-33.442
3	50	40	6	40	0.027	-31.373	12.884	-22.201	52	-34.320
4	50	50	1	50	0.015	-36.478	10.006	-20.005	44	-32.869
5	50	50	4	40	0.025	-32.041	11.966	-21.559	50	-33.979
6	50	50	6	60	0.033	-29.63	13.288	-22.469	54	-34.648
7	50	60	1	40	0.035	-29.119	13.876	-22.845	55	-34.807
8	50	60	4	60	0.043	-27.331	14.548	-23.256	58	-35.269
9	50	60	6	50	0.054	-25.352	15.632	-23.880	61	-35.707
10	100	60	1	60	0.072	-22.853	16.998	-24.608	72	-37.147
11	100	60	4	50	0.082	-21.724	17.882	-25.048	80	-38.062
12	100	60	6	40	0.118	-18.562	20.098	-26.063	87	-38.790
13	100	40	1	50	0.062	-24.152	16.344	-24.267	67	-36.522
14	100	40	4	40	0.072	-22.853	17.000	-24.609	72	-37.147
15	100	40	6	60	0.079	-22.048	17.566	-24.894	75	-37.501
16	100	50	1	40	0.082	-21.724	17.88	-25.047	80	-38.062
17	100	50	4	60	0.091	-20.819	18.432	-25.311	82	-38.276
18	100	50	6	50	0.104	-19.659	19.002	-25.576	86	-38.690
19	150	50	1	60	0.158	-16.027	21.072	-26.474	92	-39.276
20	150	50	4	50	0.181	-14.846	22.987	-27.230	95	-39.555
21	150	50	6	40	0.228	-12.841	26.879	-28.588	98	-39.825
22	150	60	1	50	0.169	-15.442	21.611	-26.694	94	-39.463
23	150	60	4	40	0.188	-14.517	23.888	-27.564	96	-39.645
24	150	60	6	60	0.196	-14.155	24.662	-27.841	96	-39.645
25	150	40	1	40	0.168	-15.494	21.434	-26.622	94	-39.463
26	150	40	4	60	0.176	-15.09	22.144	-26.905	95	-39.555
27	150	40	6	50	0.189	-14.471	24.044	-27.620	96	-39.645

contributions of each of the factors and the error.

Using Minitab, ANOVA is performed to determine which parameter and interaction significantly affect the performance characteristics. Tables 9 and 10 show the ANOVA results for material removal rate for copper and graphite electrodes respectively. The ANOVA table shows the percentage contribution of each parameter. For MRR, the calculation of S/N ratio follows "Larger the Better" model. Therefore, pulse-on time (T_{ON}) has the maximum effect on material removal rate. Similarly, interaction between pulse-on time (T_{ON}) and discharge current (I) and pulse-on time (T_{ON}) and average machining voltage (V) have some influence on MRR of the EDM ~~of~~ process parameters.

It can be concluded that, pulse-on-time is a significant factor for material removal rate. Pulse-on-time is current discharge duration on the voltage gap. Short peak current duration or short pulse duration may cause less surface vaporization of the workpiece during machining, whereas long pulse duration may cause the plasma channel to expand in the machining gap (Wang, 1999). The increase in pulse-on-time means that the same heating flux is applied for a longer time. This will cause an increase of heat that is conducted into the workpiece as the plasma channel expands which will result in an increase in the MRR (Panda, 2008; Natsu et al., 2006).

Effect of input factors on surface roughness

Main effect plots for means and main effect plots for signal to noise ratio surface roughness for copper and graphite electrodes are shown in Figures 6 and 7, respectively. The interaction plots for surface roughness between pulse-on time and both discharge current and average machining voltage for copper and graphite electrodes are also shown in Figures 8 and 9, respectively. It is noticed from the main effects plot that the pulse-on-time parameter (A) is the most significant parameter, while the average machining voltage (D) has some contribution.

The response table for signal to noise ratio for surface roughness is shown in Table 11 and the corresponding ANOVA table is shown in Tables 12 and 13. For SR, the calculation of S/N ratio follows "Smaller the Better" model. Therefore, pulse-on-time (T_{ON}) has the maximum

Table 7. Experimental design using L_{27} orthogonal array for graphite electrodes.

Exp. no.	E	F	G	H	MRR	S/N ratio	SR	S/N ratio	GS	S/N ratio
1	20	10	1	50	0.014	-37.077	8.026	-18.090	33	-30.370
2	20	10	5	45	0.062	-24.152	10.800	-20.669	37	-31.364
3	20	10	10	40	0.088	-21.110	12.033	-21.608	41	-32.256
4	20	40	1	45	0.094	-20.537	13.756	-22.770	43	-32.669
5	20	40	5	40	0.101	-19.914	15.066	-23.560	46	-33.255
6	20	40	10	50	0.108	-19.332	16.989	-24.603	50	-33.979
7	20	60	1	40	0.121	-18.344	18.222	-25.212	52	-34.320
8	20	60	5	50	0.129	-17.788	19.011	-25.580	54	-34.648
9	20	60	10	45	0.144	-16.833	19.889	-25.972	55	-34.807
10	60	60	1	50	0.377	-8.4732	23.632	-27.470	73	-37.267
11	60	60	5	45	0.408	-7.7868	24.092	-27.638	76	-37.616
12	60	60	10	40	0.442	-7.0916	24.881	-27.917	80	-38.062
13	60	10	1	45	0.152	-16.363	20.688	-26.314	58	-35.269
14	60	10	5	40	0.173	-15.239	21.020	-26.453	62	-35.848
15	60	10	10	50	0.198	-14.067	21.872	-26.798	65	-36.258
16	60	40	1	40	0.221	-13.112	22.078	-26.879	67	-36.522
17	60	40	5	50	0.265	-11.535	22.874	-27.187	69	-36.777
18	60	40	10	45	0.352	-9.0692	23.122	-27.281	72	-37.147
19	100	40	1	50	0.588	-4.6125	28.678	-29.151	90	-39.085
20	100	40	5	45	0.608	-4.3219	29.065	-29.267	92	-39.276
21	100	40	10	40	0.648	-3.7685	29.867	-29.504	94	-39.463
22	100	60	1	45	0.692	-3.1979	30.602	-29.715	96	-39.645
23	100	60	5	40	0.712	-2.9504	31.089	-29.852	98	-39.825
24	100	60	10	50	0.800	-1.9382	32.224	-30.164	101	-40.086
25	100	10	1	40	0.488	-6.2316	25.438	-28.110	83	-38.382
26	100	10	5	50	0.501	-6.0033	26.534	-28.476	85	-38.588
27	100	10	10	45	0.528	-5.5473	27.843	-28.894	87	-38.790

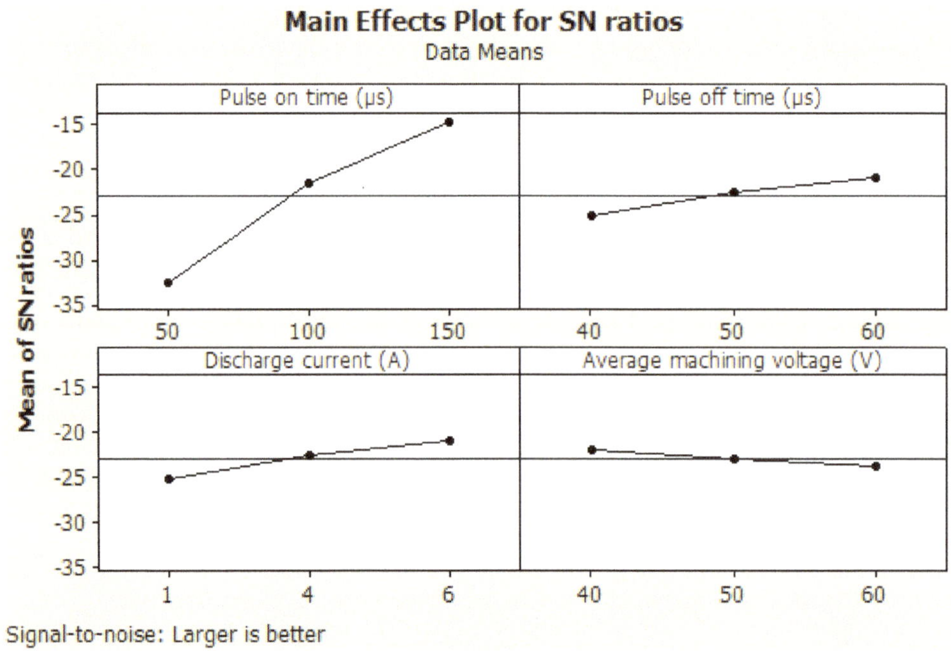

Figure 2. Main effects plot for S/N ratios of each factor on MRR with Cu electrode.

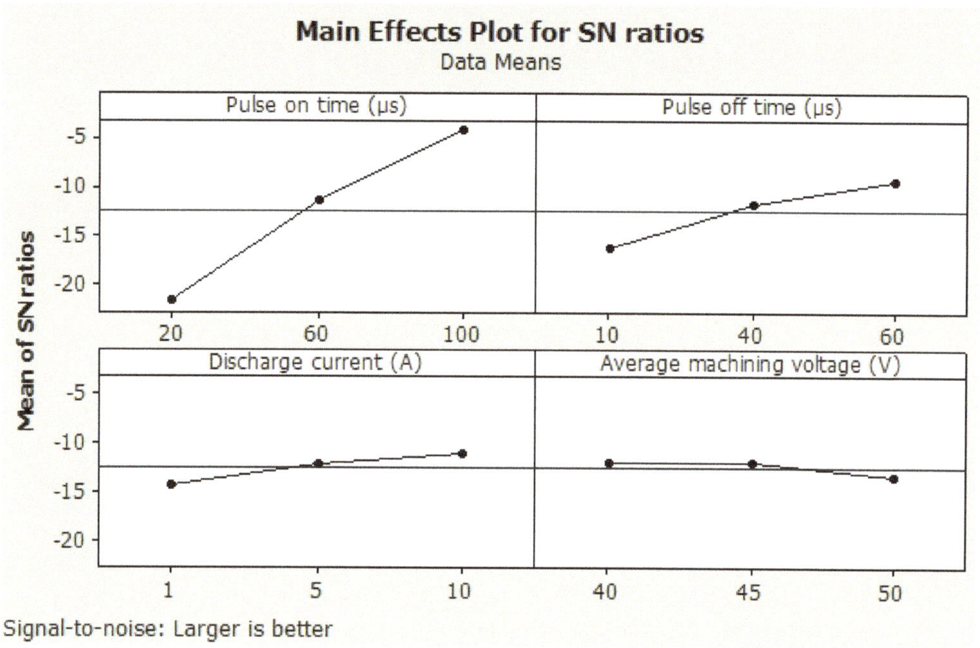

Figure 3. Main effects plot for S/N ratios of each factor on MRR with graphite electrode.

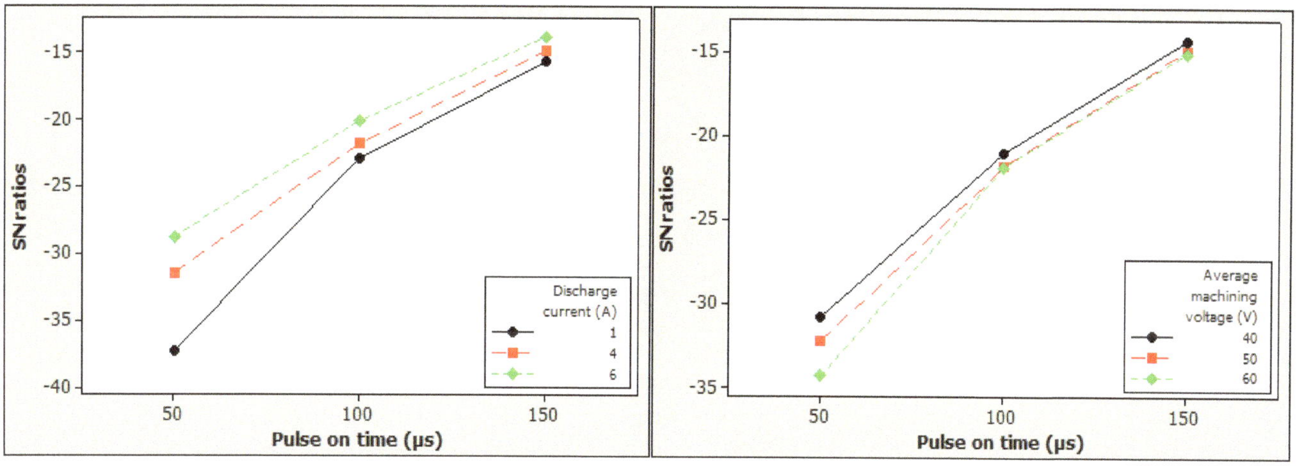

Figure 4. Interaction graphs for MRR between pulse-on time and both discharge current and average machining voltage for Cu electrode.

effect on surface roughness.

When using copper electrode, pulse-on-time (T_{ON}) with a contribution of 86.33% has the greatest effect on the machining output characteristics. Parameter C that is,. discharge current (I) with a 7.51% share is the next most significant influence on the output parameters, followed by parameter B that is, machine's pulse-off-time, (T_{OFF}) 4.7%. However, for graphite electrode, pulse-on-time with a contribution of 80.12% has the greatest effect on the machining output characteristics. Parameter B i.e. machine's pulse-off-time, with 15.16% share is the next most significant influence parameter, followed by parameter C, that is, discharge current with a 2.1%. Surface finish quality was better when applying smaller pulse-on-time. This is because of small particle size and crater depths formed by electrical discharge. As a result, the best surface finish will be produced. The selection of these machining parameters for EDM of any material should be used for a higher surface quality is required.

It was observed that when discharge current and particularly pulse-on time increased with pulse-off time, machined work piece surface exhibited a higher surface roughness due to irregular topography. Discharge current had an effect on surface roughness at low pulse time, but the influence of pulse-on time was more significant than discharge current at higher pulse times. It was noticed

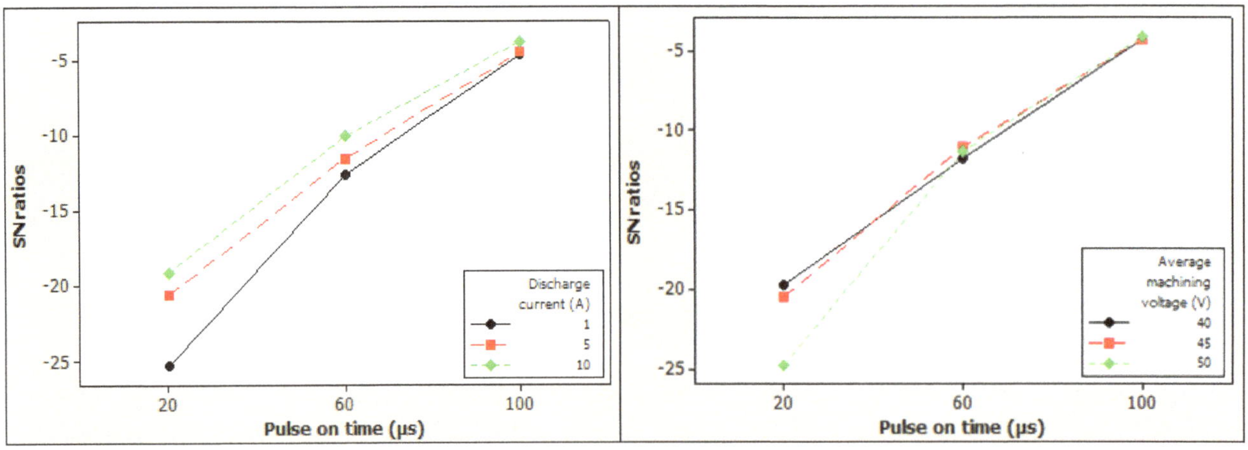

Figure 5. Interaction graphs for MRR between pulse-on time and both discharge current and average machining voltage for Graphite electrode.

Table 8.S/N ratio response table for material removal rate.

Level	Cu electrode				Gr electrode			
	T_{ON}	T_{OFF}	I	V	T_{ON}	T_{OFF}	I	V
1	-32.47	-25.16	-22.06	-25.26	-21.676	-16.199	-14.217	-11.974
2	-21.60	-22.67	-22.68	-23.00	-11.415	-11.800	-12.188	-11.979
3	-14.76	-21.01	-20.90	-23.77	-4.286	-9.378	-10.973	-13.425
Delta	17.71	4.15	4.36	1.72	17.391	6.821	3.244	1.452
Rank	1	3	2	4	1	2	3	4

Table 9. ANOVA results for material removal rate using Cu electrode.

Sequence of Variation	Degrees of freedom (DF)	Sum of squares (SS)	Mean square (variance) MS = SS/DF	Contribution (%)
T_{ON}	2	1435.23	717.616	86.49
T_{OFF}	2	78.45	39.227	4.73
I	2	86.40	43.202	5.21
V	2	13.30	6.651	0.8
T_{ON} * I	4	42.07	10.517	1.27
T_{ON} * V	4	7.59	1.899	0.23
Residual Error	10	105.91	10.591	1.28
Total	26	1768.96		

that high discharge current and pulse times will produce a poor surface finish due to deeper and wider crates on the machined surface. Excellent machined surface quality could be obtained by setting machining parameters at a low short pulse-on-time and pulse-off-time.

Effect of input factors on gap size

Main effect plots for means and main effect plots for signal to noise ratio of gap size for copper and graphite electrodes are shown in Figures 10 and 11, respectively. The interaction plots for gap size between pulse-on time and both discharge current and average machining voltage for copper and graphite electrodes are also shown in Figures 12 and 13, respectively. It was observed that, the gap size has highly affected with pulse-on-time and being slightly affected with average machining voltage. This may be due to that in EDM process the values of gap size are proportional to

Table 10. ANOVA results for material removal rate using graphite electrode.

Sequence of variation	Degrees of freedom (DF)	Sum of squares (SS)	Mean square (variance) MS = SS/DF	Contribution (%)
T_{ON}	2	1375.67	687.837	81.1
T_{OFF}	2	215.22	107.610	12.68
I	2	48.34	24.171	2.85
V	2	12.60	6.299	0.74
T_{ON} * I	4	26.25	6.563	0.77
T_{ON} * V	4	31.09	7.772	0.92
Residual Error	10	81.95	8.195	0.97
Total	26	1791.13		

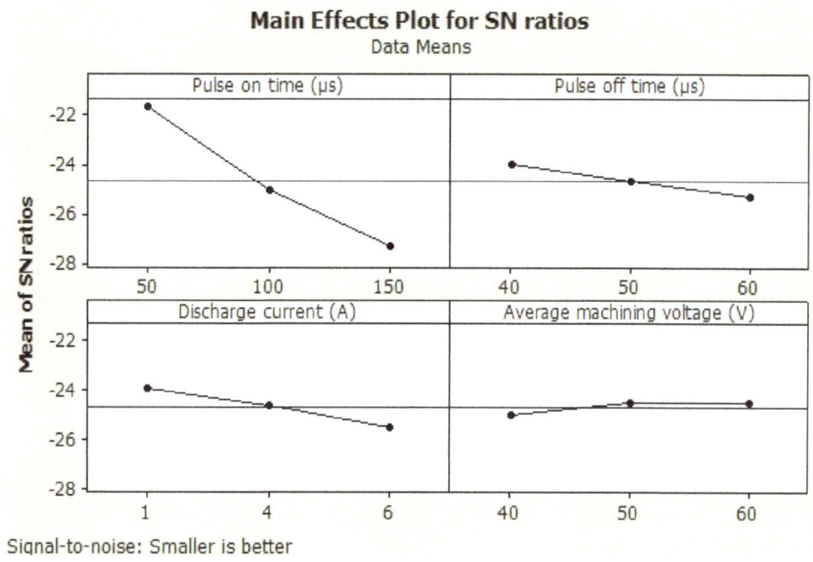

Figure 6. Main effects plot for SN ratios of each factor on SR with Cu electrode.

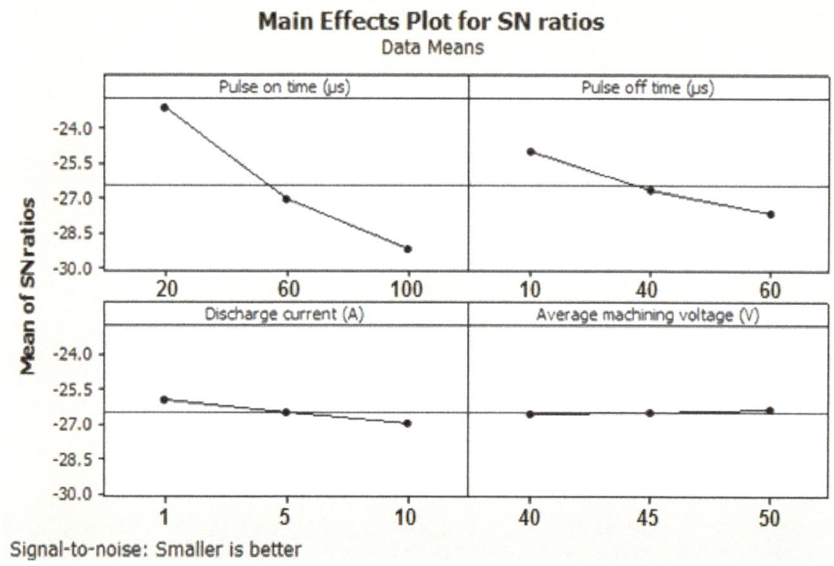

Figure 7. Main effects plot for SN ratios of each factor on SR with graphite electrode.

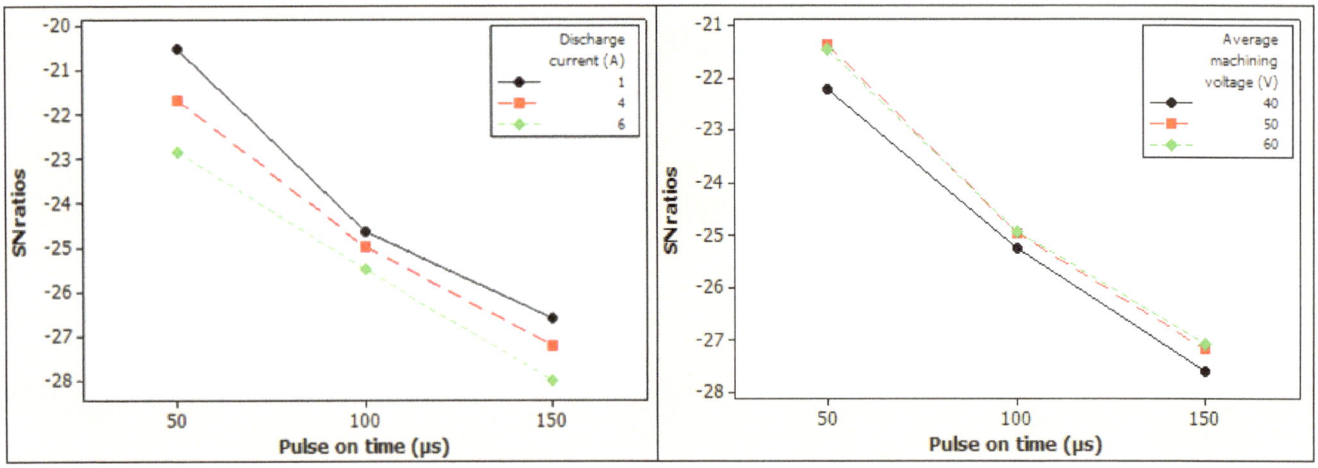

Figure 8. Interaction graphs for SR between pulse-on time and both discharge current and average machining voltage for Cu electrode.

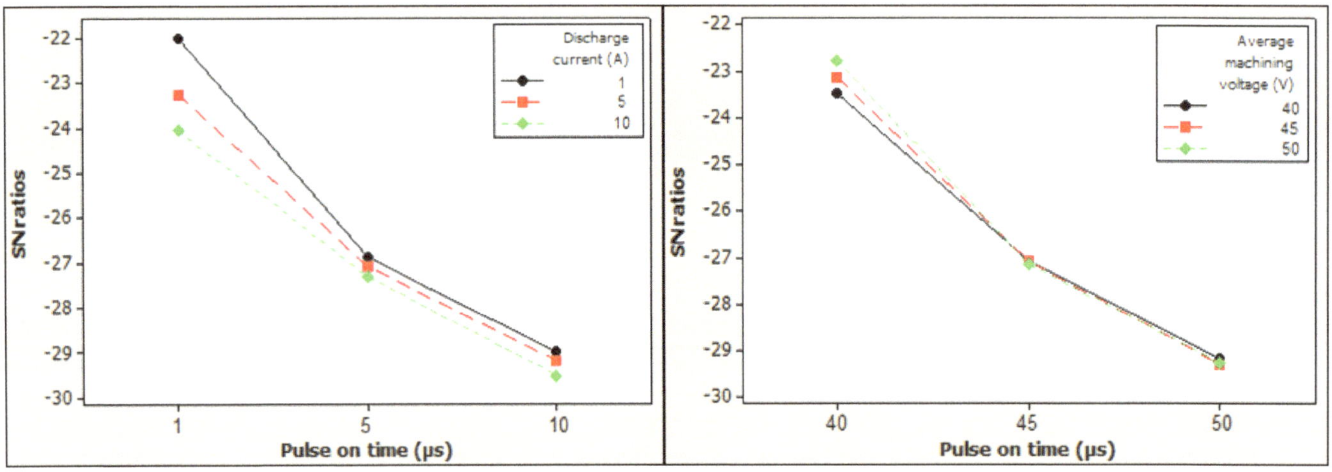

Figure 9. Interaction graphs for SR between pulse-on time and both discharge current and average machining voltage for graphite electrode.

Table 11. S/N ratio response table for surface roughness.

Level	Cu electrode				Gr electrode			
	T_{ON}	T_{OFF}	I	V	T_{ON}	T_{OFF}	I	V
1	-21.67	-24.00	-23.91	-25.01	-23.12	-25.05	-25.97	-26.57
2	-25.05	-24.70	-24.63	-24.50	-27.10	-26.69	-26.52	-26.50
3	-27.28	-25.31	-25.46	-24.49	-29.24	-27.72	-26.97	-26.39
Delta	5.61	1.32	1.55	0.52	6.12	2.68	1.00	0.18
Rank	1	3	2	4	1	2	3	4

material removal rate. Long pulse-on-time may cause the plasma channel to expand in the machining gap. This will cause an increase of heat that is conducted into the workpiece (Wang, 1999; Panda, 2008; Natsu et al., 2006) which leads to increase in the material removal rate and gap size.

The response table for signal to noise ratio for gap size is shown in Table 14 and the corresponding ANOVA analysis is shown in Tables 15 and 16 for the two types of electrodes. For gap size, the calculation of S/N ratio

Table 12. ANOVA results for surface roughness using Cu electrode.

Sequence of Variation	Degrees of freedom (DF)	Sum of squares (SS)	Mean square (variance) MS = SS/DF	Contribution (%)
T_{ON}	2	143.526	71.7631	86.33
T_{OFF}	2	7.801	3.9005	4.7
I	2	10.777	5.3886	7.51
V	2	1.595	0.7976	0.96
T_{ON} * I	4	1.683	0.4208	0.51
T_{ON} * V	4	0.296	0.0740	0.09
Residual Error	10	7.826	0.7826	0.94
Total	26	173.505	83.1272	

Table 13. ANOVA results for surface roughness using graphite electrode.

Sequence of variation	Degrees of freedom (DF)	Sum of squares (SS)	Mean square (variance) MS = SS/DF	Contribution (%)
T_{ON}	2	173.631	86.8155	80.12
T_{OFF}	2	32.846	16.4231	15.16
I	2	4.545	2.2725	2.1
V	2	0.141	0.0706	0.07
T_{ON} * I	4	2.505	0.6263	0.58
T_{ON} * V	4	0.641	0.1603	0.15
Residual Error	10	19.823	1.9823	1.83
Total	26	234.132		

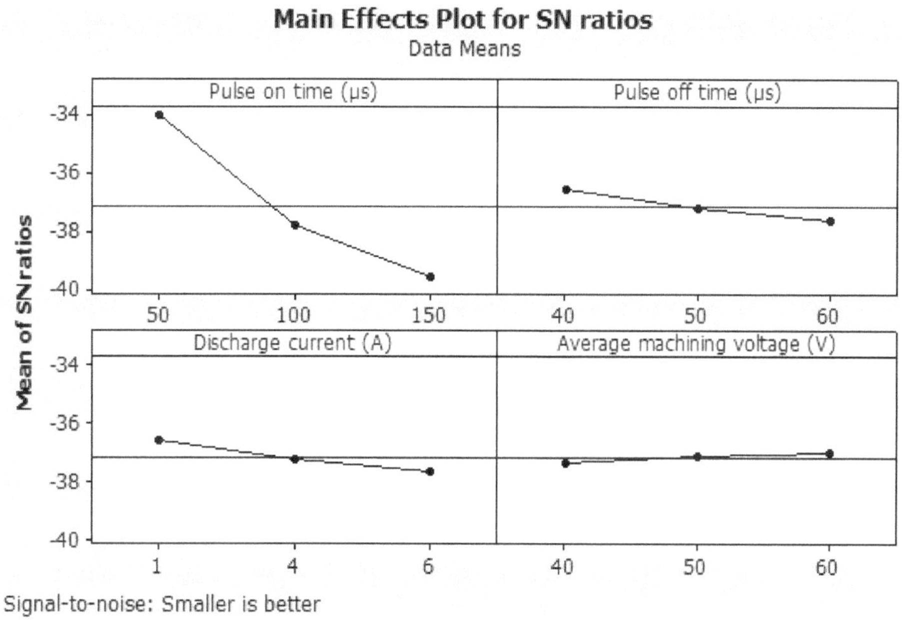

Figure 10. Main effects plot for SN ratios of each factor on GS with Cu electrode.

follows "Smaller the Better" model. Therefore, pulse-on-time (T_{ON}) has the maximum effect on gap size. Also, Tables 15 and 16 show that, the contribution amounts of pulse-on-time for copper and graphite electrodes are 91.4

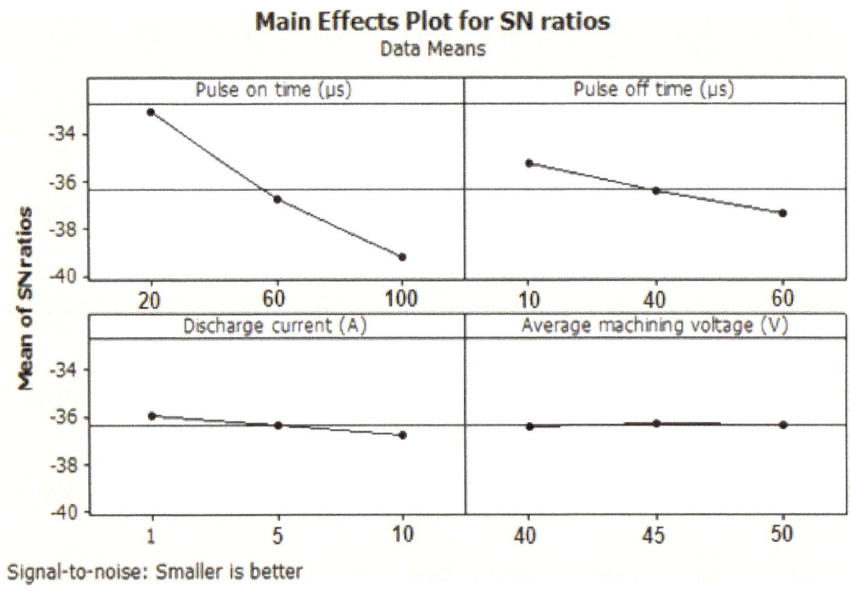

Figure 11. Main effects plot for SN ratios of each factor on GS with graphite electrode.

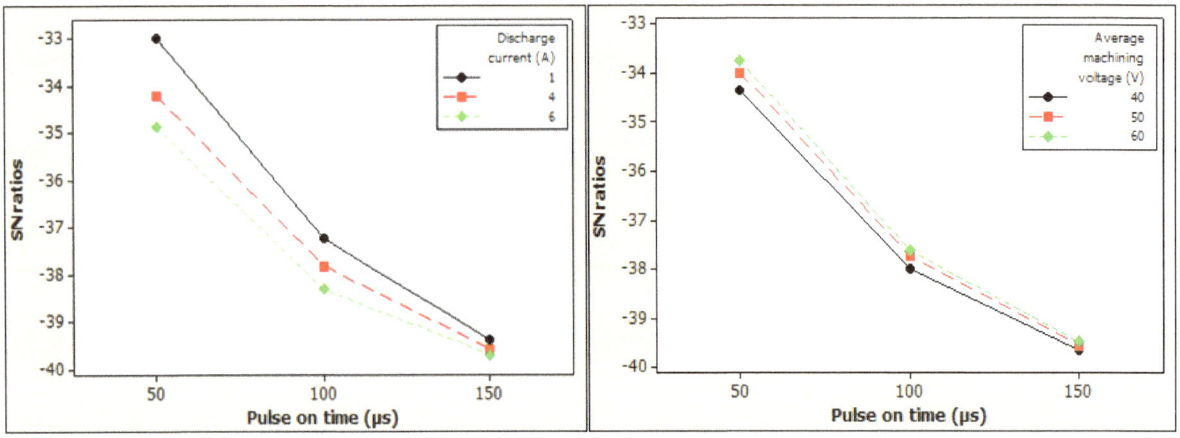

Figure 12. Interaction graphs for GS between pulse-on time and both discharge current and average machining voltage for Cu electrode.

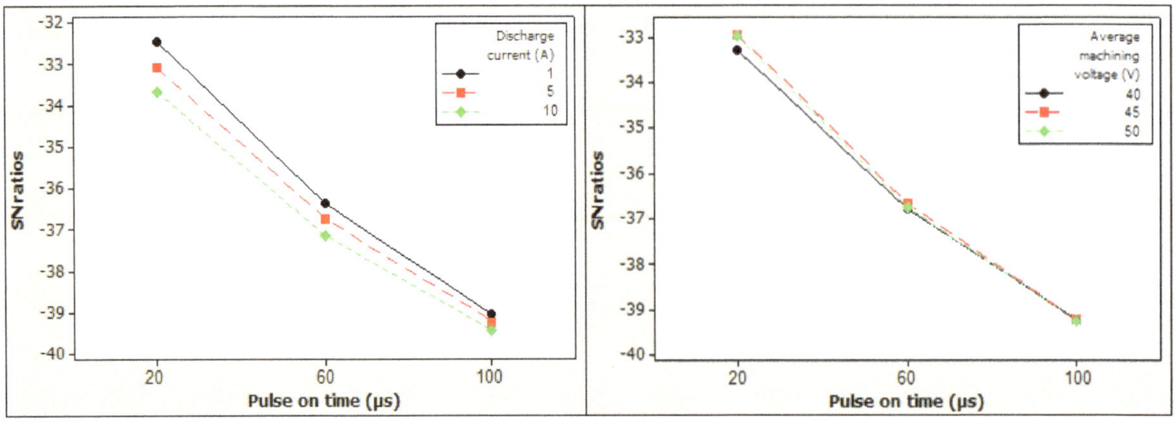

Figure 13. Interaction graphs for GS between pulse-on time and both discharge current and average machining voltage for graphite electrode.

Table 14. S/N ratio response table for gap size.

Level	Cu electrode				Graphite electrode			
	T_{ON}	T_{OFF}	I	V	T_{ON}	T_{OFF}	I	V
1	-34.04	-36.55	-36.55	-37.34	-33.07	-35.24	-35.95	-36.44
2	-37.80	-37.24	-37.21	-37.11	-36.75	-36.46	-36.36	-36.29
3	-39.56	-37.61	-37.64	-36.96	-39.24	-37.36	-36.76	-36.34
Delta	5.52	1.06	1.09	0.37	6.16	2.13	0.81	0.15
Rank	1	3	2	4	1	2	3	4

Table 15. ANOVA results for gap size using Cu electrode.

Sequence of variation	Degrees of freedom (DF)	Sum of squares (SS)	Mean square MS = SS/DF	Contribution (%)
T_{ON}	2	142.983	71.4917	91.4
T_{OFF}	2	5.247	2.6237	3.35
I	2	5.419	2.7094	3.46
V	2	0.639	0.3193	0.41
T_{ON} * I	4	1.933	0.4833	0.62
T_{ON} * V	4	0.160	0.0399	0.05
Residual Error	10	5.725	0.5725	0.73
Total	26	162.106	78.2398	

Table 16. ANOVA results for gap size using Graphite electrode.

Sequence of variation	Degrees of freedom (DF)	Sum of squares (SS)	Mean square MS = SS/DF	Contribution (%)
T_{ON}	2	173.073	86.5363	87.54
T_{OFF}	2	20.536	10.2681	10.39
I	2	2.977	1.4885	1.51
V	2	0.104	0.0519	0.05
T_{ON} * I	4	0.504	0.1260	0.13
T_{ON} * V	4	0.114	0.0286	0.03
Residual Error	10	3.578	0.3578	0.36
Total	26	200.886		

and 87.54% respectively.

Experimental verification

After performing the statistical analysis on the experimental data, it has been observed that there is one particular level for each factor for which the responses are either maximum (in case of material removal rate) or minimum (in case of surface roughness and gap size). The signal to noise ratio (S/N ratio) of each responses corresponding to each factor level also has a maximum and a minimum value. The optimal parameter setting for copper and graphite electrodes have been evaluated from the Figures 2 and 3 for material removal rate, Figures 6 and 7 for surface roughness and Figures 10 and 11 for gap size. The optimal setting comes as shown in Table 17. The optimal process parameters that have been identified to yield the best combination of process variables are A3B3C3D1 (highest level of pulse-on time, highest level of pulse-off time, highest level of discharge current and lowest level of average machining voltage) for copper and graphite electrodes. Using these optimum parameter settings, verification experiments have been carried out and the experimental results are shown in Table 18.

Conclusion

This research investigates the parameter optimization of electrical discharge machining on tool steel 2714

Table 17. Optimal parameter settings of input factors.

Physical requirement	Optimal combination Cu-electrode				Optimal combination Gr-electrode			
	T_{ON}	T_{OFF}	I	V	T_{ON}	T_{OFF}	I	V
Max. MRR	150	60	6	40	100	60	10	40
Min. SF	50	40	1	60	20	10	1	50
Min. GS	50	40	1	60	20	10	1	50

Table 18. Verification experimental results and calculation of various response factors.

Verification exp. for	Cu-electrode			Graphite-electrode		
	MRR	SF	GS	MRR	SF	GS
Max. MRR	**0.428**	27.956	99	**0.82**	32.441	101
Min. SF	0.005	**8.564**	37	0.014	**8.026**	33
Min. GS	0.005	8.564	**37**	0.014	8.026	**33**

workpiece with copper and graphite tool electrodes using Taguchi approach. The main conclusions of this research are as follows:

(1) Machining performance of the electrical discharge machining process can be improved effectively by using optimum factors as determined within this work.
(2) Pulse-on-time (T_{ON}) has the most significant influence on material removal rate, surface roughness and gap size within the specific test range for copper and graphite electrodes.
(3) Material removal rate, surface roughness and gap size are slightly affected with average machining voltage for both copper and graphite electrodes.
(4) The mean S/N ratio for each level of the cutting parameters reveals that, when using copper electrodes, the effect of cutting parameters can be ranked as follows: pulse-on-time, discharge current, pulse-off-time and average machining voltage.
(5) The effect of cutting parameters can be ranked as follows: pulse-on-time, pulse-off-time, discharge current and average machining voltage, when using graphite electrodes.
(6) The optimal process parameters that have been identified to yield the best combination of process variables are A3B3C3D1 for copper electrode and E3F3G3H1 for graphite electrode (highest level of pulse-on-time, highest level of pulse-off-time, highest level of discharge current and lowest level of average machining voltage).

Conflict of Interests

The author(s) have not declared any conflict of interests.

REFERENCES

Amit J, Pradeep K (2012). Investigating effects of process variables on MRR in EDM by using Taguchi parameter design approach. Int. J. Theor. Appl. Res. Mech. Eng. (IJTARME), 1(2):2319-3182.

Lin JL, Wang KS, Yan BH, Tarng YS (2000). Optimization of the electrical discharge machining process based on the Taguchi method with fuzzy logics. J. Mater. Process. Technol. 102:48-55.

Lorelei G, Laurenţiu S, Ionuţ P (2012). Taguchi modeling of brass electroerosion by single pulse machining: Nonconventional Technologies Review. Romanian Association of Nonconventional Technologies Romania.

Mahapatra SS, Amar P (2006). Optimization of wire electrical discharge machining (WEDM) process parameters using Taguchi method. Int. J. Adv. Manuf. Technol. 34(9-10):911-925.

Marafona J, Wykes C (2000). A new method of optimizing material removal rate using EDM with Copper–Tungsten electrodes. Int. J. Mach. Tools Manuf. 40 (2):153-164.

McGeough JA (1998). Advanced methods of machining. Chapman and Hall, New York.

Mohd AL, Mohd Radzi, HCD, Nurul Amin AKM (2009). The implementation of Taguchi method on EDM process of Tungsten carbide. Eur. J. Sci. Res. 26(4):609-617.

Natsu W, Shimoyamada M, Kunieda M (2006). Study on expansion process of EDM arc plasma. JSME Int. J. 49:600-605.

Nipanikar SR (2012). Parameter optimization of electro discharge machining of AISI D3 steel material by using Taguchi method. J. Eng. Res. Stud. III(III):07-10.

Panda DK (2008). Study of thermal stresses induced surface damage under growing plasma channel in electro-discharge machining. J. Mater. Process. Technol. 202:86-95.

Pandit SM, Mueller TM (1987). Verification of on-line computer control of EDM by data dependent systems. J. Eng. Ind. 109:109-121.

Peace GS (1993). Taguchi Methods: A Hands-on Approach, Addison-Wesley Publishing Co., Inc.

Ranjit KR (1996). A premier on the Taguchi method, Van Nostrand Reinhold, New York.

Ranjit KR (2001). Design of Experiments Using the Taguchi Approach. John Wiley & Sons, New York.

Renjie J, Yonghong L, Yanzhen Z, Baoping C, Hang L, Jianmin M (2010a). Optimizing machining parameters of silicon carbide ceramics with ED milling and mechanical grinding combined process. Int. J. Adv. Manuf. Technol. 51(1-4):195-204. DOI 10.1007/s00170-010-2628-0.

Renjie J, Yonghong L, Yanzhen Z, Baoping C, Rong S (2010b). Electric discharge milling and mechanical grinding compound machining of silicon carbide ceramic. Mach. Sci. Technol. 14(4):502-521. DOI: 10.1080/10910344.2010.533623.

Renjie J, Yonghong L, Yanzhen Z, Fei W, Baoping C, Xin D (2012). Machining performance optimization in end ED milling and mechanical grinding compound process. Mater. Manuf. Proces. 27(2):221-228. DOI: 10.1080/10426914.2011.568569.

Singh J, Walia RS, Satsangi PS, Singh VP (2011). Parametric optimization of hybrid electric discharge machining process with continuous and discontinuous ultrasonic vibrations on workpiece. National Conference on Advancements and Futuristic Trends in Mechanical and Materials Engineering, Punjab Technical University, Jalandhar.

Tzeng YF, Chen FC (2003). A simple approach for robust design of high-speed electrical discharge machining technology. Int. J. Mach. Tools Manuf. 43 (3): 217-227.

Biing HY, Wang CC (1999). The machining characteristics of Al_2O_3/6061 Al composite using rotary electro-discharge machining with a tube electrode. J. Mater. Process. Technol. 95: 222-231.

Yan BH, Wang CC, Liu WD, Huang FY (2000). Machining characteristics of Al_2O_3/6061Al composite using rotary EDM with a disklike electrode. Int. J. Adv. Manuf. Technol. 16: 322-333.

Yan-Cherng L, A-Cheng W, Der-An W, Chih-Cherng C (2009). Machining performance and optimizing machining parameters of Al_2O_3–TiC ceramics using EDM based on the Taguchi method. Mater. Manuf. Process. 24:667-674.

Performance of reactive powder concrete slabs with different curing conditions

Mohammed Mansour Kadhum Alkafaji

Civil Department, College of Engineering, Babylon University/ Babil, Iraq.

In this paper, the material performance of Reactive Powder Concrete (RPC) with two different curing conditions, water-curing of 25°C and heat-treatment of 90°C and 95% relative humidity was experimentally studied. An experimental investigation is carried out to study the influence of using steel fiber and heat-treatment on RPC slab specimens. All the slabs are simply supported along the four edges and loaded concentrically by a square plate of dimensions (70×70×25 mm). The test results showed adequate improvements by increasing adding steel fiber, using of heat-treatment on the compressive strength, indirect tensile strength and modulus of elasticity. It also showed a great positive effect on the flexural strength. The 28-day compressive strength varied between 163.4 and 194.6 MPa for the non-heat-treated and heat-treated specimens. Test results proved that the increase slab thickness leads to significant increase in ultimate load. The ultimate load of RPC slabs that have 20, 40 and 60 mm thickness are 17.5, 13.0 and 12.6% respectively compared to RPC slab specimens non-heat-treated when steel fibers are used at 2% by volume and the percentage increment when thickness decrement. It was found that the deflection at ultimate load, showed a significant decrease when the slab thickness was increased.

Key words: Reactive powder concrete; mechanical properties; curing conditions; steel fibers.

INTRODUCTION

Reactive Powder Concrete (RPC), a cement-based composite material well known for its ultra-high strength, high-durability and low-porosity made its international debut in 1994 (Richard and Cheyrezy, 1994). The advance mechanical and physical properties of RPC are obtained by optimizing packing density of concrete mixture with precise gradation of all mix particles, and by using highly refined silica fume to improve the microstructure of hydrated cement pastes through the pozzolanic reaction. To produce a very high compressive strength of RPC, applications of pressure before and during setting and heat-treating after setting are normally required. Compressive strengths of 200 to 800 MPa, modules of elasticity of 50 to 60 GPa and flexural strength of 6 to 13 MPa have been achieved with RPC (Richard and Cheyrezy, 1995).

On the other hand, maximum compressive strengths for the commonly used normal weight concrete in most concrete structures are in the range of 30 to 60 MPa with the corresponding modules of elasticity of 14 to 35 GPa

Table 1. Chemical composition of the cement.

Chemical composition	**Component**	**CaO**	**SiO_2**	**Al_2O_3**	**Fe_2O_3**	**MgO**	**SO_3**	**Na_2O**
	Content [%]	63.2	20.2	4.1	3.4	2.8	2.6	0.3
Physical properties	Fineness using Blaine (m^2/kg)				3270			
	Soundness using autoclave method				0.19 %			

(Mindess et al., 2003). Due to advantage of the super material properties, RPC has been applied to many real constructions where the substantial weight savings of structure can be appreciated and where its distinguished features of material property can be fully utilized (Adeline and Cheyrezy, 1998; Etienne et al., 2001; Rebentrost and Cavill, 2006). In a typical RPC mix proportioning, in order to enhance the homogeneity and optimize the density of concrete granular mixture, the coarse aggregates of conventional concrete are totally eliminated such that the amounts of fine aggregate, Portland cement and silica fume need to be substantially increased.

Twelve square RPC slabs of dimensions (45×45) inch were tested to investigate the boundary between a flexural failure and a punching shear failure. The variables considered were the slab thickness (2, 2.5, and 3 in) and loading plate dimensions (from (1×1) inch to (3×3) inch. All the slabs are supported on the fully restrained end and loaded concentrically (Harris, 2004).

It was concluded that for small plate tests, a very small loading area was required to force a punching shear failure in the RPC slabs and based on this assessment, a (2 inch) slab thickness should provide sufficient punching shear capacity for bridge applications. The actual slab thickness selected for design should also consider factors such as flexural capacity and deflection criteria. Also, the RPC is more suited for precast operations due to the controlled environment and special formwork required. He also concluded that the casting method has a strong influence on the orientation of the steel fibers, which in turn influences the flexural strength in orthogonal directions in the slabs, The brittle punching shear failure mechanism exhibits limited warning prior to failure, but is still significantly less abrupt than for a reinforced concrete punching shear failure (Harris, 2004).

The effect of subjecting UHPC to a temperature of 90°C during its early ages (for 24 to 48 h) was studied. It was shown that such heat treatment accelerated the pozzolanic reaction of silica fume with $Ca(OH)_2$ and resulted in high strength development of UHPC. The compressive strength of a heat treated UHPC was observed to increase from 220 to 280 MPa over 8 years' storage in water. The chain length of the C-S-H phases increased from values between 5 and 6 to 9 during this period. A distinct increase in compressive strength of UHPC not subjected to heat treatment was also observed at high ages. The compressive strength was 250 MPa at an age of six years which was 58% more than the 28 day strength of 160MPa (Schachinger et al., 2008).

Cwirzen studied the influence of the curing regime on the mechanical properties and microstructure of ultra-high-strength mortar. Nine different curing procedures were applied in which the start and duration of the heat treatment were varied. The studied mortars had a water-binder ratio of (w/b) 0.17 and additions of amorphous silica fume and fine quartz filler. The microstructure and microchemistry were investigated by electron scanning microscopy and mercury intrusion porosimetry. The results revealed that longer heating times increased the degree of hydration, refined the microstructure and resulted in higher ultimate compressive strength. Very late and very early application of the heat treatment caused a lower degree of hydration and a smaller long-term increase of compressive strength (Cwirzen, 2007).

Up to now, all previous studies concise on the flexural behavior of the simply supported slab under normal water curing. While in this paper, the first object of project is focused on the behavior of square RPC reduced scale slab under heat treatment and the second object of this research was to determine the mechanical properties of the produced materials. Additionally, mode of failure of RPC slab specimens was evaluated.

EXPERIMENTAL PLAN

Materials

The RPC considered here is prepared by the following ingredients: ASTM Type I Portland cement produced in Iraq of (TASLUJA) and taken from local markets; natural sand (0-2.0 mm, with a specific gravity of 2.7); a polycarboxylate-based superplasticizer (SP) SikaViscocrete-5930 and brass-coated steel micro-fibers, with a density of 7850 kg/m³, a length of 13 mm, a diameter of 0.18 mm and an l/d ratio of 72. The cross section of the fiber was circular. The chemical composition and physical properties of the cement and silica fume used are presented in Tables 1 and 2 respectively. The composition of RPC with a water to cement ratio (W/C) of 0.18 used in this study is given in Table 3.

Concrete mixing and properties

All the dry constituents of the RPC were batched by an electronic balance. The dry constituents were first mixed for about 4 min at low and high speed in Pan Mixer. Water and superplasticizer were added gradually and re-mixed for about 5 min at high speed until the materials were uniformly mixed. Lastly the fibers were introduced, dispersed uniformly by using a sieve and additional mixing was applied for about 2 min.

The mixing procedures proved satisfactory in that the dispersion of fibers were found to be uniform and there was no evidence of fiber balling. Flow table tests, as per ASTM C230, were undertaken

Table 2. Chemical composition and physical properties of silica fume.

Component	Content [%]
Chemical composition	
SiO_2	90
Al_2O_3	1
Fe_2O_3	1
CaO	0.4
MgO	1
Cl	< 0.3
SO_3	< 2
Loss o n ignition, LOI	3
Physical properties	
Specific gravity	2.2
Percent passing 45 μm (%)	Approximately 90-100
Bulk density [kg/m^3]	640
Form	Amorphous

Table 3. Composition of RPC.

Mix ID	Slab thickness (mm)	Amount kg/m^3					
		Cement	Water	SF	SP	FS	Steel fiber (%)
RPC1-20	20	860	195	215	43	980	1
RPC2-20	20	860	195	215	43	980	2
RPC1-40	40	860	195	215	43	980	1
RPC2-40	40	860	195	215	43	980	2
RPC1-60	60	860	195	215	43	980	1
RPC2-60	60	860	195	215	43	980	2

SF, Silica fume; FS, Fine Sand; SP, Superplasticizers (quantity of SP represented in percentage by weight of cement material), Steel Fibers (quantity is represented in percentage by volume of the total mixture).

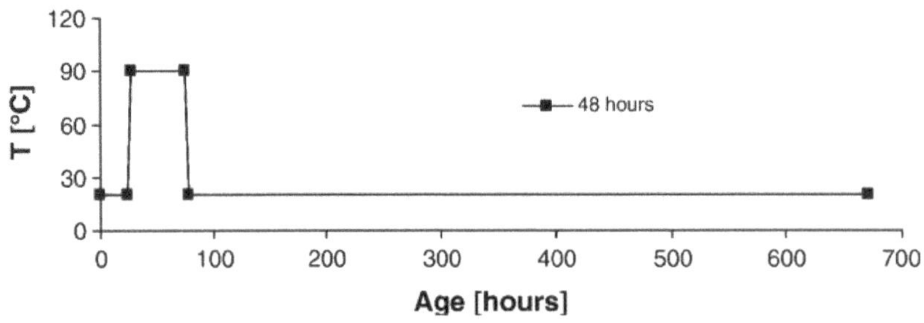

Figure 1. Curing regime adopted.

before casting of the specimens to assure that the fiber reinforced concrete mix had achieved a flow between 160 and 250 mm for the selected mix.

The hardened specimens were demolded after 24 h, and cured by two procedures were applied. The first procedure included demoulding of the specimens 24 h after casting and their subsequent storage at 95% RH and 25°C. The second procedure included a heat treatment at 90°C which was followed by storage at 95% RH. Based on an extensive study whose results are described by Cwirzen (2007), the heat treatment started after 24 h from casting and lasted 48 h. The curing protocol adopted in this study is indicated in Figure 1. Heating rate of treatment was 12°C /h. This extended (48 h) high temperature (90°C) heat treatment which is different from conventional curing process was preferred due to the high amount of reactive cementitious materials in RPC. Studies showed that high mechanical properties can be achieved under

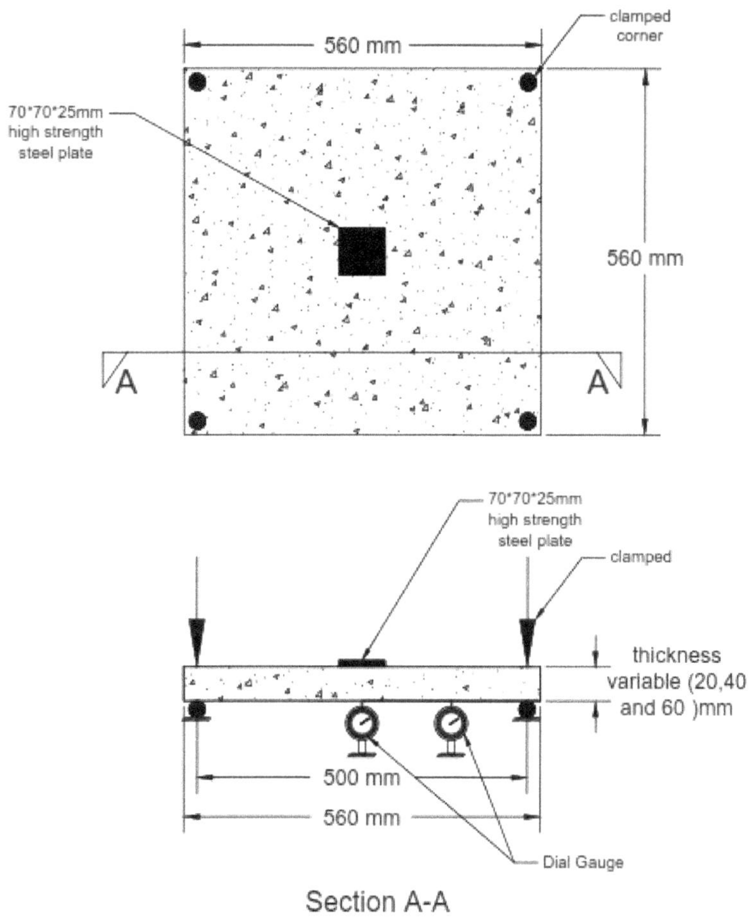

Figure 2. Slab test schematic.

these conditions at early ages. Heat treatment densified the microstructure of the RPC matrix. The specimens, which were subjected to heat treatment, were kept in laboratory conditions for cooling before testing in this study.

Test specimens

In the present work, twelve RPC square slabs with full side length of 560 mm, and center to center span between supports of 500 mm with a different thickness of 20, 40 and 60 mm. The slab was simply supported along its four edges and loaded by a central steel plate of dimension 70×70×25 mm. Three experimental variables are investigated in this study to show their effects on the behavior of the reinforced reactive powder concrete slab specimens. These variables are percentage of steel fibers, types of curing and thickness of the RPC slab.

Concrete tests

The compressive strength test was carried out according to B.S: 1881: part 116 (B.S. 1881:1983) using a digital testing machine with a capacity of (2000 kN). For each measurement, a set of three specimens 5×5×5 cm^3 was used. The tests were performed at the following ages: 1, 7 and 28 days. The loading was applied at a rate of 0.9 kN s^{-1}. Splitting tensile strength has been determined by testing six standard cylinders of 100×200 mm for every mix depending on ASTM C496-04 specification (ASTM C 496/C 496M, 2005). The cylinders were cast, demoulded and cured in a similar way as for slab specimens. Prismatic specimens (50×50×300 mm) were used to determine the flexural behaviors in conformity with ASTM C78-02 (ASTM C78-02, 2002). Each prism was simply supported and subjected to a two point loading using an electrical testing machine with a capacity of (2000 kN). The tests were performed at the age of 28 days. Elastic modulus test was carried out on 100×200 mm cylindrical specimens. The 40% of ultimate compressive strength of concrete specimen was applied on the concrete cylinders to perform the elastic modulus test as specified by ASTM C469 (ASTM C469-02, 2002). In each taken slab, each data presented here are the average of test results of the three specimens.

Testing procedure

Before the testing day, the slab was taken out from the curing container, cleaned and painted with white paint on both surfaces, to achieve clear visibility of cracks during testing. The slab was labeled and carefully placed along the edges on simple supports. The point load was applied at the center of the slab and the dial gauges were positioned in their places, so that a precise setup of the testing equipment was achieved. Figure 2 shows the testing of slab specimens mentioned above. The slab specimens were tested

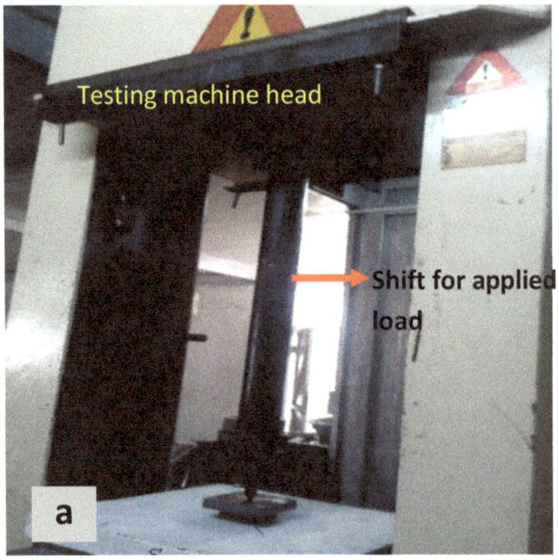

Photograph showing the testing machine.

Figure 3. Photograph showing the details of slabs testing setup.

used a hydraulic testing machine with a capacity of (20000 kN) available in the structural Laboratory in Civil Engineering Department; College of Engineering is illustrated in Figure 3a.

The load was applied gradually in increments of 4 kN until failure occurs. This amount of incremental loading allows sufficient number of loads and corresponding value of deflections to be taken during the test which gave a good picture for the structural behavior of the slab specimens. Tow dial gauges having accuracy of 0.001 mm per deviation and capacity of 10 mm and a maximum sensor length of 50 mm were used to measure the mid-span deflection and 125 mm from mid-span of slab specimen see Figure 3b and c).

It is worth pointing out here that the 560 mm length of the square slabs of the present investigation was larger than the maximum space of 450 mm offered by the MFL system testing machine. Therefore a special supporting frame was manufactured and used inside the testing machine, as shown in Figure 4, to provide the required span of the slab. This supporting frame was made using four steel beams of the type W-shape (W4×13) welded and arranged to form a square shape. Each of which had a 25 mm steel bar welded on its top face to provide a simple support for the slab edge.

RESULTS AND DISCUSSION

Engineering properties of RPC

The engineering properties of RPC under two different curing conditions are shown in Table 4. The highest measured 28-day compressive strength values for the mixes of the heat-treated and non-heat-treated specimens are shown in Figure 5. The highest compressive strength value of 194.6 MPa was obtained in case of the heat-treatment incorporating a volume fraction of 2% steel fibers. On the other hand, the 28-day

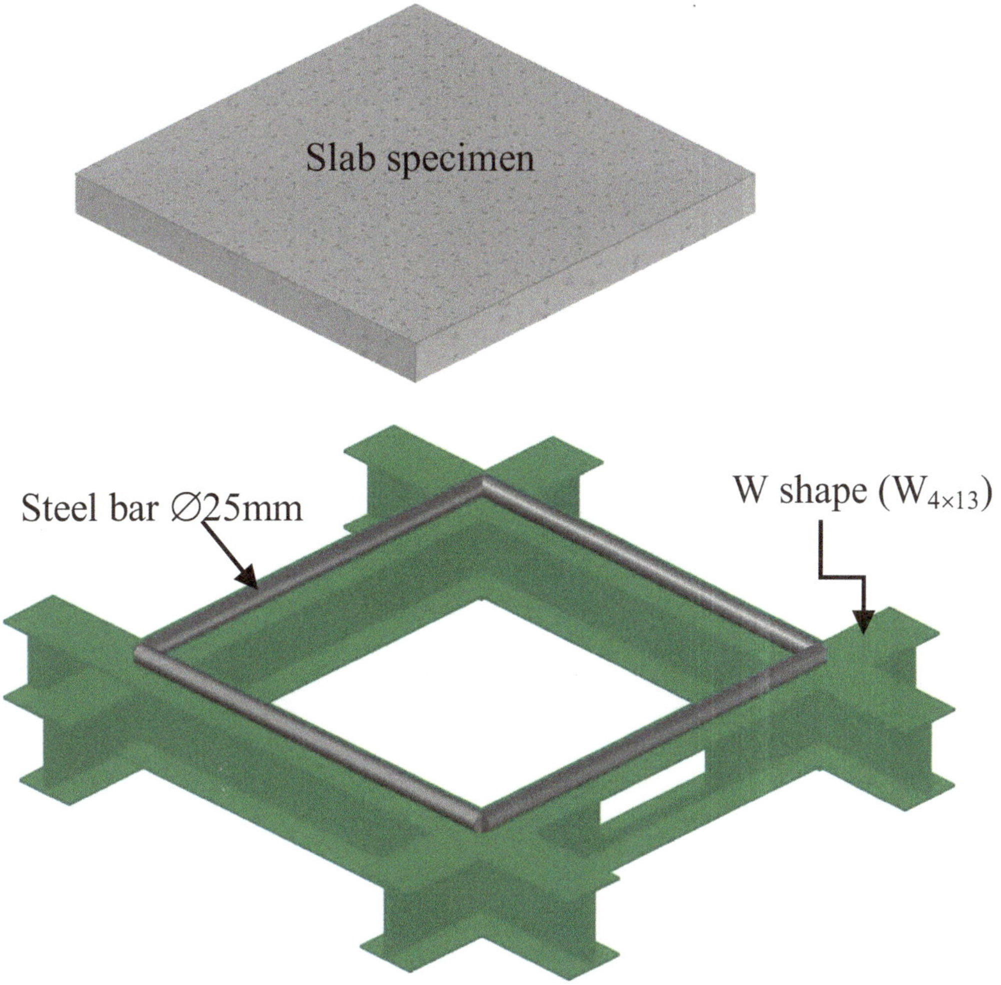

Figure 4. Supporting frame of slab specimens.

Table 4. Engineering properties of RPC under water curing of 25°C and heat treated.

Mix	Compressive strength [MPa]			Flexural 28 day [MPa]	Splitting tensile 28 day [MPa]	Modulus of elasticity 28 day [GPa]
	1 day	7 day	28 day			
RPC1/N	119.2	139.0	163.4	24.0	13.2	42.01
RPC2/N	128.6	152.3	173.0	28.7	17.0	45.15
RPC1/H	143.0	166.8	181.5	30.8	15.4	44.56
RPC2/H	154.5	178.0	194.6	36.0	19.7	48.30

compressive strength value of the non-heat-specimens reached 173 MPa and was measured in mix RPC2/N. Because the pozzolanic reaction resulting from the ingredients of silica fume, in the RPC mixture will be activated energetically by the high temperature and moisture of curing. This pozzolanic reaction causes adenser microstructure of C-S-H cement hydrate and results in a faster development of strength gain.

The splliting tensile strength of RPC for all mixes is much lower than the compressive strength, because of the ease with cracks which propagate under tensile loading, and is usually not considered in design. However, it is an important property, since cracking in concrete is most generally due to the tensile stresses that occur under load, or due to environmental changes. Referring to Table 4, the splitting tensile strength of RPC cylinders can be increased up to (27.9 and 28.8%) as the micro fine steel fibers volume ratio increases from 1 to

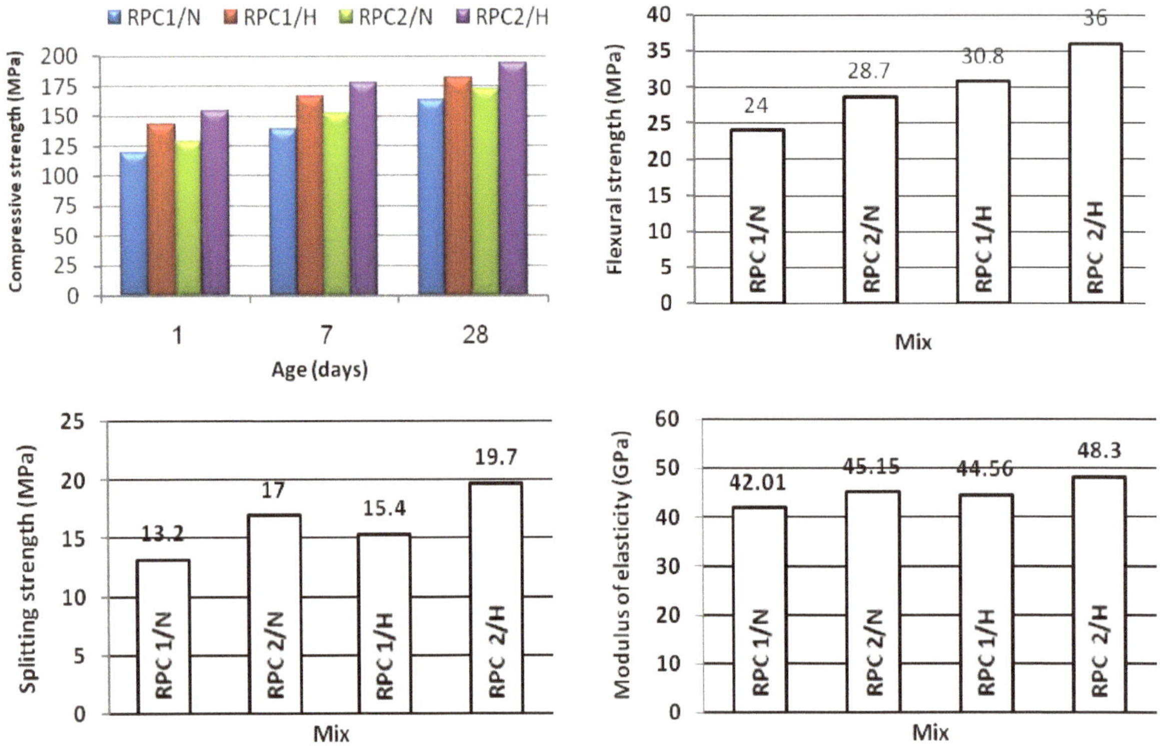

Figure 5. Engineering properties results of the studied RPC.

2% for heat-treated and non-heat-treated RPC mixes respectively. This is due to the fact that fibers bridge tensile cracks and retard their propagation (Graybeal and Hartman, 2002). The steel fibers improve the characteristics of cement-based matrices in the hardened state; they are able to bridge cracks, transmit stress across a crack and counteract crack growth (Mahdi, 2009). As shown in Figure 5, the presence of micro steel fibers influenced the flexural strength of the studied mixes. The results showed an increase of the flexural strength from 24 to 28 MPa after the addition of 2% of steel fibers and up to 36 MPa in the case of the heat-treated mixes.

The splitting tensile strength is related to the failure mechanism in tension due to the concentration of stresses at the tip of an elliptical crack. According to this theory the tensile strength will increase with shorter cracks, higher total modulus of elasticity and lower porosity (Penttala et al., 1995). The modulus itself is influenced by three concrete phases; binder matrix, aggregates and ITZ. Consequently, the wide and porous ITZ or weak and cracked aggregates will lower the tensile strength. Furthermore, the larger maximum aggregate size the greater their ultimate influence on the mechanical properties. Thus, it can be concluded that the main function of the short steel fibers is to stop the opening and the propagation of the microcracks. A negligible increase of the flexural strength of the RPC after the addition of steel fibers can be explained by the unfavorable combination of at least two effects. The first effect is due to de-bonding of the binder matrix from the aggregate surface in tension. The second is due to described possible increase of the internal stresses caused by the presence of micro cracked aggregates. In order to overcome these effects significantly, higher amount of the steel fibers would have to be incorporated into the binder matrix.

With respect to these curves, it can be noticed that the highest value of the modulus of elasticity of 48.3 GPa was measured in the RPC2/H mix. The lowest value of the modulus of elasticity which equaled 42.01 GPa was measured in case of the RPC1/N mix. The heat treatment did not have significant effect on the measured values.

Typical failure modes of RPC specimens from the compressive and splitting tensile tests are shown in Figure 6a and b respectively. It is noted that, unlike the normal weight concrete specimens with the failure modes of either in crushed state or two separated pieces, these failed RPC specimens are still kept together by the steel fibers.

First crack load, ultimate load and load versus deflection for slab specimens

In this study, twelve slabs were tested. The test results of these slabs included the first crack load, peak load, load-deflection curves and crack pattern as well as modes of

Figure 6. Typical failure modes of RPC specimen after testing: (a) Compressive test; (b) Splitting tensile test.

Table 5. Test results of the first crack load, ultimate load and mid-span deflection for RPC slab specimens.

Slab Identification	First crack load (kN)	Peak load (kN)	Mid-span deflection at first crack (mm)	Mid-span deflection at peak load (mm)	Mode of failure
RPC1-20/N	42.2	79.0	2.6	9.1	Flexural
RPC1-40/N	45.8	94.6	2.1	6.9	Flexural
RPC1-60/N	47.7	109.2	1.5	4.2	Flexural
RPC2-20/N	43.0	89.6	4.4	13.6	Flexural
RPC2-40/N	76.2	103.5	3.8	11.4	Flexural
RPC2-60/N	85.3	114.0	2.5	8.9	Flexural
RPC1-20/H	49.8	93.2	3.0	10.2	Flexural
RPC1-40/H	53.0	101.0	2.5	7.8	Flexural
RPC1-60/H	54.8	117.3	1.8	5.3	Flexural
RPC2-20/H	50.1	105.3	5.1	14.6	Flexural
RPC2-40/H	86.0	117.0	4.6	12.1	Flexural
RPC2-60/H	97.2	128.4	2.9	10.3	Flexural

failure data are presented in Table 5. In all slabs, flexure failure occurred and the cracks initially formed near the center of the slab on the tension face and radiated to the edges. As loading continued, the cracks along the diagonals on the tension face widened and extended with additional cracks forming at the central region of the tension face. At higher loads, further flexural cracks were formed progressively and widened as the loading increases. As the flexural cracks widened, the fiber pullout reached its final stages and the slab failed. The test revealed that the increasing micro steel fibers content usually leads to an increase in the peak load and ultimate deflection with delays the propagation of cracks and controls their growth in the slab as shown in Figure 7. This can be attributed to the fact that the efficiency of steel fibers in arresting the propagation and controlling the growth of cracks within the slab, and hence, steel fibers maintain the slab integrity throughout the post cracking stages of behavior. The slab, hence could withstand greater loads and deflection before failure. It is evident from Table 5 that the deflection at ultimate load is considerably increased by the presence of steel fibers and using heat treated.

Figure 8 shows clearly that the fiber volume fraction has a significant effect on the ultimate load carrying capacity of slab specimens. Heat-treatment curing seems very effective ways to increase the peak load of RPC slab specimens. This can be attributed to the improvement of hydration process under these curing regimes. The RPC slabs under heat-treatment at volume fraction (1 and 2%) results in an increase in the mid-span deflection at peak load of 26.2 and 15.7%, respectively in compared to that of non-heat slabs.

The load versus deflection relationship of slab specimens is presented in Figures 9 to 11. Deflections of these slab specimens were measured by using two dial gages, one located at the center of the slab and the other at 125 mm away from slab center. In all RPC slab specimens, the deflections were recorded immediately after the application of the load. At peak load, the rate of increase in deflection was so fast that no reliable deflection value could be measured. Accordingly, all load-

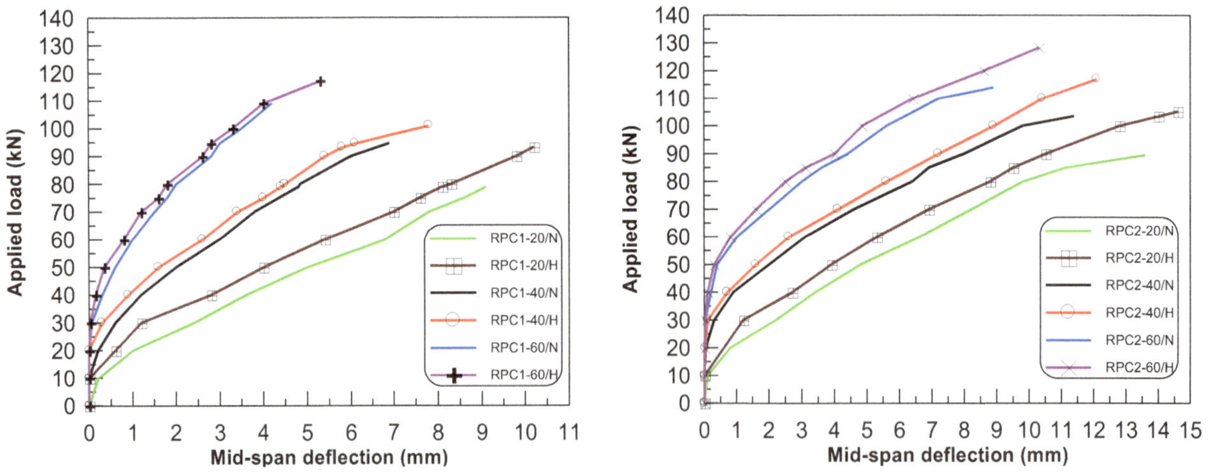

Figure 7. Load - mid-span deflection curves of RPC slab specimens.

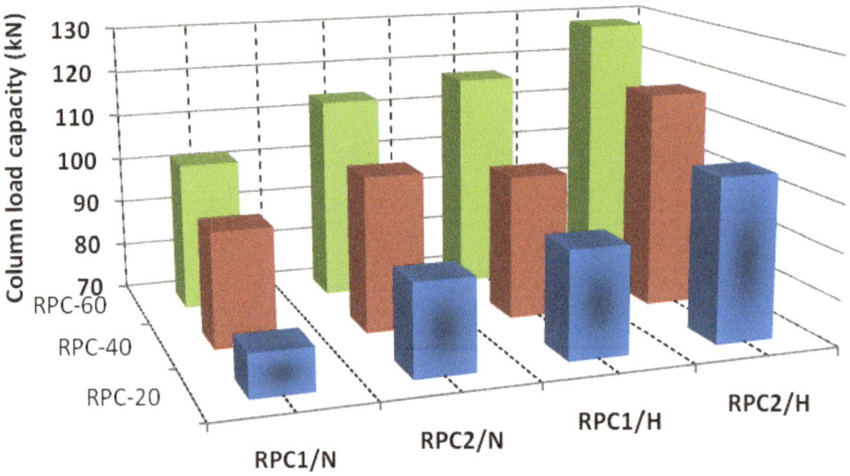

Figure 8. Effect of steel fiber percentage and curing conditions on the ultimate load for slab specimens.

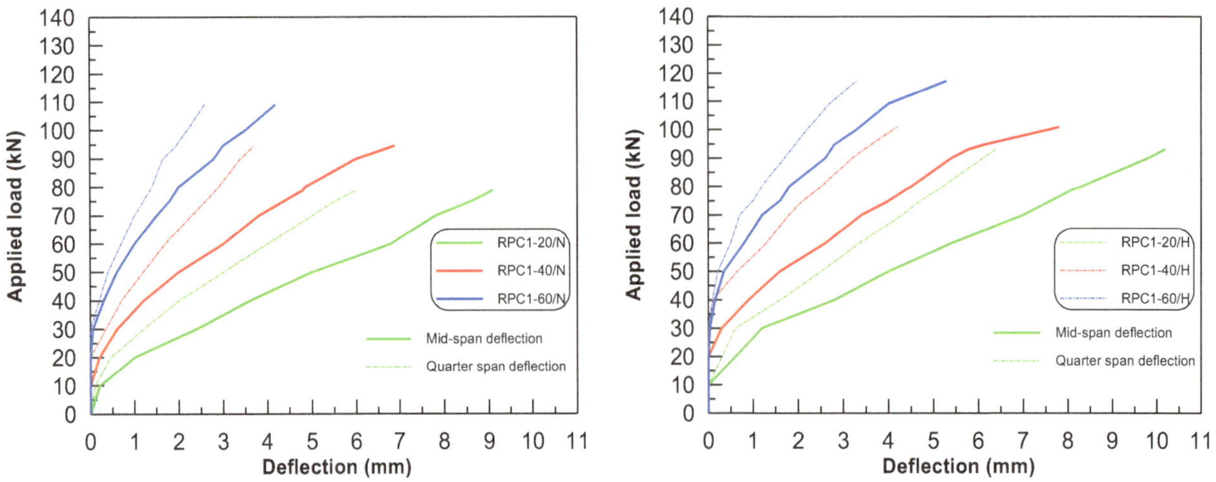

Figure 9. Load - deflection curves of RPC slab specimens with 1% steel fibers.

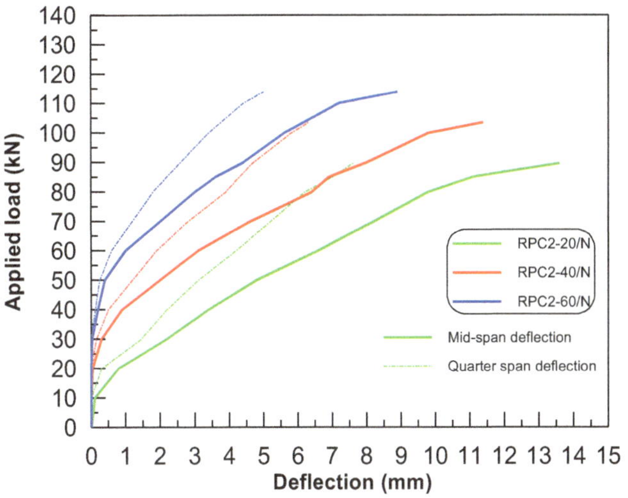

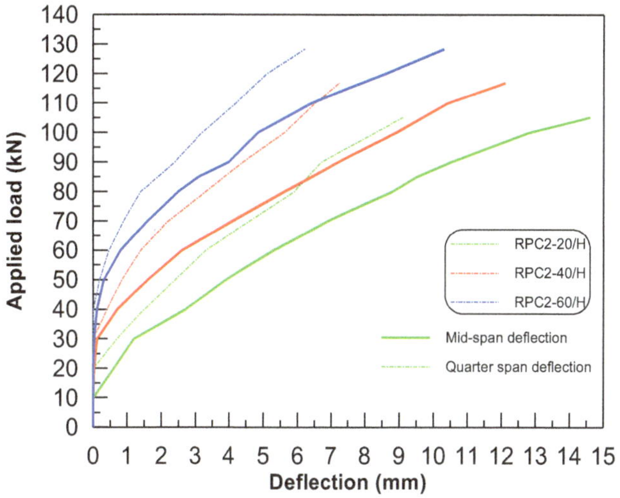

Figure 10. Load - deflection curves of RPC slab specimens with 2% steel fibers.

deflection curves were terminated at their peak load value. In addition, it can be concluded that the increase in the percentage of fibers with using heat-treatment increases the load carrying capacity and increases the deflection in slab specimens. This can be attributed to the fact that the efficiency of steel fibers in arresting the propagation and controlling the growth of cracks within the slab, and hence, the steel fibers maintain the slab integrity throughout the post cracking stages of behavior. The slab, hence could withstand greater loads and deflection before failure. For all slabs that fibers tend to align in the direction of the flow of the material and also with the formwork. For RPC mixes containing 1% micro steel fibers, it is obvious from the results that the percentage of peak load increases when the thickness of slab specimens increases from 20 to 60 mm, the peak load increased to 38.2 and 25.9% for non-heat specimens and heat-treated respectively and mixes containing 1% micro steel fibers. This result ensures that the slab thickness has a significant effect on shear strength of RPC slabs under heat or non-heat-treated. Also for the same mixes, it can be noticed that in increasing the slab thickness, the deflection at peak load shows significant decreases from 9.1 mm (RPC1-20/N) to 4.2 mm (RPC1-60/N) and 10.2 mm (RPC1-20/H) to 5.3 mm (RPC1-60/H) for non-heat treated and heat-treated respectively.

It is clear from the above Table that the thin slab with 20 mm thickness seems clearer in ductility behavior in comparison to the slab specimens. From the test results, it is clear that the role of slab thickness on the behavior of slab specimens, the values of mid-span deflection of slabs increases with decreasing of slab thickness. This indicates that the use of RPC allows for smaller, thinner, lighter sections to be designed while strength is still maintained or even improved and taking advantage by minimizing material usage and cost in addition to reducing the weight of slabs. The appearance of the RPC slab specimens after failure is shown in Figure 12.

Conclusions

The following observations and conclusions can be made on the basis of the current experimental results.

1. Test results showed that RPC has very high mechanical performance. Heat-treatment method seems very effective way to increase the compressive strength of RPC. This can be attributed to the improvement of hydration process under these curing regimes. Compressive strength of RPC with water-cementitious material ratio of 0.18 exceeded 190 MPa after heat-treatment curing in this study.
2. The inclusion of steel fibers leads to a considerable increase in splitting tensile strength and flexural tensile strength while the increase in compressive strength and modulus of elasticity is relatively lower. Therefore, it is recommended to use steel fibers as an enhancing material to RPC.
3. It was observed that fibers tend to align in the direction of the flow of the material and also with the formwork.
4. In this study, it was observed that the amount of ultra fine steel fibers in RPC slabs has an influence on the value of the peak load and deflection.
5. The experimental results clearly indicate that the presence of ultra fine steel fibers in RPC slabs resulted in an enhanced stiffness, reduced crack width, reduced rate of crack propagation.
6. In thin RPC slabs, the presence of steel fibers provides sufficient tensile strength and ductility even if conventional reinforcement is avoided. This allows the designer to reduce the amount of reinforcement needed to resist tension for many structural applications with

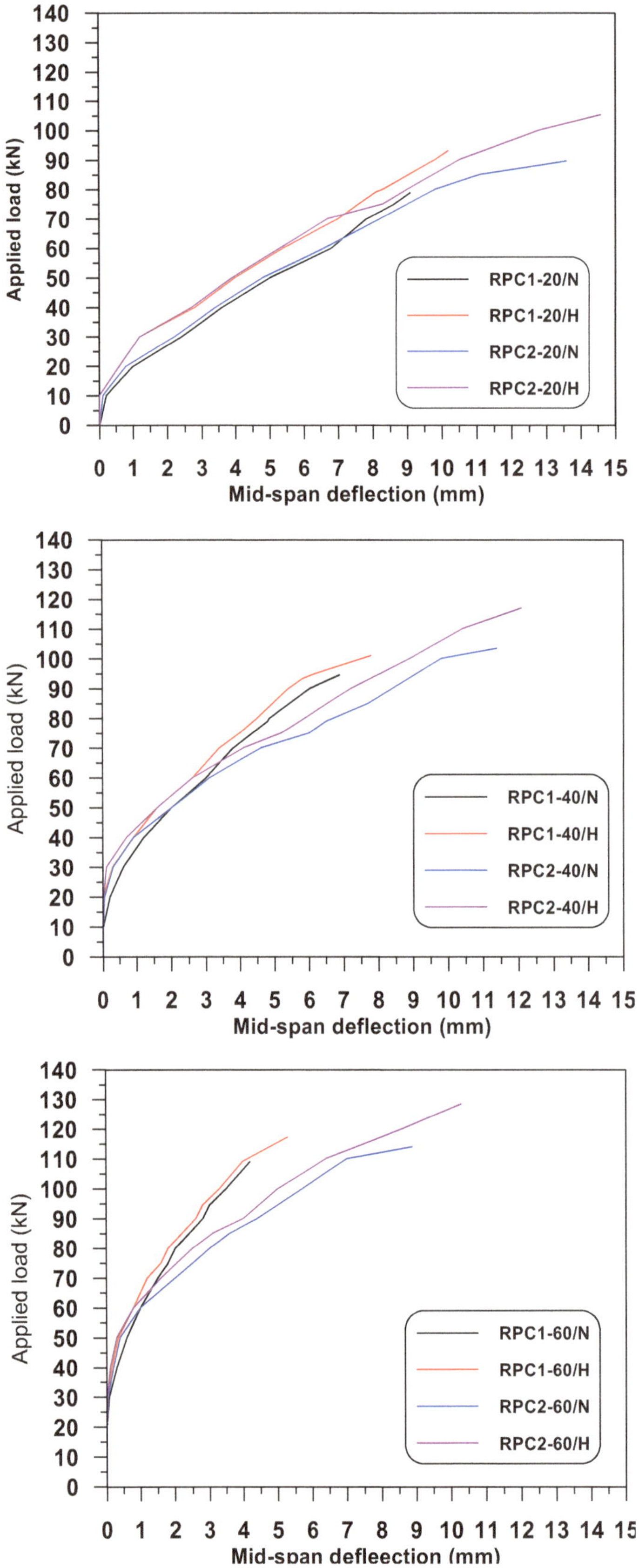

Figure 11. Effect of curing condition on load - mid-span deflection curves of RPC slab.

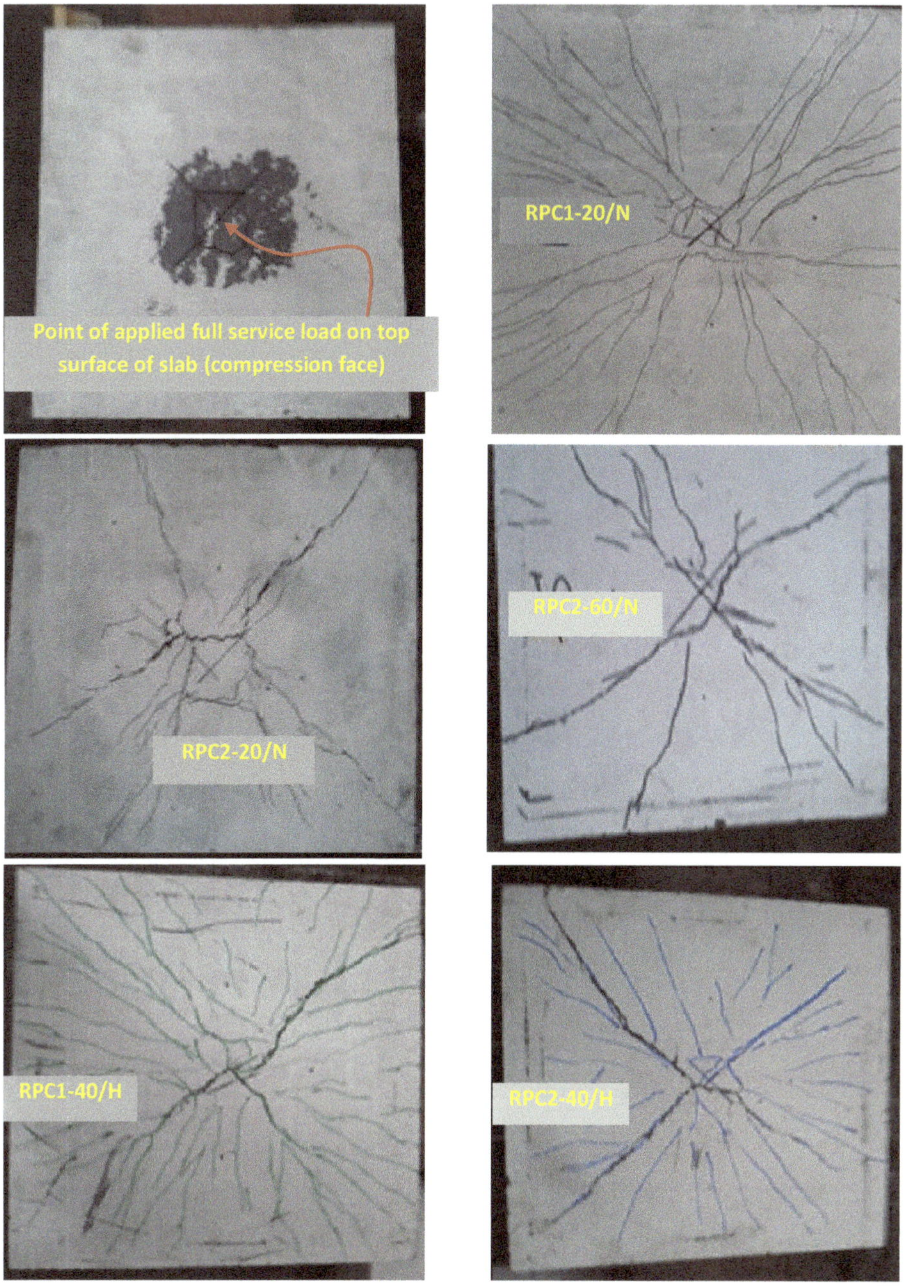

Figure 12. Typical crack patterns of slab specimens after failure at bottom view of failure (tension face).

beneficial effects on the cost and self weight and make it possible to create thinner sections, longer spans, and taller structures.

7. All the slabs behave similar in the aspect of failure patterns under load regardless of the variables. The using of heat-treatment method increases the first crack and ultimate loads of slab specimens.

8. Due to presence of fibres ductile failure is observed. In fact crack get arrested by fibres does not get fractured but pulling out of fibres from matrix is observed.

Conflict of Interest

The author have not declared any conflict of interest.

REFERENCES

Adeline R, Cheyrezy M (1998). The Sherbrooke Foot- bridge: the First RPC Structure. FIP 98, Amsterdam, May.

ASTM C 496/C 496M (2005). Standard Test Method for Splitting Tensile Strength of Cylindrical Concrete Specimens. Vol. 04.02.

ASTM C78-02 (2002). Standard Test Method for Flexural Strength of Hydraulic-Cement Mortars. Vol. 04(02):6.

ASTM C469-02 (2002). Standard Test Method for Static Modulus of Elasticity and Poisson's Ratio of Concrete in Compression.

BS 1881-116 (1983). Method of Determination of Compressive Strength of Concrete Cubes. British Standards Institution, P. 3.

Cwirzen A (2007). The effect of the heat-treatment regime on the properties of reactive powder concrete. Adv. Cement Res.19(1):25-33.

Etienne D, Causse M, Behloul M (2001). Design and Building of Seoul Peace Footbridge. Third International Arch Bridge Conference, September. Paris, pp. 865-876.

Graybeal B, Hartman J (2002). Ultra high performance concrete material properties. Transportation Research Board Conference, USA, November 15:1-14.

Harris DK (2004). Characterization of Punching Shear Capacity of Thin UHPC Plates. Faculty of the Virginia Polytechnic Institute and State University.

Mahdi BS (2009). Properties of self compacted reactive powder concrete exposed to saline solution. Ph.D. Thesis, Building and Construction Engineering Department, University of Technology, Baghdad, P. 223.

Mindess S, Young JF, Darwin D (2003). Concrete. 2nd Ed., Prentice Hall.

Penttala V, Hannonen P, Jarvinen M Komonen J (1995). Effects of aggregates and microfillers on the flexural properties of concrete. Report 5, Helsinki University of Technology, Espoo, P. 113.

Richard P, Cheyrezy M (1994). Reactive Powder Concretes With High Ductility and 200–800 MPa Compressive Strength. ACI Spring Convention, San Francisco, April.

Richard P, Cheyrezy M (1995). Composition of reactive powder concretes. Cement Concrete Res. 25(7):1501–1511.

Rebentrost M, Cavill B (2006). Reactive Powder Concrete Bridges. 6th Austroads Bridge Conferences, 12–15 September, Perth, Australia.

Schachinger I, Hilbig H, Stengel T (2008). Effect of Curing Temperature at an Early Age on Long Term Strength Development of UHPC. 2nd International Symposium on Ultra High Performance Concrete, Kassel, 05-07 March. pp. 205-212.

Electrochemical properties of metals in cassava fluid

O. R. Adetunji[1]*, P. O. Aiyedun[1], O. J. Alamu[2] and A. S. Surakat[1]

[1]Mechanical Engineering Department, College of Engineering, University of Agriculture, Abeokuta P. M. B. 2240, Abeokuta, Nigeria.
[2]Mechanical Engineering Department, Osun State University, Oshogbo, Osun State, Nigeria.

This study was carried out to investigate and evaluate the electrochemical properties of mild steel (0.6, 1.0 and 1.46 mm), galvanized steel (0.36, 0.6 and 1.24 mm) and aluminium (0.9 and 1.3 mm) sheets exposed to cassava fluid. Corrosion rates of mild steel (10.595 mm/yr), galvanized steel (11.26 mm/yr) and aluminium (9.556 mm/yr) sheets exposed to cassava fluid were got using weight loss technique. The weight loss of each sample for specific periods of immersion (120,168,216, 264 and 312 h) was determined on average of two samples exposed under same cassava fluid in different plastic containers. It was found that aluminium samples had highest corrosion resistance followed by mild steel then galvanized steel. Also corrosion rates for all metallic samples studied decreased over the immersion durations.

Key words: Electrochemical, properties, metals, cassava, fluid.

INTRODUCTION

The wastage of metal due to corrosion has become an important engineering problem. Probably, no other source of waste except that affecting human life is of greater concern to all. It has well been said that only through the elimination of waste caused by corrosion and the increase in our national efficiency we can hope to lower the cost of living on one hand, and raise our standard of living on the other. The elimination of waste is a total asset. It has no liabilities (Fontana, 1986).

Corrosion is a chemical or electrochemical reaction between a metal and its environment which involves removal of the metal or its conversion to an oxide or other compound (Walker, 1980). Corrosion occurs in different forms in metallic structure such as in the construction industries, industrial process equipment and allied industries (Abiola and Oforkar, 2002). Corrosion was also represented in global perspective as a tremendous economic loss and much as to be done to combat it (Adetunji and Adewoye, (2010).

Cassava, *Manihot esculenta* Crantz, is a perennial woody shrub with an edible root, which grows in tropical and sub- tropical areas of the world (Burrelli, 2003). Cassava is one of the most important food crops in the tropics. Cassava is a tropical root crop that serves as a food security and income generation crop for many millions of people in the developing world (Scott et al., 2002). Cassava is grown widely in Nigeria and in many regions of the tropics, where it serves as one of the basic food source for about 200 to 300 million people. In 1999, Nigeria produced 33 million tonnes making it the world's largest producer (Sobowale et al., 2007). During fermentation of grated cassava tuber (fufu - local food popular among South Western Nigerians made from cassava tuber), lactic acid bacteria, yeast and other bacteria contribute significantly to starch breakdown, acidification, detoxification and flavour development (Oyewole, 1991). Lactic acid bacteria are found to be useful in flavouring foods, in inhibiting spoilage bacteria and pathogens, in intestinal health and other health benefits related to blood cholesterol levels, immune competence and antibiotics production.

Wider utilization of cassava products can be a catalyst for rural industrial development and raise the incomes of producers, processors and traders. It can also contribute to the food security status of its producing and consuming households (Fontana, 1986). Besides starch, the cassava tuber contains some soluble carbohydrates, that is glucose and sugar, which can be inverted poisonous prussic acid. The nutritive value for cassava tuber is estimated at 127 calories per 100 g to contain protein

*Corresponding author. E-mail: adetunjiolayide@yahoo.co.uk.

(0.8 to 1.0%), fat (0.2 to 0.5%), carbohydrate (32%), ash (0.3 to 0.5%), moisture (65%) and fibre (0.8%). Cassava is considered to reflect different level of cyanogenic glucosides, the precursor of the highly toxic cyanic acid. Common synonyms are hydrocyanic acid and prussic acid.

Mild steel which accounts for a great deal of metallic material in the construction industries, industrial process equipment and allied industries is susceptible to various forms of corrosion such as pitting corrosion, uniform corrosion, intergranular corrosion, stress corrosion cracking and fatigue corrosion (Afolabi, 2005).

Sacrificial protection involves attaching blocks of zinc to the iron or steel. Sacrificial protection is sometimes used on ships' hulls, below waterline. As zinc is above iron in the reactivity series, it will corrode in preference to iron – the zinc blocks are sacrificed to protect the iron, hence the name "sacrificial protection".

Aluminium is one of the most abundant metals in the earth's crust and its supply is limited by the economics of mining, extracting and processing. Because of their light weights and mechanical strength, aluminium and its alloys are very active material for engineering applications. Since the 1950's, galvanized steel has remained the principal material of construction for factory assembled cooling towers. This fact attests to the cost-effectiveness of galvanized steel, and when properly maintained this material can offer 20 years or more life expectancy in cooling applications.

A galvanized steel sheet includes a galvanized coating layer having a coating weight of about 20 to 60 g/m^2 formed on at least one surface of the steel sheet, and a zinc phosphate coating layer having a coating weight of about 0.5 to 3.0 g/m^2 formed on the galvanized coating layer.

This steel sheet exhibits superior perforative corrosion resistance. The steel sheet also exhibits superior press workability by further controlling the magnesium, nickel, and manganese contents. Zinc is a highly reactive metal. It exhibits a low corrosion rate only if a continuous passive film forms on the surface. A key requirement of corrosion control with zinc is that the surface needs to remain largely dry and in contact with the air in order to develop and maintain this passive film. Storage stain (white rust) is simply the chemical compound, zinc hydroxide/carbonate zinc/oxide zinc, which forms when zinc is kept in contact with moisture during storage or transportation. Zinc corrosion products are typically white, but under certain conditions may also take the form of a grey or black deposit on the metal surface. Accelerated corrosion of galvanized steel pole, white rust and storage stain (tiger striping) can occur when galvanized surfaces are held for extended periods in wet conditions immediately after the hot dip process. The surfaces may become wet either by rain fall or by condensation. The corrosion products form after zinc reacts with moisture. Corrosive compounds such as chlorides from marine and sulphur containing atmospheres accelerate the formation of white rust. The corrosion of galvanized steel cooling towers may be referred to as white rust and the consequence of white rust can be premature failure of galvanized steel components (Amuda et al., 2006).

The only impending factor to achieving the laudable objective of fabrication and installation of machineries for processing of cassava and its product by agro-processing equipment fabrication factories is the frequency at which the machineries break down due to the corrosion of the parts resulting from its contact with cassava juice during processing. It is important to study and identify the corrosion rate of mild steel, galvanized steel and aluminium when used as machine parts of agro-processing equipment fabrication factories with reference to time variation and pH values of contact fluid. Also, providing possible guide lines to selection of material for fabrication of processing machineries to minimize cost of production, maintenance (replacement) and, durability of machine.

Selection of appropriate material for the fabrication of cassava processing machineries was the major objective of this study. The specific objectives were as follows:

(1) Comparison of the corrosion rates of mild steel; galvanized steel and aluminium sheets in cassava fluid.
(2) Evaluation of corrosion performance of mild steel; galvanized steel and aluminium sheets in cassava fluid at different time of immersion.
(3) Effect of pH of cassava fluid on the corrosion performance of mild steel, galvanized steel and aluminium sheets.

This study was carried out with limitations to three test specimens or samples that is, mild steel, galvanized steel and aluminium.

MATERIALS AND METHODS

The materials used for this study were mild steel sheet samples (0.6, 1.0 and 1.46 mm), aluminium sheet samples (0.9 and 1.3 mm), galvanized steel sheet samples (0.36, 0.6 and 1.24 mm), undiluted cassava fluid, plastic containers, dilute sulphuric acid (H_2SO_4), distilled water, alcohol (acetone), serviette papers, paper tape, plastic bowls; apparatus- sensitive weighing balance, pH meter, quant meter, fissions instrument (applied research laboratory 2460), bench shearing machine.

The weights of the samples were recorded using the sensitive weighing balance. Each weighed sample was separately and fully immersed into plastic container filled with cassava fluid (test medium). The fluid was sourced by grinding pilled cassava tuber and squeezing out the fluid using cloth sieve. The containers were first washed with detergent, rinsed in distilled water, and cleaned and dried with towel and tissue paper.

The experimental procedure adopted for this study was total immersion test method because of good reproducibility. Weighed test samples were separately and fully immersed in the test medium of each of the seventy eight plastic containers and were studied for periods of 120, 168, 216, 264 and 312 h at ambient temperature. After the lapse of each period, two of test specimens designated to each of the periods were washed with distilled water, rinsed with acetone, dried with the serviette paper, re-weighed with

Table 1. Mild steel chemical composition in %.

MS sample (mm)	C	Si	S	P	Mn	Ni	Cr	Mo	Cu	Fe
0.6	0.07112	0.10080	0.0358	0.0168	0.35353	0.04995	0.10808	0.04773	0.11297	99.1033
1.0	0.07364	0.09610	0.0228	0.0055	0.34188	0.03923	0.12260	0.04771	0.08184	99.1687
1.46	0.04855	0.08184	0.0328	0.0145	0.30990	0.03700	0.11043	0.04618	0.06563	99.2531

Table 2. Aluminium chemical composition in %.

Al sample (mm)	Fe	Si	Mn	Cu	Zn	Ti	Mg	Pb	Sn	Al
0.9	0.493	0.194	0.005	0.004	0.003	0.015	0.000	0.001	0.010	99.77
1.3	0.454	0.183	0.032	0.035	0.052	0.011	0.002	0.006	0.003	99.17

Table 3. Corrosion rates for mild steel (0.6, 1.0 and 1.46 mm).

Sample	Mild steel (0.6 mm)			Mild steel (1.0 mm)			Mild steel (1.46 mm)		
Time (h)	CR (mm/yr)	pH	W/A (kg/m^2)	CR (mm/yr)	pH	W/A (kg/m^2)	CR (mm/yr)	pH	W/A (kg/m^2)
120	4.3	4.0	45.9	6.2	4.2	67.4	10.6	4.0	114.2
168	3.1	4.0	46.4	5.3	3.9	80.4	7.8	3.9	117.8
216	2.4	3.9	46.8	4.2	3.8	82.4	6.2	4.3	119.9
264	2.0	3.9	47.6	3.4	4.1	81.0	5.1	4.5	121.1
312	1.7	4.1	48.5	2.9	4.6	82.1	4.2	4.0	118.4

the sensitive weighing balance and the pH values of the test medium were also measured with a pH meter. The weight recorded was the average of two reading. The weights of the specimens after soaking and the pH readings were recorded.

Calculation

The weight loss of each designated sample at respective immersion period was determined and the corresponding corrosion rate in millimetre per year (mm/yr) was calculated according to the relationship.

The corrosion rate (mm/yr) =KW / DAT

Where: K = 87.6 which is universal constant based on dimension,
W = weight loss (mg),
D = density of test specimen (g/cm^3),
A = area of test specimen (cm^2),
T = time of immersion (h).

RESULTS

Comparative study of galvanized steel, aluminium, and mild steel samples corrosion rates in cassava fluid

The corrosion rates of mild steel (0.6 mm) ranged from 4.263 to 1.729 mm/yr for immersion time of 120 to 312 h while that of galvanized steel ranged from 4.905 to 2.234 mm/yr for the same period. Corrosion rates of 1.0 mm aluminum sample ranged from 6.582 to 2.506 mm/yr.

The weight losses per specimen surface area of the mild steel samples 0.6 mm, 1.0 mm and 1.46 mm thickness as a function of the time of immersion had an average trend (Tables 1 to 3).

This can be related to formation of lactic and formic acid during the fermentation of cassava fluid. This increases the acidity of the cassava fluid and in turn causes the mild steel in contact with it to corrode. The pH of the cassava fluid varied over the times of immersion that implies that the corrosion products of the mild steel had impacts on the pH of the cassava fluid due to formation of iron cyanide.

The weight losses per specimen surface area of these samples with reference to time of immersion had an averagely decreasing trend. The performance of the aluminium was observed to be consistent (Tables 3 and 4). This can be attributed to the formation of aluminium oxide which protects the surface of the aluminium samples and the stability of the oxide in environment with pH between 4 and 9 according to findings in the literature of this study.

DISCUSSION

The decreasing trend of corrosion rates of samples with immersion time (Figures 1 to 4) can be related to formation

Table 4. Corrosion rates for aluminium sheet (0.9 and 1.3 mm).

Sample	Aluminium I (0.9 mm)			Aluminium (1.3 mm)		
Time (h)	CR (mm/yr)	pH	W/A (kg/m^2)	CR (mm/yr)	pH	W/A (kg/m^2)
120	6.6	4.1	24.3	9.6	4.1	35.3
168	4.9	4.2	25.6	6.8	4.0	35.4
216	3.6	4.0	24.2	5.3	4.0	35.3
264	3.0	4.0	24.3	4.3	3.9	35.2
312	2.5	3.9	24.1	3.6	3.9	34.8

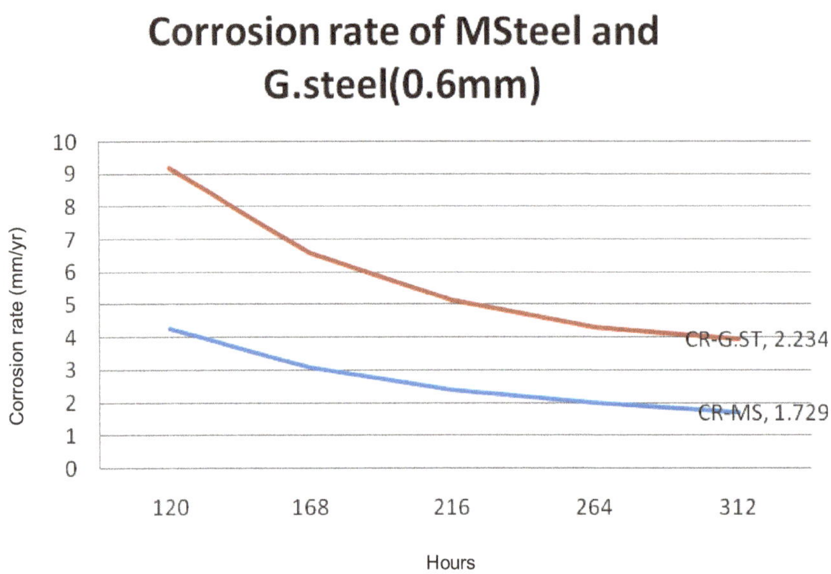

Figure 1. Corrosion rate of mild steel vs. galvanised steel.

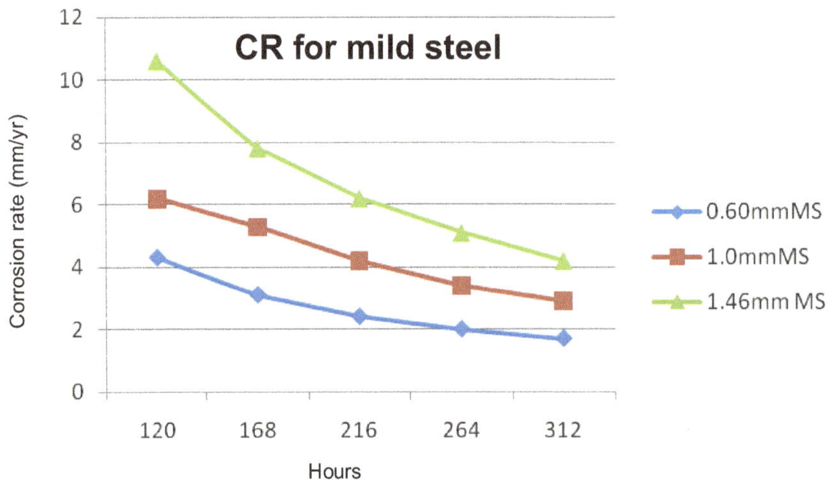

Figure 2. Corrosion rates of mild steel.

of lactic and formic acid during the fermentation of cassava fluid and metallic composition of individual metal as contained in Tables 1 and 2. But the influence of corrosion products which covered the samples lowered the corrosion rates with increase in soaking time. The difference in the corrosion resistance of these samples

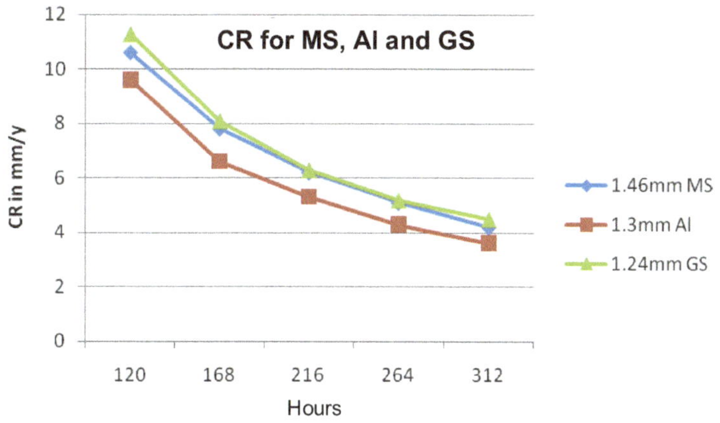

Figure 3. Corrosion rates of mild steel, galvanized steel and aluminium.

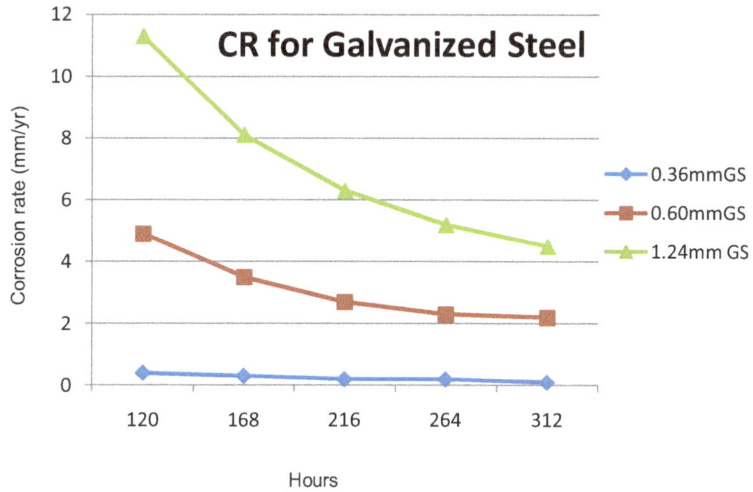

Figure 4. Corrosion rates of galvanized steel in relation to soaking time.

Table 5. Corrosion rates for galvanized steel (0.36, 0.60 and 1.24 mm).

Time (h)	GS (0.36 mm)			GS (0.60 mm)			GS (1.24 mm)		
	CR (mm/yr)	pH	W/A (kg/m^2)	CR (mm/yr)	pH	W/A (kg/m^2)	CR (mm/yr)	pH	W/A (kg/m^2)
120	0.4	4.6	35.6	4.9	5.6	47.9	11.3	4.9	149.9
168	0.3	4.1	37.9	3.5	4.2	47.9	8.1	4.9	110.4
216	0.2	4.4	36.5	2.7	4.3	48.0	6.3	4.1	110.3
264	0.2	4.4	36.1	2.3	4.4	49.9	5.2	4.7	112.1
312	0.1	4.5	35.4	2.2	4.6	56.7	4.5	5.0	113.8

of mild steel (0.6, 1.0 and 1.46 mm thickness) can be explained from the point of view of composition. The variation in the content of elements such as carbon, manganese, silicon, copper, nickel and chromium in mild steel affects the corrosion behaviour of the steel to some degree. The corrosion rates decreased with increasing contents of carbon, manganese and silicon. Both silicon and manganese contents of mild steel 0.6 mm thickness are higher than those of 1.0 mm thickness while carbon, silicon and manganese contents of both mild steel 0.6 and 1.0 mm thickness are higher than those of 1.46 mm thickness.

The purity of 0.90 mm aluminium was higher than that of 1.3 mm. Thus the corrosion rates of the former were higher than those of latter. The corrosion rates of galvanized steel increased with increase in thickness as observed in Table 5. This might be a coincidence as effect of pH which showed the strength of the corrodent

Plate 1. Photomicrograph of aluminium sheet 0.9 mm immersed in cassava fluid (×400; Etched in 2 % Nital and 98% Ethly alcohol).

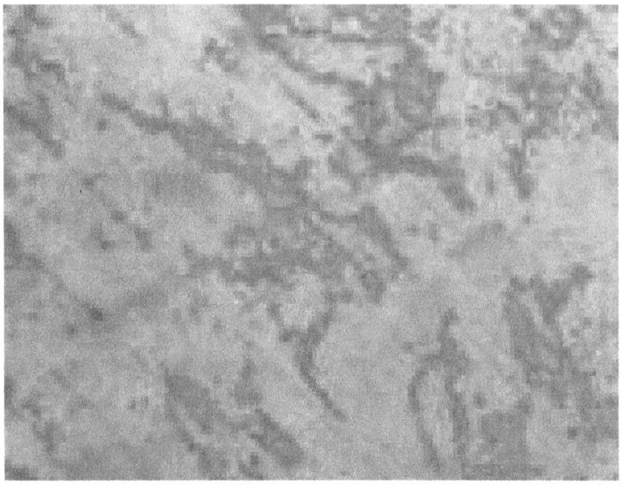

Plate 3. Photomicrograph of 1.0 mm mild steel immersed in cassava fluid (×400: Etched in 2% Nital and 98% Ethly alcohol).

Plate 2. Photomicrograph of galvanized sheet (0.6 mm) immersed in cassava fluid (×400: Etched in 2 % Nital and 98% Ethly alcohol).

(cassava fluid) was responsible for the increase in corrosion rate. Plates 1 to 3 showed the effect of corrodents on the microstructures of the samples investigated

In summary, the electrochemical properties of the metals – mild steel (0.6, 1.0 and 1.46 mm thickness), galvanized steel (0.36, 0.6 and 1.24 mm thickness) and aluminium (0.9 and 1.3 mm) were investigated in undiluted cassava fluid using weight loss technique. Considering all conditions of same test medium, time of immersion and removal rate, the following conclusions were drawn from the result of the study:

(1) Aluminium had the highest resistance to corrosion followed by mild steel then galvanized steel in cassava fluid.

(2) The corrosion rates of the aluminium, galvanized steel and mild steel all had decreasing trend with increase in immersion time in cassava fluid.

REFERENCES

Abiola OK, Oforkar NC (2002).Inhibition of Corrosion of Mild Steel in Hydrochloric Acid. J. Corros. Sci. Eng., 3: 21-26.

Adetunji O, Adewoye O (2010). Mechanical Properties of Mild Steel Rod in Cassava Fluid. Pac. J. Sci. Technol., 11(2): 22-26.

Afolabi AS (2005). Effect of Chemical Inhibiting Admixtures on Corrosion Protection of Mild Steel Reinforced Concrete in a Chloride Environment. Int. Res. J. Eng. Sci. Technol., 2(1): 156–160.

Burrell MM (2003). Starch the Need for Improved Quality or Quantity and Overview. J. Exp. Bot., 218: 4574–4762.

Fontana MG (1986). Corrosion Engineering, 3rd Edition. McGraw-Hill: New York, NY, pp. 240-250.

Amuda MOH, Lawal GI, Soremekun OO (2006). "Inhibitive Strength of Some Organic Compounds on the Corrosion Behavior of Mild Steel in Cassava Fluid. West Indian J. Eng., Trinidad and Tobago, pp. 228-234.

Scott GJ, Rosegrant MW, Ringler MW (2002). Roots and Tubers. For the 21st Century: Trends, Projections and Policy Options. Food, Agric. Environ., 31: 1-71.

Sobowale AO, Olurin TO, Oyewole OB (2007). Effect of Lactic Acid Bacteria Starter Culture Fermentation of Cassava on Chemical and Sensory Characteristics of Fufu Flour. Afr. J. Biotechnol., 6(16): 1954-1958.

Walker R (1980). A Corrosion Inhibitor for Metals. J. Chem. Educ., 57(11): 789-791.

Analysis and parametric optimization of flux cored arc welding process for IS 2062 mild steel plates using Taguchi method and utility concept

P. Sreeraj[1] , T. Kannan[2] and S. Maji[3]

[1]Department of Mechanical Engineering, Younus College of Engineering and Technology, Kerala, 691010 India.
[2]SVS College of Engineering, Coimbatore, Tamilnadu, 642109 India.
[3]Department of Mechanical Engineering, IGNOU, Delhi, 110068, India.

Optimization of flux cored arc welding (FCAW) process parameters were carried out to obtain optimum weld bead geometry in mild steel plates of IS 2062 by using an alternate Taguchi method. Design of experiments based on multi-objective Taguchi approach was employed for development of mathematical model correlating important process parameters like welding voltage (V), welding current (I), stick out (N) and wire speed (T), with weld bead parameters like penetration (P) and percentage of dilution as responses. The optimum welding parameters were investigated using Taguchi method with L_{16} orthogonal array (OA). The developed models have checked for adequacy and significance based on analysis of variance (ANOVA) test. Because of high reliability, easiness in operation, high penetration good surface finish and high productivity FCAW became a natural choice for fabrication industries. Based on confirmation test, the proposed model can be effectively used to optimize the welding performance. The utility concept was used for optimization of bead geometry and results are presented.

Key words: Flux cored arc welding (FCAW), optimization, analysis of variance (ANOVA), Taguchi approach, utility concept.

INTRODUCTION

Quality is a vital factor in today's manufacturing world. Quality can be defined as the degree of customer satisfaction. Quality of a product depends on how it performs in desired circumstances. Quality is avery vital factor in the field of welding. The quality of a weld depends on mechanical properties of the weld metal which in turn depends on metallurgical characteristics and chemical composition of the weld. The mechanical and metallurgical feature of weld depends on bead geometry which is directly related to welding process parameters. In other words, quality of weld depends on in process parameters. Flux cored arc welding (FCAW) welding is a multi-objective and multifactor metal fabrication technique. The process parameters have a direct influence on bead geometry.

Figure 1 shows the weld bead geometry. Mechanical strength of weld metal is highly influenced by the composition of metal but also by weld bead shape.

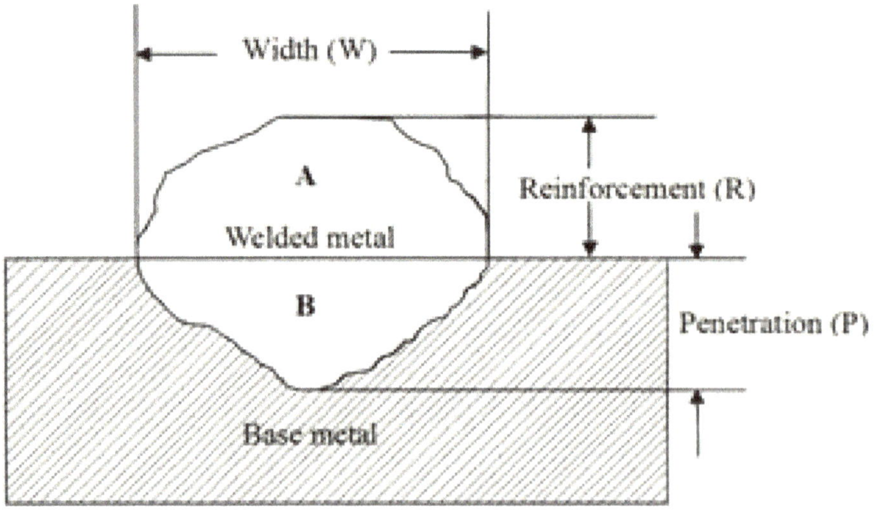

Figure 1. Weld bead geometry. Percentage dilution (D) = [B / (A + B)] × 100.

(Kannan et al., 2013). This is an indication of bead geo geometry. It mainly depends on wire speed, welding current, voltage and stick out etc. Therefore, it is necessary to study the relationship between in process parameters and bead parameters to study weld bead geometry (Lijo and Somashekher, 2013). This paper highlights the study carried out to develop mathematical models to optimize weld bead geometry, in bead on plate welding deposited by FCAW.

The multi characteristics optimization model based on Taguchi method and utility concept has been employed to determine the optimal combination of welding parameters to attain minimum percentage of dilution and maximum penetration simultaneously. The confirmation test was also conducted to verify the results.

TAGUCHI METHOD

Taguchi method uses a special type of design of orthogonal arrays (OA) to study the entire parameter space with smaller number of experiments. The OA is a small set from all possibilities which helps to determine the least number of experiments, which will further helps to determine the optimum level for each process parameters and establish the relative importance on individual process parameters. Appropriate selection of OA is the first step of Taguchi approach. The minimum number of experimental trials performed in an OA is given by N_{min} = (L-1) F + 1; where F is the number of factors and L is the number of levels. Taguchi method utilizes a statistical measure of performance called signal-to-noise (S/N) ratio. The ratio depends on the quality characteristics of product/process to be optimized. The experimental results are then transferred to S/N ratio. This ratio can be used to measure the quality characteristics deviating from desired values. Usually, there are three categories in the analysis of the S/N ratio, that is, the lower-the-better, higher-the-better and nominal-the-better. Regardless of category of quality characteristics larger S/N ratio corresponds to the better quality characteristics (Jagannatha et al., 2012). The optimal process parameters are the levels with highest S/N ratio. Analysis of variance (ANOVA) tests are performed to see that the process parameters are statically significant. Finally, a confirmation experimented is conducted to verify the optimal process parameters.

The multi-response methodology based on Taguchi's robust design technique and utility concept was used for optimizing the multi responses like dilution and penetration. Taguchi's standard S/N ratios were selected to obtain the optimum parameters. They were the larger-the-better type for penetration and dilution as calculated by Equation 1, and Equation 2 represents smaller-the-better type, respectively.

$$\eta = -10log_{10}\left|\frac{1}{p^2}\right| \quad (1)$$

$$\eta = -10log_{10}\left|D^2\right| \quad (2)$$

UTILITY CONCEPT

Welding process is a multi-factor multi-objective problem; the optimum solution is a compromise. The models developed were used for optimization of FCAW process parameters to obtain good bead geometry. Utility can be defined as the usefulness of a product or service into the levels of expectations to the consumers. The usefulness of a process can be represented by a unified index termed as utility which is the summation of the individual utilities of various quality characteristics (Norasiah et al., 2012). It is difficult to obtain the best combination process parameters, when there are multi responses to be optimized. If x_i represents the measure of effectiveness and n represents the number of responses, then overall utility function can be defined as:

$$U(x_1, x_2 .... x_n) = f\,[U_1(x_1), U_2(x_2) .... U_n(x_n)] \quad (3)$$

Where $U(x_1, x_2 .... x_n)$ is the overall utility of n process response characteristics and $U_1(x_1)$ is the utility of I_{th} response. Assignment of weights is based on the requirements and priorities among the various responses. Therefore, weighted form can be expresses as:

Table 1. Chemical composition of base metal and filler wire.

	Elements weight (%)								
Material	**C**	**SI**	**Mn**	**P**	**S**	**Al**	**Cr**	**Mo**	**Ni**
IS 2062	0.150	0.160	0.870	0.015	0.016	0.031	-	-	-
E7 IT- 1C	1.12	0.90	1.75	0.030	0.030	-		0.30	0.50

Table 2. Welding parameters and their levels.

Parameter			**Factor levels**			
	Unit	**Notation**	**1**	**2**	**3**	**4**
Welding voltage	V	V	20	22	24	25
Welding current	A	I	87	123	138	155
Stick out	mm	N	15	20	25	30
Wire speed	mm/min	T	25	40	50	53

$$U(x_1, x_2...............x_n) = \sum_{i=1}^{n} WiUi(Xi) \qquad (4)$$

Where $\sum_{i=1}^{n} Wi = 1$

Where W_i is the weight assigned to *i* th response. The utility concept employs the weighing factors to each S/N ratio for each trial and multi response S/N ratio calculated using Equation 5:

$$\eta = \eta_1 W_1 + \eta_2 W_2 \qquad (5)$$

Where W_1 and W_2 are the weight factors associated with S/N ratios. It is taken as 0.5. The overall mean of η associated with *k* number of trials is computed as:

$$M = \frac{1}{k}\sum_{1}^{n} \eta k \qquad (6)$$

These weighing factors are determined based on priorities among the various responses to be simultaneously optimized. In this experimental work, it is taken as 0.5. This gives priorities to all the responses for simultaneous maximization and minimization (Cochran and Coxz, 1987). The overall mean of η associated with *k* number of trials is computed from Equation 6.

EXPERIMENTATION

Test plates of size 300 × 200 × 6 mm were cut from mild steel plate of grade IS - 2062 and one of the surfaces is cleaned to remove oxide and dirt before welding. E7 IT-1C wire of 1.2 mm diameter was used for bead on plate welding. The properties of base metal and filler wire are shown in Table 1.

The selection of the welding electrode wire was based on the matching, the mechanical properties and physical characteristics of the base metal, weld size and existing electrode inventory (Norasiah et al., 2012). These have good surface appearance, good radiographic standard quality and minimum electrode wastage.

PLAN OF INVESTIGATION

The research work is carried out in the following steps (Tarng and Yang, 1998): Identification of factors, finding the limit of process variables, development of OA, conducting experiments as per OA, recording responses, development of mathematical models and checking adequacy of developed models. Utility concept applied further for optimization.

Identification of factors and responses

The percentage of dilution has got an important influence in welding. The properties of the welding are significantly influenced by dilution obtained. Hence, control of dilution is important in welding where a high dilution is highly desirable. When dilution is quite low, the final deposit composition will be closer to that of filler material and hence corrosion resistant properties of welding will be greatly improved. The chosen factors have been selected on the basis to get optimal dilution and optimal weld bead geometry (Kannan et al., 2013). These are wire speed rate (T), welding voltage (V), welding current (I) and stick out (N). The responses chosen were depth of penetration (P) and percentage of dilution (D). The responses were chosen based on the impact of parameters on final composite model.

Finding the limits of process variables

Working ranges of all selected factors are fixed by conducting trial run. This was carried out by varying one of the factors while keeping the rest of them as constant values. Working range of each process parameters was decided upon by inspecting the bead for smooth appearance without any visible defects. The chosen level of the parameters with their units and notation are given in Table 2.

Development of orthogonal array

In full factorial design, number of experimental runs may be high. This results increase in cost and time. In order to avoid this,

Table 3. Design matrix and observed values of weld bead geometry.

Trial No.	Design matrix				Bead parameter	
	V	I	N	T	P (mm)	D (%)
1	1	1	1	1	1.235	17.623
2	1	2	2	2	1.347	17.462
3	1	3	3	3	1.388	17.842
4	1	4	4	4	1.425	17.442
5	2	1	2	3	1.657	18.332
6	2	2	1	4	1.586	16.692
7	2	3	4	1	1.456	17.823
8	2	4	3	2	1.738	20.424
9	3	1	3	4	1.416	17.912
10	3	2	4	3	1.537	18.182
11	3	3	1	2	1.465	18.218
12	3	4	2	1	1.368	17.512
13	4	1	4	2	1.487	18.221
14	4	2	3	1	1.398	17.943
15	4	3	2	4	1.457	17.841
16	4	4	1	3	1.868	21.512

P, Penetration; D, dilution %.

Taguchi's L_{16} (4^4) OA consisting 16 sets of data has been selected to optimize the experimental data. This is shown in Table 3.

Figure 2. Welded specimen

Conducting experiments as per orthogonal array

In this work, 16 experimental runs were allowed as per the OA as shown Table 3 at random (Saurav et al., 2008). At each run settings, all parameters were disturbed and reset for next deposit. This is very essential to introduce variability caused by errors in experimental set up.

Recording of responses

For measuring the clad bead geometry, the transverse section of each weld overlays was cut using band saw from mid length. Position of the weld and end faces were machined and grinded. The specimen and faces were polished and etched using a 5% nital solution to display bead dimensions. The weld bead profiles were traced using a reflective type optical profile projector at a magnification of ×10 (Gunaraj and Murugan, 1999). Then the bead dimension such as depth of penetration and dilution were measured (Montgomery, 2003). The profiles traced using AUTO CAD software. The welded specimen is shown in Figure 2 (Kannan and Yoganath, 2010; Dawei et al., 2013). The measured weld bead dimensions and percentage of dilution is shown in Table 3.

RESULTS AND DISCUSSION

Analysis of single response

Experiments were conducted on IS 2062 mild steel plate to study the performance of bead on plate welding using the FCAW method by L_{16} OA. The values of single response S/N ratios of penetration and dilution are calculated using larger-the-better criteria, calculated using Equation 1. The combined multi-response S/N ratio is calculated using Equation 5, respectively. The individual values of S/N ratios are shown in Table 6. It can be found that the optimal combination of $V_2I_4N_1T_3$ is

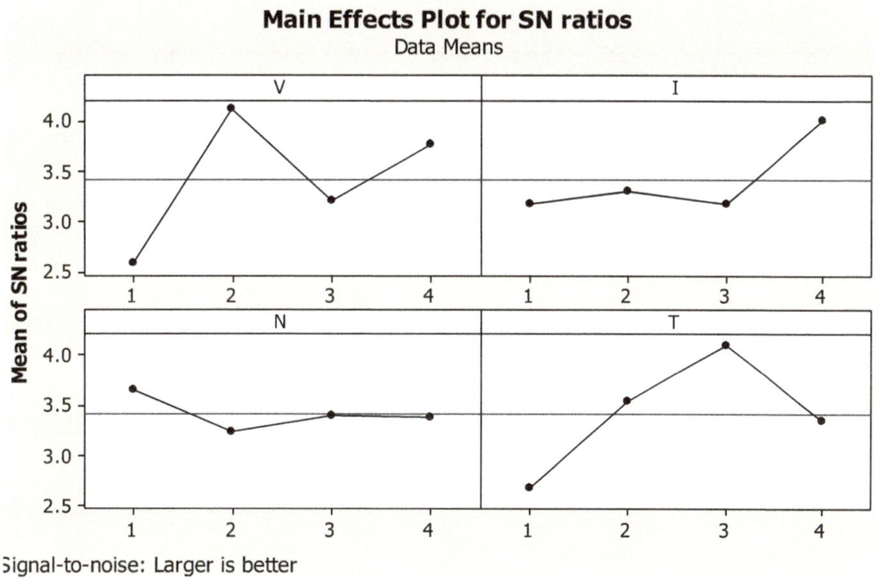

Figure 3. Main effects of plots based on S/N ratio of penetration.

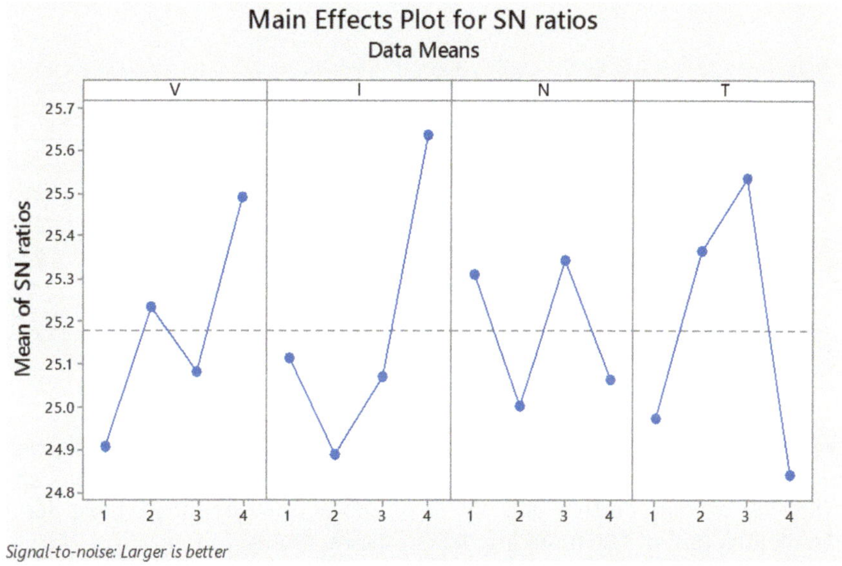

Figure 4. Main effects of plots based on S/N ratio of dilution.

the largest value of S/N ratio of penetration and for dilution it is $V_4I_4N_1T_3$, respectively. Therefore, $V_2I_4N_1T_3$ is the optimal combination of responses for penetration and for dilution $V_4I_4N_1T_3$ is the optimal combination. The main effect plots (Figures 3 and 4) shows that the optimum condition for penetration and dilution are at level, 22 V, 155 Amps, 15 mm and 53 mm/min for penetration 25V, 155 Amps, 15 mm and 53 mm/min, respectively. Table 4 and 6 show the response table for signal to noise ratios for penetration and dilution.

The statistical software MINITAB 15 is used to determine the ANOVA. The purpose of ANOVA is to investigate which welding parameter significantly affects the performance contribution. In addition, F-test was used to determine which welding factor has significant effect on performance. When F value is large then that factor has significant effect on performance. The results of ANOVA for single-response S/N ratio of dilution and penetration are shown in Tables 5 and 7, respectively. It can be seen that welding current has highest contribution

Table 4. Response table for signal to noise ratios penetration.

Level	V	I	N	T
1	2.586	3.172	3.646	2.682
2	4.114	3.072	3.241	3.538
3	3.198	3.174	3.395	4.099
4	3.763	4.007	3.380	3.343
Delta	1.528	0.835	0.405	1.417
Rank	1	3	42	

Table 5. ANOVA for penetration.

Source	DF	Seq SS	Adj SS	Adj MS	F	P
V	3	0.036551	0.036551	0.0365512	1.55965	0.237640
I	3	0.005056	0.005056	0.0050562	0.21575	0.651356
N	3	0.035870	0.035870	0.0358705	1.53060	0.241782
T	3	0.040230	0.040230	0.0402304	1.71664	0.216825
Error	3	0.257791	0.257791	0.0234355		
Total	15	0.375499				

Table 6. Response table for signal to noise ratios on dilution.

Level	V	I	N	T
1	24.91	25.12	25.31	24.97
2	25.23	24.89	25.00	25.37
3	25.08	25.07	25.34	25.53
4	25.49	25.64	25.06	24.84
Delta	0.59	0.75	0.34	0.69
Rank	3	1	4	2

of about 65.1% for penetration and stick out with 69.7.7% contribution for dilution. The other parameters have less contribution. It is clear that the welding current and stick out are one of the significant factors that has more impact than any other factors on dilution and penetration.

Optimal parameter combination of Multi response

The optimal combination of process parameters for simultaneous optimization of dilution and penetration is obtained by means of values of multi-responses S/N ratio of overall utility and is shown in Table 8.

The larger values of multi-responses S/N ratio means the comparable sequence exhibiting a stronger correlation with the reference sequence. Based on the study, combination of $V_2I_4N_1T_3$ shows the largest values of multi-response S/N ratio for the factors V, I, N and T, respectively. Therefore, $V_2I_4N_1T_3$ is the optimal parameter combination of flux cored arc bead on plate welding for IS 2062 steel plates, which are shown in Figure 5. Table 9 shows the response table for multi-response signal to noise ratios.

The results of ANOVA for multi-response S/N ratio are shown in Table 10. On examining the percentage of contribution (P %) of different factors, it can be seen that welding current is about 34.7% and other parameters have significant contributions. It is clear that welding current is one of the significant factors that have more impact than any other factors. The main effect plot, (Figure 5), shows that optimum condition for multiple responses is/are at level, 22 V, 155 Amps, 15 mm and 53 mm/min, respectively.

Confirmation test

The final step is the verification experiments to validate

Table 7. Analysis of variance for dilution.

Source	DF	Adj SS	Adj MS	F	P
Regression	4	5.8337	1.45843	1.08	0.412
V	1	2.4490	2.44895	1.81	0.205
I	1	3.1407	3.14068	2.33	0.155
N	1	0.2160	0.21601	0.16	0.697
t	1	0.0281	0.02809	0.02	0.888
Error	11	14.8450	1.34955		
Total	15	20.6787			

Table 8. S/N ratios of penetration and dilution and multi response.

S/N ratio D (%)	S/N ratio P	S/N ratio multi response
24.9216	1.83334	13.37747
24.8419	2.58735	13.71463
25.0289	2.84779	13.93835
24.8319	3.07630	13.9541
25.2642	4.38645	14.82533
24.4502	4.00606	14.22813
25.0196	3.26323	14.14142
26.2028	4.80100	15.5019
25.0629	3.02127	14.04209
25.1928	3.73348	14.46314
25.2100	3.31675	14.26338
24.8667	2.72172	13.79421
25.2114	3.44622	14.32881
25.0779	2.91014	13.99402
25.0284	3.26919	14.1488
26.6536	5.42754	16.04057

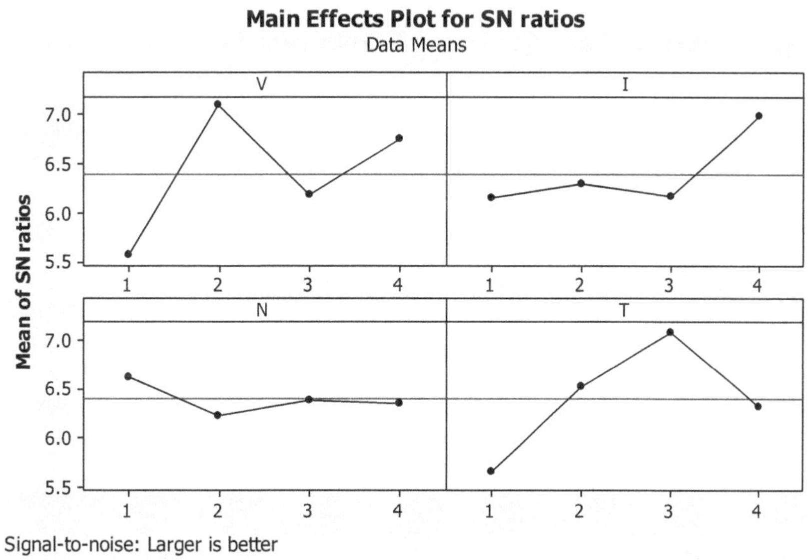

Figure 5. Main effect of plots based on S/N ratios multiple responses.

Table 9. Response table for multi-response S/N ratios.

Level	I	S	N	T
1	5.571	6.154	6.626	5.667
2	7.091	6.289	6.222	6.520
3	6.181	6.157	6.378	7.078
4	6.744	6.987	6.361	6.323
Delta	1.520	0.833	0.404	1.411
Rank	1	3	4	2

Table 10. Analysis of variance for multi response S/N ratios.

Source	DF	Seq SS	Adj SS	Adj MS	F	P
V	3	5.3253	5.3253	1.77510	25.82	0.012
I	3	0.3385	0.3385	0.11568	1.64	0.347
N	3	1.9070	1.9070	0.63658	9.25	0.050
T	3	4.0699	4.0699	1.35664	19.73	0.018
Error	3	0.2063	0.2063	0.06876		
Total	15	11.8470				

Table 11. Results of confirmation tests

Level	Performance Characteristic	Optimal setting	Predicted S/N ratio	Experimental S/N ratio
SRO	Penetration (P)	$V_2I_4N_1T3$	5.4275	7.5062
	Dilution (D)	$V_4I_4N_1T_3$	13.7874	14.7424
MRO	P and D	$V_2I_4N_1T_3$	16.0405	17.7646

optimum conditions suggested by the matrix experiments which indeed give the projected experiment. Confirmation experiments is performed by conducting a test with a specific combination of factors and levels previously evaluated. After determining the optimum conditions new experiment was conducted with optimum levels. Then predicted parameters were calculated using Equation 7.

$$\eta\_opt = \bar{\eta} + \sum_{i=1}^{p} \eta_{mi} - \bar{\eta} \quad (7)$$

Where *m* is the total mean of the responses S/N ratio at the optimal level and m_i is the S/N ratio at the optimal parameter. The predicted optimal values of single and multi-response function are shown in Table 11. The improvement in S/N ratio is found. So there is considerable improvement in quality with multi-response function.

In order to validate the experiment, four trials are conducted according to the optimal parameters level, $V_2I_4N_1T_3$ and corresponding values of performance measures are taken. Table 11 shows the predicted multi-response S/N ratio and multi-response S/N ratio obtained from experiment. It may be noted that there is a good agreement between estimated value (16.0405) and the experiment (17.7646). Therefore, the condition $V_2I_4N_1T_3$ is treated as optimal. The optimal condition, $V_2I_4N_1T_3$, (22 V, 155 Amps, 15 mm and 53 mm/min) is confirmed by ANOVA.

Conclusions

Optimization of process parameters of FCAW has been carried out using Taguchi's OA with multi response analysis as discussed in this paper. From the analysis, the following conclusions have been drawn:

1) It has been found that the combination $V_2I_4N_1T_3$ show the largest values of multi-response S/N ratio for factors V, I, N and T. Therefore, $V_2I_4N_1T_3$ is optimal parameter combination with voltage, current, stick out and wire feed

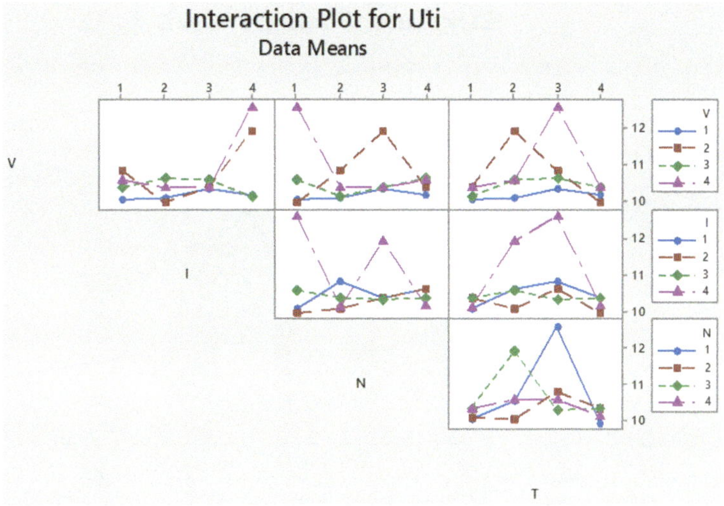

Figure 6. Interaction plot for multi-response.

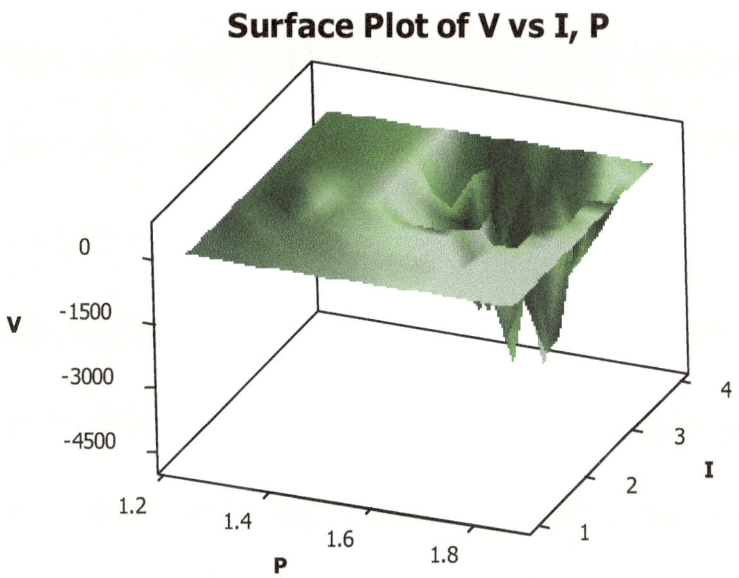

Figure 7. Interaction effect of voltage and current on penetration.

of the flux cored bead on plate welding on IS 2022 mild steel plates.

2) Through ANOVA, percentage of contribution to the stick out is more as compared to other parameters. Hence, the most significant factor for the flux cored bead on plate welding for maximization of dilution and maximization of penetration is stick out.

3) It can be found that there is a good agreement between estimated and experimental value. Therefore, in this welding process $V_2I_4N_1T_3$ is treated as optimum parameter combination.

4) From the experimental results, it can be found that welding current and stick out has largest impact on dilution and penetration.

5) Figure 6 illustrates the interaction effects of process parameters for multi-response optimization. It is observed that if the lines are parallel and not over lapping there is little or no evidence of an interaction in the parameters.

6) Figure 7 shows the interaction effects of voltage and current on penetration. Figure 8 shows surface plot voltage and current on dilution.

7) Figure 9 shows combination of process parameters with penetration.

8) Figures 10 to 13 show microstructures of various

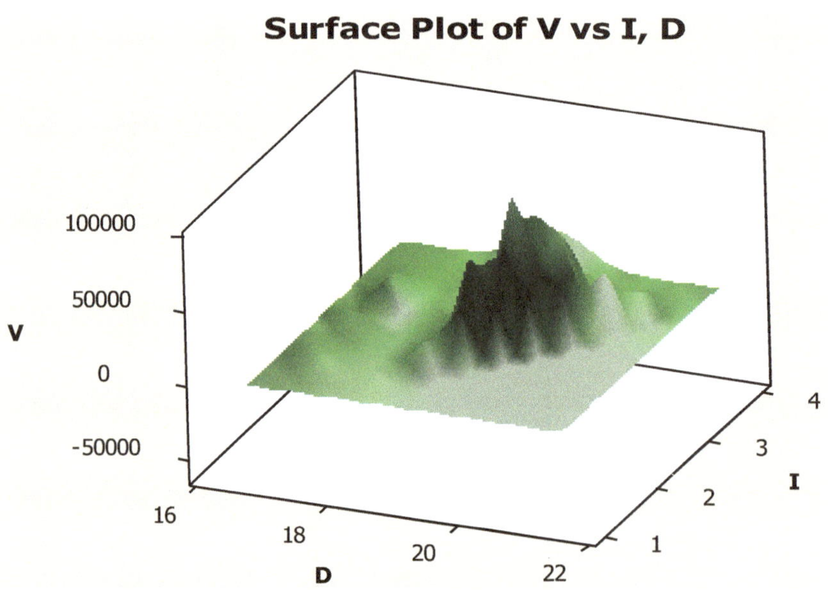

Figure 8. Interaction effects of voltage and current on dilution.

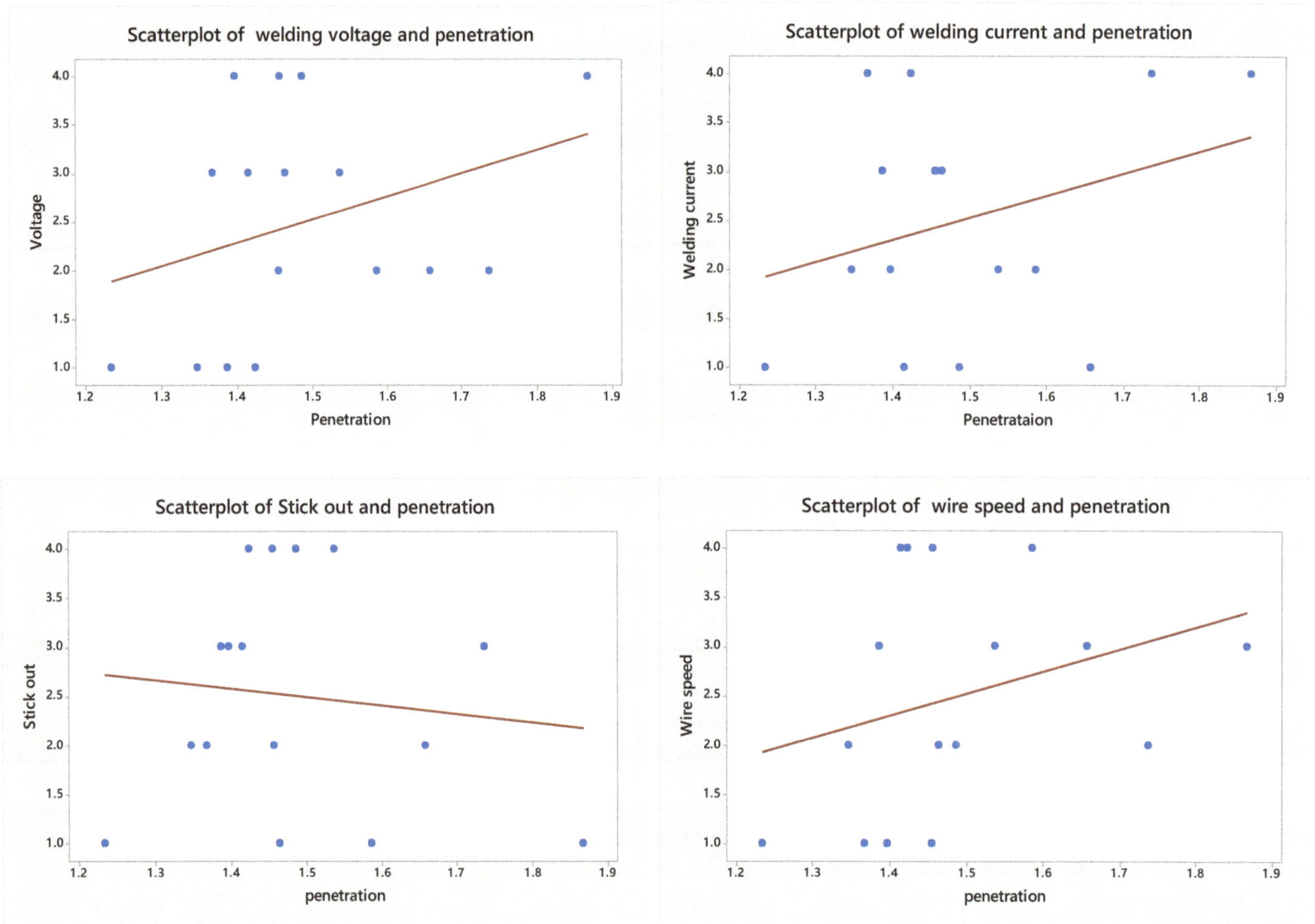

Figure 9. Combination of process parameters and penetration.

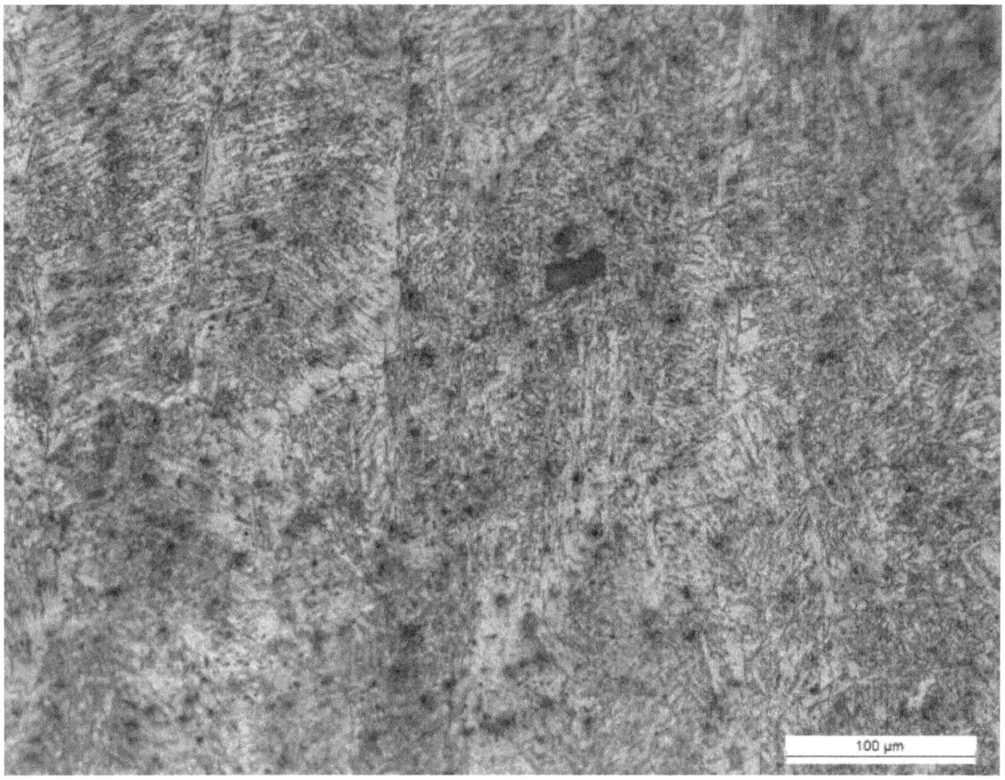

Figure 10. Microstructure of HAZ.

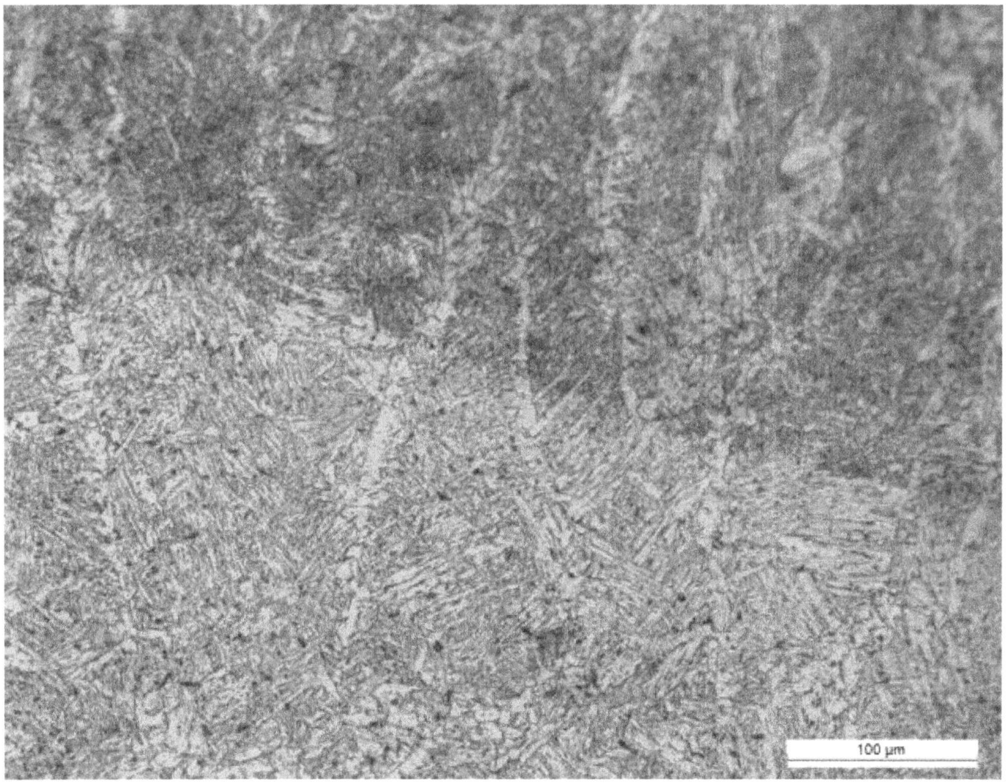

Figure 11. Microstructure of fusion line.

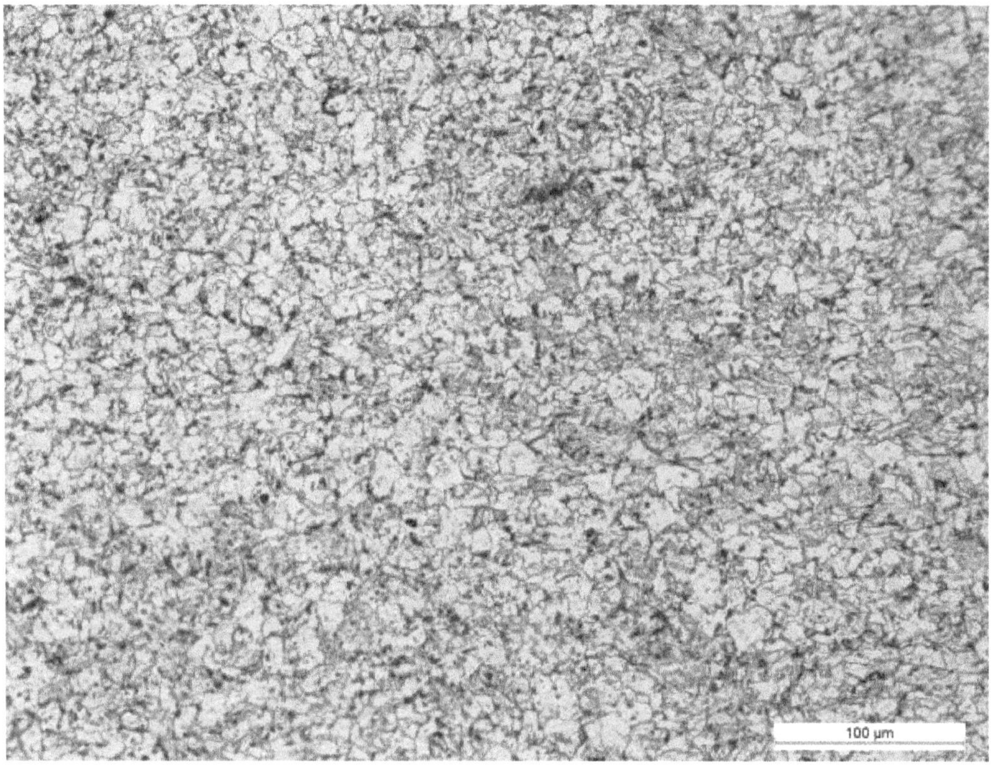

Figure 12. Microstructure of weld zone.

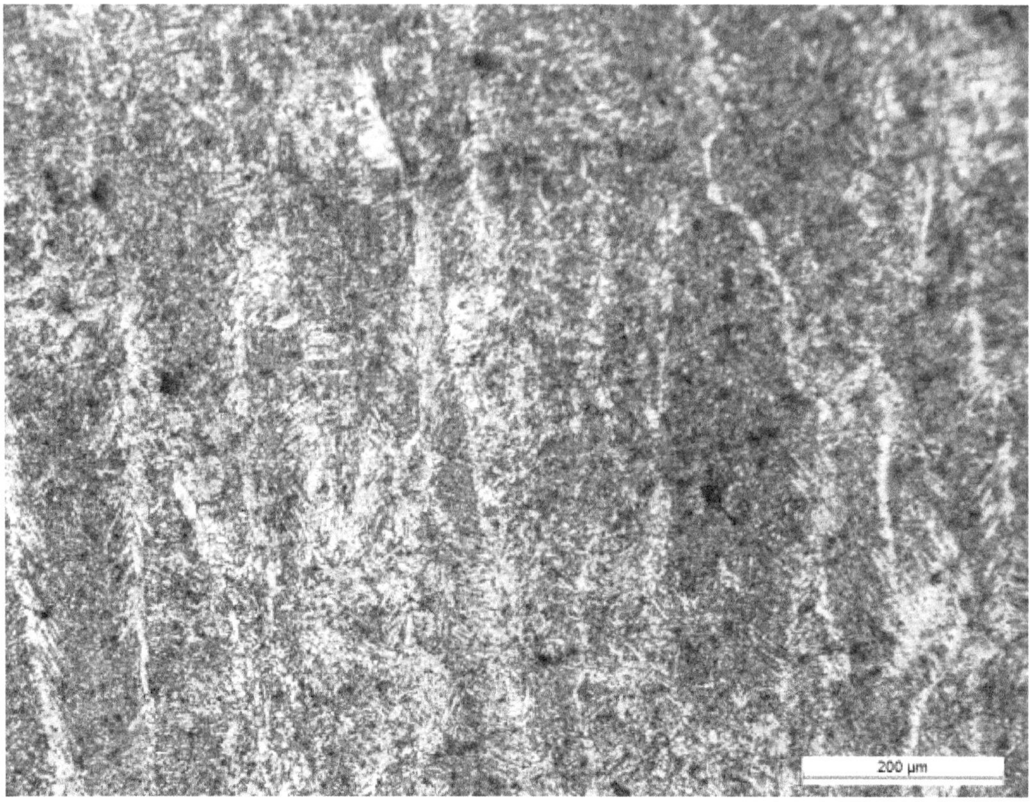

Figure 13. Microstructure of HAZ.

welding zones of specimens. From the microstructure figures, it can be seen that in the weld zone the grain structure is somewhat fine. In HAZ, it is somewhat coarse this is due to dendrite formation.

Conflict of Interest

The author(s) have not declared any conflict of interest.

ACKNOWLEDGEMENT

The authors sincerely acknowledge the help and facilities extended to them by the Department of Mechanical Engineering, Younus College of Engineering and Technology, Kollam, India.

REFERENCES

Cochran WG, Coxz GM (1987). Experimental Design., New York, John Wiley & Sons. P. 370.

Dawei Z, Yuanxun W, Suning S, Zongguo L (2013). Multi objective optimal design of small scale resistance spot welding process with principal component analysis and response surface methodology. J. intell. manuf. DOI,10.1007/s10845-013-0733-2.

Gunaraj V, Murugan N (1999). Prediction and comparison of the area of the heat effected zone for the bead on plate and bead on joint in SAW of pipes, J. Mater. Process. Technol. 95:246-261.

Kannan TY (2010). Effect of process parameters on clad bead geometry and shape relationships of stainless steel cladding deposited by GMAW. Int. J. Manuf. Technol. 47:1083-1095.

Kannan T, Murugan N, Sreeharan BN (2013). Optimization of flux cored arc welding process parameter using genetic and memetic algorithms. J. Manuf. Sci. Prod. 13(4):239-250.

Jagannatha N, Hiremath SS, Sadashivappa K (2012). Analysis and parametric optimization of abrasive hot air jet machining for glass using Taguchi method and utility concept. Int. J. Mech. Mater. Eng. 7(1):9–15.

Montgomery DC (2003). Design and analysis of Experiments, John Wiley & Sons (ASIA) Pvt. Ltd.

Norasiah M, Manurung Y, Hafidzi M, Abas S, Tham G, Haruman E (2012). Optimization and modelling of spot welding parameters with simultaneous response consideration using multi objective Taguchi method and utility concept. J. Mech. Sci. Technol. 26(8):2365-2370.

Saurav D, Ashish B, Pradip KP (2008). Grey based Taguchi method for optimization of bead geometry in submerged arc bead on plate welding. Int. J. Adv. Manuf. Technol. 39:1136-1143.

Somashekher L, Hiremath S (2013). Response surface modelling of micro holes in electro chemical discharge machining process. Proceed Engr. 64:1335-1404.

Tarng YS, Yang WH (1998) Optimization of weld bead geometry in gas tungsten arc welding by the Taguchi method. Int. J. Adv. Manuf. Technol. 14:549-554.

Seismic isolation in buildings to be a practical reality: Behavior of structure and installation technique

A. B. M. Saiful Islam*, Mohammed Jameel and Mohd Zamin Jumaat

Department of Civil Engineering, University of Malaya, Kuala Lumpur, Malaysia.

Although a great deal of research has been carried out regarding seismic isolation, there is a lack of proper research on its behavior and implementing technique in low to medium seismic region. The basic intention of seismic protection systems is to decouple the building structure from the damaging components of the earthquake input motion, that is, to prevent the superstructure of the building from absorbing the earthquake energy. This paper reviews a number of articles on base isolation incorporation in building structure. Lead rubber bearing (LRB), high damping rubber bearing (HDRB), friction pendulum system (FPS) have been critically explored. This study also addressed the detail cram on isolation system, properties, characteristics of various device categories, recognition along with its effect on building structures. Meticulous schoolwork has also been accomplished about installation technique for various site stipulations. The entire superstructure is supported on discrete isolators whose dynamic characteristics are chosen to uncouple the ground motion. Displacement and yielding are concentrated at the level of the isolation devices, and the superstructure behaves very much like a rigid body. Rigorous reckoning illustrated the isolation system as very innovative and suitable in buildings to withstand the seismic lateral forces and also contributed to safety ensuring flexibility of structures.

Key words: Seismic protection, base isolation, idealized behavior, hysteresis loop, ductility, installation technique.

INTRODUCTION

To minimize the transmission of potentially damaging earthquake ground motions into a structure is achieved by the introduction of flexibility at the base of the structure in the horizontal direction while at the same time introducing damping elements to restrict the amplitude or extent of the motion caused by the earthquake somewhat akin to shock absorbers. In recent years this relatively new technology has emerged as a practical and economic alternative to conventional seismic strengthening. This concept has received increasing academic and professional attention and is being applied to a wide range of civil engineering structures. To date there are several hundred buildings in Japan, New Zealand, United States, India which use seismic isolation principles and technology for their seismic design.

It may come as a surprise that the rubber foundation elements can actually help to minimize earthquake damage to buildings, considering the tremendous forces these buildings must endure in a major quake (Cheng et al., 2008; Islam and Ahmad, 2010). Contrasting the conventional design approach based on an increased resistance (strengthening) of the structures, the seismic isolation concept is aimed at a significant reduction of dynamic loads induced by the earthquake at the base of the structures themselves (Micheli et al., 2004, , Islam et al., 2010a, 2010b). Seismic isolation separates the structure from the harmful motions of the ground by providing flexibility and energy dissipation capability through the insertion of the isolated device so called isolators between the foundation and the building structure (Ismail et al., 2010).

Invention of lead rubber bearing (LRB, 1970's) and high damping rubber bearing (HDRB,1980's) gives a new dimension to the seismic base isolation design of BI structure (Islam, 2009; Hussain et al., 2010). ICBO (1997) included design specification for isolated structures as in Uniform building Code, UBC 1997. A

*Corresponding author. E-mail: abm.saiful@gmail.com.

significant amount of both past and recent research in the area of base isolation has focused on the use of elastomeric bearings, such as HDRB and LRB (Islam et al., 2011a, 2011b). Providakis (2008) investigated seismic responses of multi- storied buildings for near fault motion isolated by LRB. Dall'Asta and Ragni (2006, 2008), Bhuiyan et al. (2009) covered experimental tests, analytical model and nonlinear dynamic behavior of HDRB.

Although it is a relatively recent technology, seismic isolation for multi storied buildings has been well evaluated and reviewed (Barata and Corbi, 2004; Hong and Kim, 2004; Matsagar and Jangid, 2004; Komodromos, 2008; Lu and Lin, 2008; Spyrakos et al., 2009; Polycarpou and komodromos, 2010). Base isolator with hardening behavior under increasing loading has been developed for medium-rise buildings (up to four storeys) and sites with moderate earthquake risk (Pocanschi and Phocas, 2007). Nonlinear seismic response evaluation was performed by Balkaya and Kalkan (2003).

Resonant behaviour of base-isolated high-rise buildings under long-period ground motions was dealt by Ariga et al. (2006) and long period building responses by Olsen et al. (2008). Deb (2004), Dicleli and Buddaram (2007), Casciati and Hamdaoui (2008), Di Egidio and Contento (2010) have also given effort in progresses of isolated system. Komodromos et al. (2007), Kilar and Koren (2009) focused the seismic behavior and responses through dynamic analyses of isolated buildings. Wibowo et al. (2010) have done the failure and collapse modeling analysis of weak storey building. The low to medium earthquake risk region is prone to seismic hazard in the United States too. Many places at the Midwestern areas in US (Southern Illinois, Kentucky, Southern Indiana etc) experience low to moderate seismicity. However, for building construction in these zones, seismic base isolation can be a suitable alternative as it ensures flexibility of building and reduces the lateral forces in a drastic manner. Apart from this as the additional cost of using the system is favorable; these sites can also benefit from implementation of this base-isolation system, especially in such buildings as museums, data centers etc. Though the application of isolator is going to be very familiar all over the world, there is a lack of proper research to implement the device practically in low to medium risk seismic zone especially for local buildings in Dhaka, Bangladesh region as per the local requirements. So, this concern is a very burning matter for this study. Site specific earthquake data are also very important in seismic design. This study focused on the detail revise on isolation system, characteristics of various device categories, recognition along with its effect on building structures. Thorough schoolwork has also been accomplished about installation technique for various site stipulations.

SEISMIC ISOLATION COMING TO REALITY

Seismic isolation is intended to prevent earthquake damage to structures, buildings and building contents. One type of seismic isolation system employs load bearing pads, called isolators. They are located strategically between the foundation and the building structure and are designed to lower the magnitude and frequency of seismic shock permitted to enter the building. They provide both spring and energy absorbing characteristics. Figure 1 illustrates the behavior change of structure without isolator and with isolator incorporation.

The first seismic isolation system was proposed by Dr. Johannes Calantarients, an English medical doctor, in 1909. His diagrams show a building separated from its foundation by a layer of talc which would isolate the main structure from seismic shock. The oldest base isolated structure of the world, Mausoleum of Cyrus, is shown in Figure 2.

This technology can be used both for new structural design and seismic retrofit. In process of seismic retrofit, some of the most prominent U.S. monuments like Pasadena City Hall, San Francisco City Hall, Salt Lake City and County Building or LA City Hall. The seismic rehabilitation of the Los Angeles City Hall is a landmark event in the City's history. For Los Angeles City Hall (Figure 3a), in process of seismic upgrading, this high-rise building was placed atop a mechanical system of isolators, sliders and dampers called base isolation technology. Later on a few famous isolated buildings have also been shown in Figure 3. Bhuj Hospital (Figure 3b), New Zealand Assembly Library (Figure 3c) and New Zealand Parliament (Figure 3d) have been erected on Lead Rubber bearing type base isolator. Isolation system was also inserted at Te Papa Museum of New Zealand (Figure 3e). Figure 3f shows the practical construction of inserting seismic isolation system in Te Papa Museum building.

Early concerns were focused on the fear of uncontrolled displacements at the isolation interface, but these have been largely overcome with the successful development of mechanical energy dissipaters. When used in combination with a flexible device, an energy dissipater can control the response of an isolated structure by limiting both the displacements and the forces. Interest in base isolation as an effective means of protecting structures from earthquakes has therefore, been revived in recent years.

The following are the advance developments (Kelly, 1998) that have enabled base isolation to be a practical reality.

(a) The design and manufacture of high quality isolation bearings that are used to support the weight of the structure and at the same time, release it from earthquake induced forces.

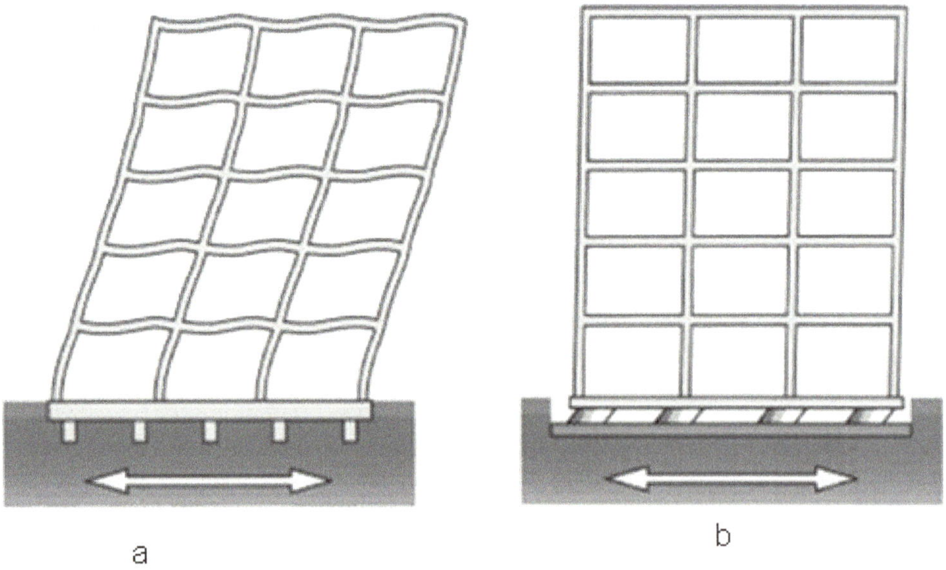

Figure 1. Behavior change while using isolator. (a) Conventional structure (b) base -isolated structure.

Figure 2. Mausoleum of Cyrus, the oldest base-isolated structure in the world.

(b) The design and manufacture of mechanical energy dissipaters (absorbers) that are used to reduce the movement across the bearings to practical and acceptable levels (4 to 6 inches) and to resist wind loads

(c) The development and acceptance of computer software for the analysis of base-isolated structure which includes nonlinear material properties and the time-varying nature of the earthquake loads.

(d) The ability to perform shaking table tests using real recorded earthquake ground motions to evaluate the performance of structures and provide results to validate computer modeling techniques.

(e) The development and acceptance procedures for estimating site-specific earthquake ground-motions for different return periods.

BASE ISOLATION SYSTEM

Basic elements of base isolation

Seismic Isolation increases the fundamental period of vibration so that the structure is subjected to lower earthquake forces. However, the reduction in force is accompanied by an increase in displacement demand which must be accommodated within the flexible

Figure 3. Some Renowned Base-isolated Structures, (a) Los Angeles City Hall, USA, (b) Bhuj Hospital, Gujrat, India, (c) New Zealand Assembly Library, (d) New Zealand Parliament, (e) Te Papa Museum of New Zealand, (f) Te Papa during Incorporation of Isolator.

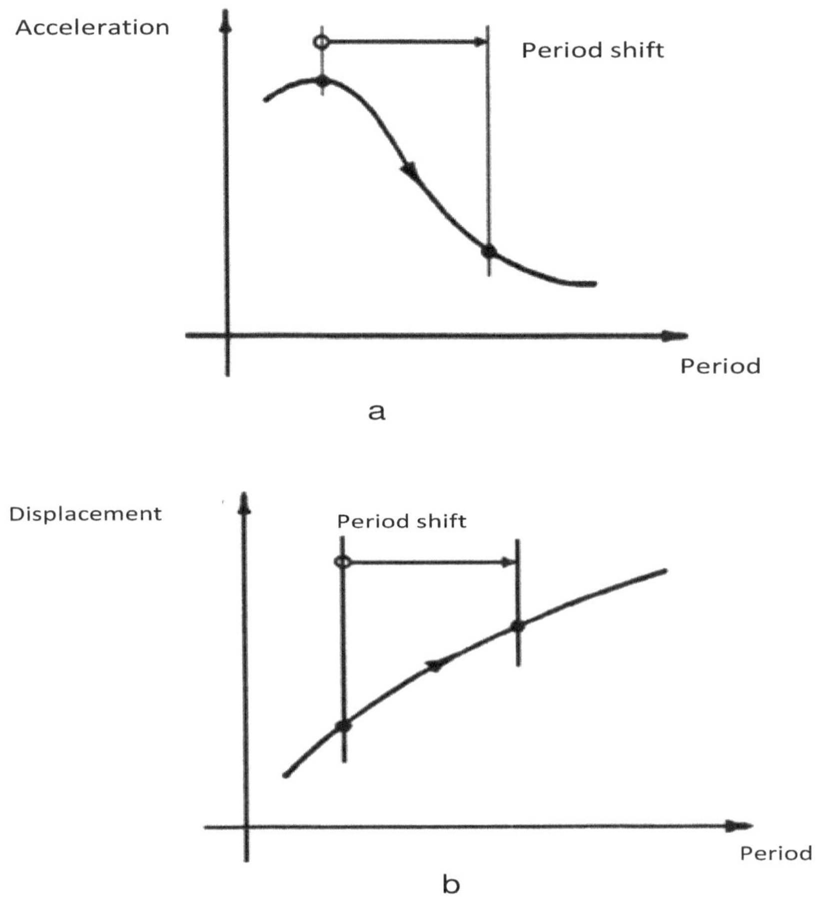

Figure 4. Impact of period elongation obtained by seismic isolation on accelerations. (a) Acceleration response spectrum, (b) displacement response spectrum and displacements of a structure.

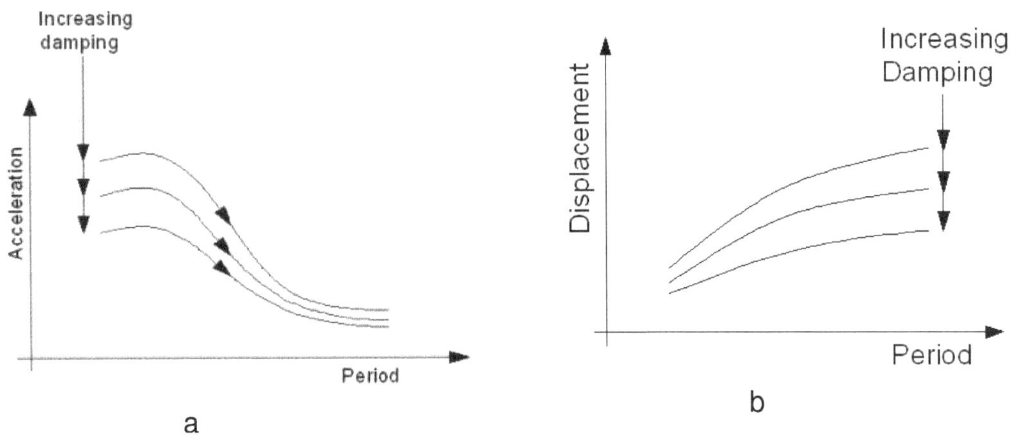

Figure 5. Acceleration and displacement response Spectrum for increasing damping. (a) Acceleration RS, (b) Displacement RS.

mount, Furthermore, longer period buildings can be lively under service loads. The following are three basic elements in any practical isolation system (Skinner et al., 1993), they are:

1. A flexible mounting so that the period of vibration of the building is lengthened sufficiently to reduce the force response.
2. A damper of energy dissipater so that the relative

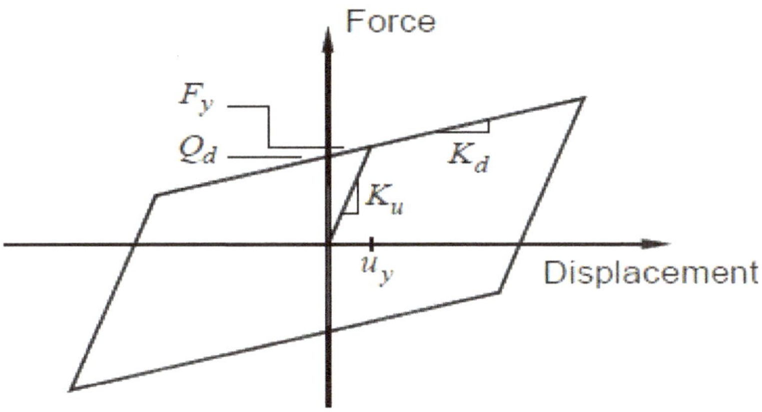

Figure 6. Idealized force-displacement (Hysteresis) Loop.

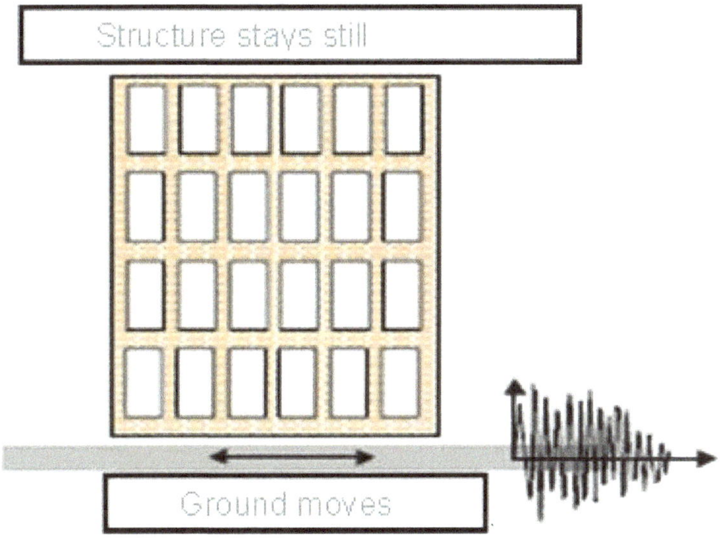

Figure 7. Base isolation strategy.

deflections across the flexible mounting can be limited to a practical design level.
3. A means of providing rigidity under low (service) load levels such as wind and braking force

Flexibility: Due to additional flexibility the period of structure is elongated. From the acceleration response curve shown in Figure 4a, it may be observed that reductions in base shear occur as the period of vibration of the structure is lengthened. The extent to which these forces are reduced is primarily dependent on the nature of the earthquake ground motion and the period of the non-isolated structure.

Energy dissipation: Additional flexibility needed to lengthen the period of the structure will give rise to large relative displacement across the flexible mount. Figure 4b shows an idealized displacement response curve from which displacements are seen to increase period (flexibility). Large relative displacements can be controlled if substantial additional damping is introduced into the structure at the isolation level. This is shown schematically in Figure 5. Also shown schematically in this figure is the smoothening effect of higher damping.

One of the most effective means of providing a substantial level of damping is through hysteric energy dissipation. The hysteric refers to the offset between the loading and unloading curves under cyclic loading. Figure 6 shows an idealized force-displacement loop where the enclosed area is a measure of the energy dissipated during one cycle of motion.

Rigidity under low lateral loads: While lateral flexibility is highly desirable for high seismic loads, it is clearly undesirable to have a structural system which will vibrate perceptibly under frequently occurring loads such as wind loads or braking loads. Mechanical energy dissipaters may be used to provide rigidity at these service

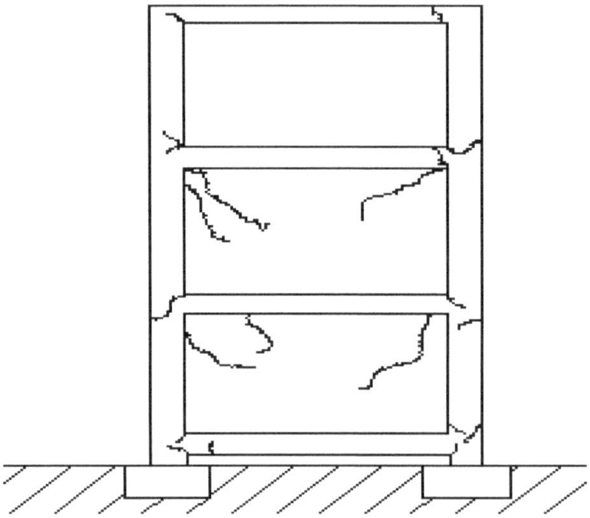

Figure 8. Failure pattern of a fixed based structure due to lateral seismic loading.

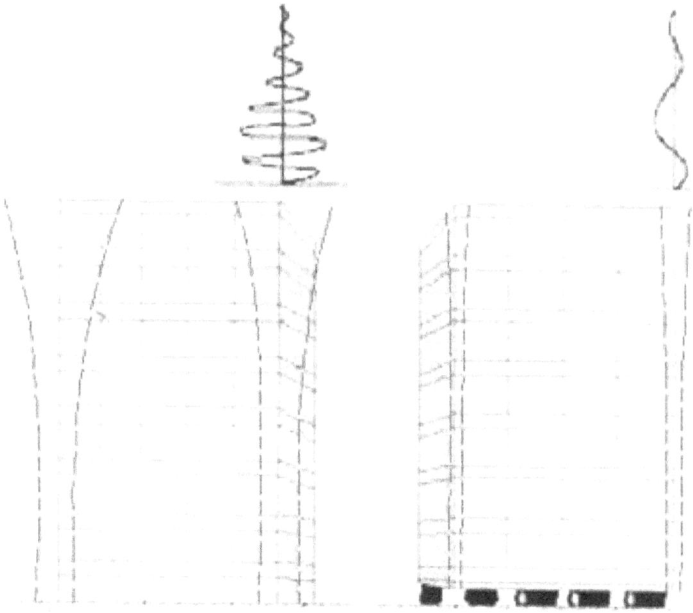

Figure 9. Fixed base and isolated base.

loads by virtue of their high initial elastic stiffness.

Fundamental concepts of base isolation

The term base isolation uses the word a) isolation in its meaning of the state of being separated and b) base as a part that supports from beneath or serves as a foundation for an object or structure. As suggested in the literal sense, the structure (a building, bridge or piece of equipment) is separated from its foundation. The original terminology of base isolation is more commonly replaced with seismic isolation nowadays, reflecting that in some cases the separation is somewhere above the base – for example, in a building the superstructure may be separated from substructure columns. In another sense, the term seismic isolation is more accurate anyway in that the structure is separated from the effects of the seism, or earthquake (Kelly, 2001).

The only way a structure can be supported under gravity is to rest on the ground. Isolation conflicts with this fundamental structural engineering requirement.

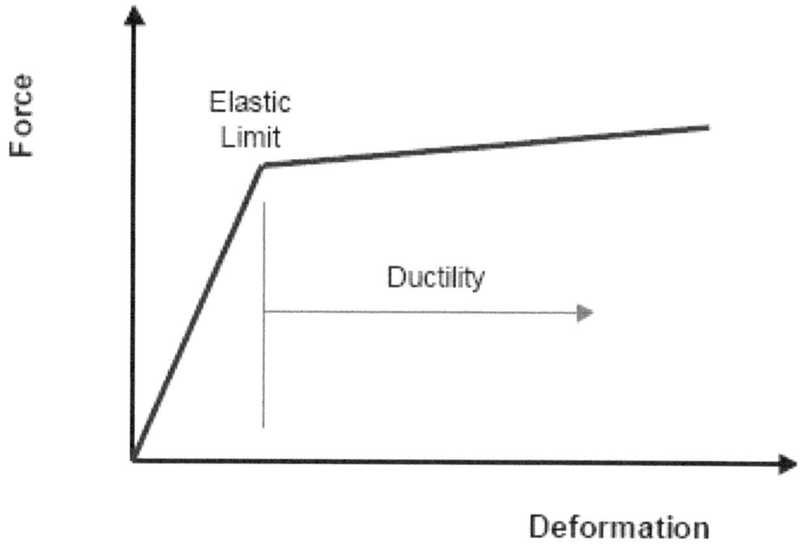

Figure 10. Ductility: deformation beyond elastic limit.

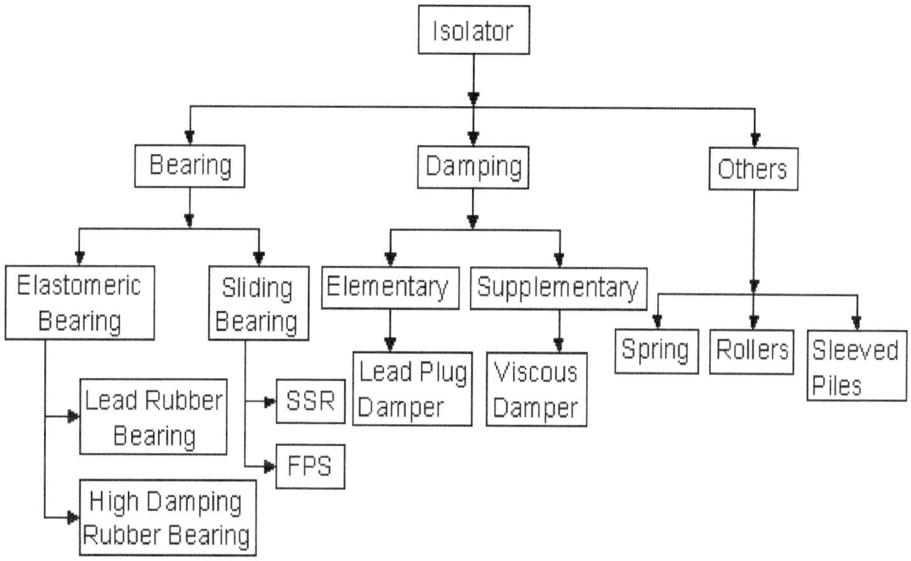

Figure 11. Schematic Diagram showing various types of Isolators used through out the world.

How can the structure be separated from the ground for earthquake loads but still resist gravity? It is practical isolation systems that provide a compromise between attachment to the ground to resist gravity and separation from the ground to resist earthquakes (Figure 7). Seismic isolation is a means of reducing the seismic demand on the structure:

Action of base isolation

Hence, "base isolation" or, seismic isolation separates upper structure from base or, from down structure by changing of fix joint with flexible one (Figure 10). Increasing of flexibility is done by the insertion of additional elements in structure, known as isolators. Usually, these isolators are inserted between upper structure and foundation (Figure 12). Seismic isolation system absorbs larger part of seismic energy. Therefore, vibration effects of soil to upper structure are drastically reduced. Figure 8 shows the failure pattern of a "fixed based" structure due to seismic loading.

But in case of isolated buildings as the ground moves, inertia tends to keep structures in place resulting in the imposition of structure with large displacements in different stories (Figure 9: left figure: dashed portion indicating displacements due to seismic loading).

For base isolated structure the situation is quite

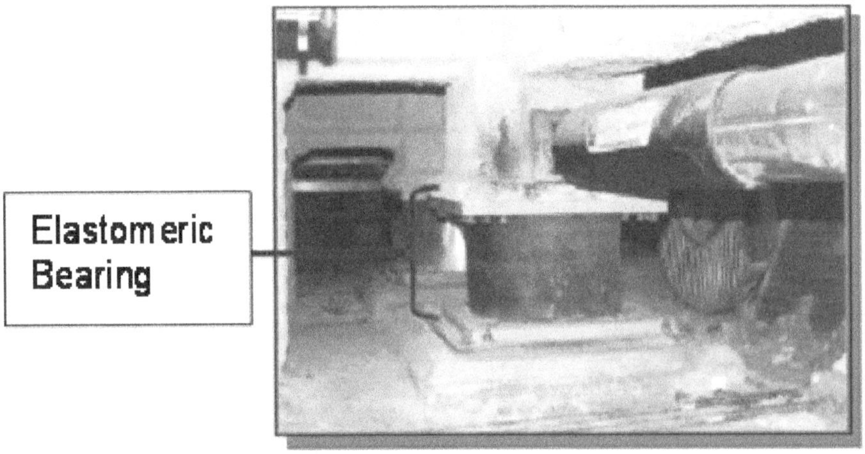

Figure 12. Installed elastomeric bearing.

different. In such cases, the whole upper structure gets a displacement (which naturally remains in limits) and the relative displacement of different stories is so small that the structure can withstand a comparatively high seismic tremor with a low seismic loading in a safe, efficient and economic manner.

Goal of base isolation

A high proportion of the world is subjected to earthquakes and society expects that structural engineers will design our buildings so that they can survive the effects of these earthquakes. As for all the load cases encountered in the design process, such as gravity and wind, should work to meet a single basic equation: CAPACITY > DEMAND. Earthquakes happen and are uncontrollable. So, in that sense, we have to accept the demand and make sure that the capacity exceeds it. The earthquake causes inertia forces proportional to the product of the building mass and the earthquake ground accelerations. As the ground accelerations increases, the strength of the building, the capacity, must be increased to avoid structural damage. But it is not practical to continue to increase the strength of the building indefinitely.

In high seismic zones the accelerations causing forces in the building may exceed one or even two times the acceleration due to gravity, g. It is easy to visualize the strength needed for this level of load – strength to resist 1 g means than the building could resist gravity applied sideways, which means that the building could be tipped on its side and held horizontal without damage.

Designing for this level of strength is not easy, nor cheap. So, most codes allow engineers to use ductility to achieve the capacity. Ductility is a concept of allowing the structural elements to deform beyond their elastic limit in a controlled manner (Figure 10). Beyond this limit, the structural elements soften and the displacements increase with only a small increase in force.

The elastic limit is the load point up to which the effects of loads are non- permanent; that is, when the load is removed the material returns to its initial condition. Once this elastic limit is exceeded changes occur. These changes are permanent and non-reversible when the load is removed.

A design philosophy focused on capacity leads to a choice of two evils:

1. Continue to increase the elastic strength. This is expensive and for buildings leads to higher floor accelerations. Mitigation of structural damage by further strengthening may cause more damage to the contents than would occur in a building with less strength.
2. Limit the elastic strength and detail for ductility. This approach accepts damage to structural components, which may not be repairable.

Base isolation takes the opposite approach, it attempts to reduce the demand rather than increase the capacity. We cannot control the earthquake itself but we can modify the demand it makes on the structure by preventing the motions being transmitted from the foundation into the structure above.

So, the primary reason to use isolation is to mitigate earthquake effects. Naturally, there is a cost associated with isolation and so it only makes sense to use it when the benefits exceed this cost. And, of course, the cost benefit ratio must be more attractive than that available from alternative measures of providing earthquake resistance.

Nowadays Base Isolation is the most powerful tool of the earthquake engineering pertaining to the passive structural vibration control technologies. It is meant to enable a building or non-building structure to survive a potentially devastating seismic impact through a proper initial design or subsequent modifications. In some cases, application of Base Isolation can raise both a structure's seismic performance and its seismic sustainability

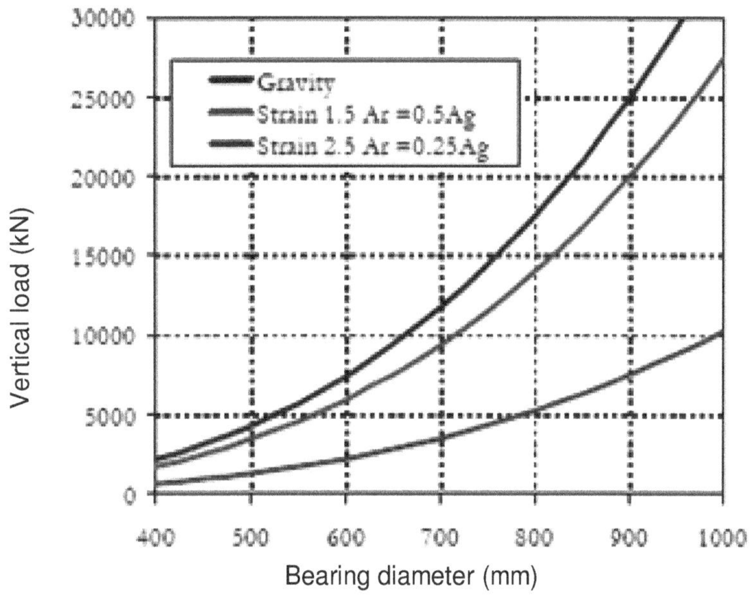

Figure 13. Load capacity of elastomeric bearings.

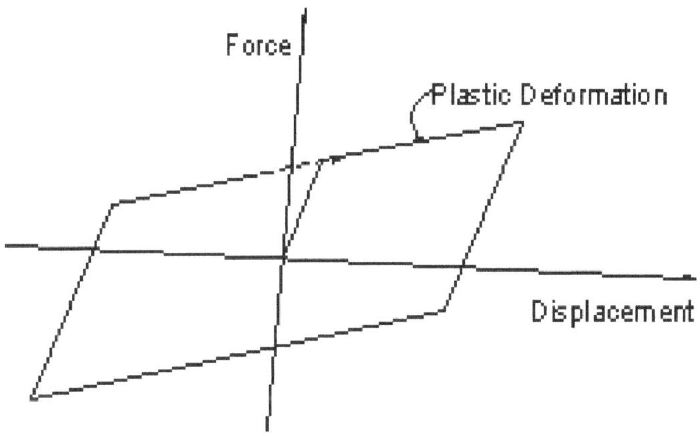

Figure 14. Area denoting dissipated energy.

considerably.

RECOGNITION OF ISOLATION TYPES

Many types of isolation system have been proposed and have been developed to varying stages, with some remaining no more than concepts and others having a long list of installed projects. A discussion of generic types of system and focus on different types (especially rubber bearing) of isolators along with their characteristics is provided subsequently.

Types of isolator

The development of isolators ensured the properties required for the achievement of perfect base isolation. The chart (Figure 11) details the various types of Isolators used through out the world. A brief description along with their basic functions and advantages is also included just after the chart. As the present research is mainly highlighting the use of LRB and HDRB type of isolator, so special attention is given to their characteristics.

Bearing type

Elastometric (rubber) bearings

Rubber bearings are formed of horizontal layers of natural or synthetic rubber in thin layers bonded between steel plates. The steel plates prevent the rubber layers from blown up or busting. In such mechanism the bearing is capable to support higher vertical loads with only smaller deflection (typically 1 to 3 mm under full gravity load).

The internal steel layers do not restrict horizontal deformations of the rubber layers in shear. So, the bearings are much more flexible under lateral loads than vertical loads. This is why; the bearing works as a flexible unit.

Characteristics of rubber isolator

As described earlier, isolation system works with the principle that a rigid mass is isolated from a flexible supporting structure.

Optimum isolation of a building from ground may be achieved by choosing a rubber bearing isolator based on the knowledge of its static and dynamic characteristics determined from laboratory experiments. For this reason, understanding the properties of rubber isolators is necessary for the vibration analysis. Some of the important properties are as follows.

Load capacity and size of rubber bearings

For most bearing types the plan size required increases as vertical load increases but the height or radius is constant regardless of vertical loads. This is because all bearings at such stages are subjected to the same displacement. Therefore, the bearing can be sized according to the vertical loads they support. Figure 13 (Kelly et al., 2006) gives a typical relationship between vertical load and bearing diameter where (vertical load absorbed as per bearing diameter). Three types of curves are seen in the plot. All the curves are showing that the lower the vertical load, the lower the required bearing diameter.

Absorption

Shocks originating due to the occurrence of an earthquake can be controlled if substantial additional damping is introduced into the isolation system. A high damping rubber isolator provides a substantial level of damping through hysteretic energy dissipation. Hysteretic refers to the offset between the loading and unloading curves under cyclic loading.

Figure 14 shows an idealized force- displacement loop where the enclosed area is measure of the energy dissipated during one cycle of motion.

Durability under cyclic loading

Rubber isolator remains more or less stable under cyclic loading. Results of cyclic displacement test applied, to a rubber isolator shows that at a speed equivalent to an actual seismic event the friction factor remains stable. Figure 15 shows a typical friction factor versus number of cycles in a cyclic loading test of rubber isolator. The figure represents that rubber isolator is durable. There are mainly two types of Rubber Bearing. They are LRB and HDRB.

Lead rubber bearing (LRB)

This type of elastomeric bearings consist of thin layers of low damping natural rubber and steel plates built in alternate layers and a lead cylinder plug firmly fitted in a hole at its centre to deform in pure shear as shown in Figure 16. The LRB was invented in New Zealand in 1975 and has been used extensively in New Zealand, Japan and United States. The steel plates in the bearing force the lead plug to deform in shear.

This bearing provides an elastic restoring force and also, by selection of the appropriate size of lead plug, produces required amount of damping. The force deformation behavior of the bearing is shown in Figure 16. Performance of LRB is maintained during repeated strong earthquakes, with proper durability and reliability.

Basic functions of LRB

(1) Load supporting function: Rubber reinforced with steel plates provides stable support for structures. Multilayer construction rather than single layer rubber pads provides better vertical rigidity for supporting a building.

(2) Horizontal elasticity function (prolonged oscillation period): With the help of LRB, earthquake vibration is converted to low speed motion. As horizontal stiffness of the multi- layer rubber bearing is low, strong earthquake vibration is lightened and the oscillation period of the building is increased.

(3) Restoration function: Horizontal elasticity of LRB returns the building to its original position. In a LRB, elasticity mainly comes from restoring force of the rubber layers. After an earthquake this restoring force returns the building to the original position.

(4) Damping function: Provides required amount of damping necessary.

LRB mainly are of two shapes. One is conventional round and the other type is square. Though their basic function remains same, yet changes in shapes are advantageous in many occasions as economy concern, reduced size, stability and capacity for large deformation.

High damping rubber bearing (HDRB)

HDRB is one type of elastomeric bearing. This type of bearing consist of thin layers of high damping rubber and steel plates built in alternate layers as shown in Figure 17. The vertical stiffness of the bearing is several hundred times the horizontal stiffness due to the presence of internal steel plates. Horizontal stiffness of the bearing is controlled by the low shear modulus of elastomer while steel plates provides high vertical stiffness as well as prevent bulging of rubber. High vertical stiffness of the bearing has no effect on the horizontal stiffness. The damping in the bearing is increased by adding extra-fine carbon block, oils or resins

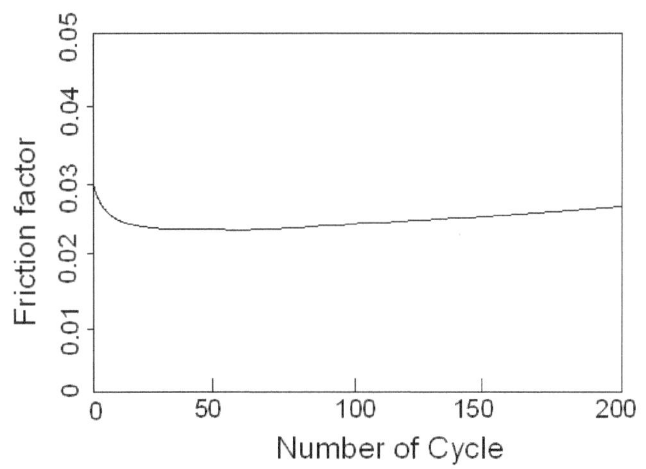

Figure 15. Typical cyclic durability test graph.

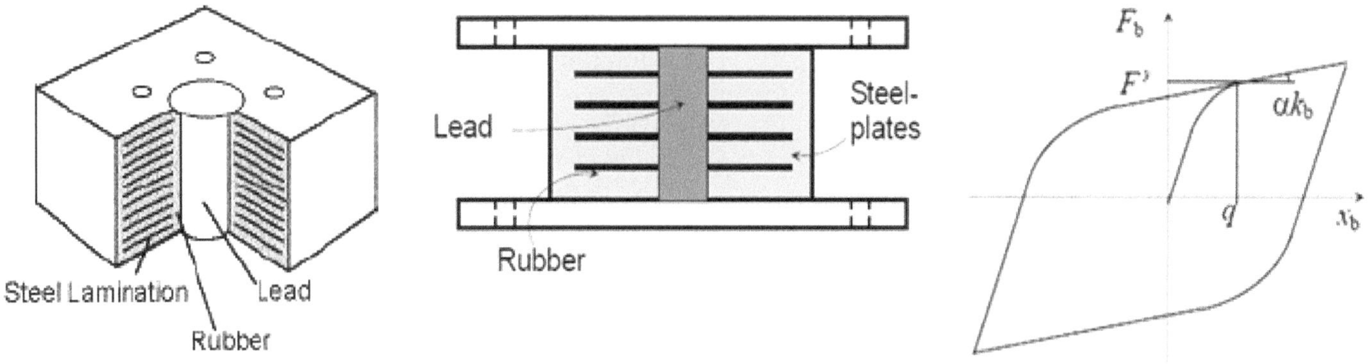

Figure 16. Geometry, schematic diagram and ideal force –deformation behavior of LRB.

and other proprietary fillers. The dominant features of HDRB system are the parallel action of linear spring and viscous damping. The damping in the isolator is neither viscous nor hysteretic, but some what in between. The ideal force deformation behavior of the bearing is shown in Figure 17.

Basic functions of HDRB

(1) Vertical load bearing function: Rubber reinforced with steel plates provides stable support for structures. Multilayer construction rather than single layer rubber pads provides better vertical rigidity for supporting a building.

(2) Horizontal elasticity function (prolonged oscillation period): With the help of HDRB earthquake vibration is converted to low speed motion. As horizontal stiffness of the multi- layer rubber bearing is low, strong earthquake vibration is lightened and the oscillation period of the building is increased.

(3) Restoration function: Horizontal elasticity of HDRB returns the building to its original position. In a HDRB, elasticity mainly comes from restoring force of the rubber layers. After an earthquake this restoring force returns the building to the original position.

(4) Damping function: Provides required amount of damping up to a higher value.

Sliding bearings

Sliding isolation system (Figure 18) is simple in concept and it has a theoretical appeal. A layer with a defined coefficient of friction will limit the acceleration to this value and the forces, which can be transmitted, will also be limited to the coefficient of friction multiplied by weight. Following are some of the utilities of using sliders:

(1) Sliding movement provides flexibility and the force-displacement traces a rectangular shape that is the optimum for equivalent viscous damping.

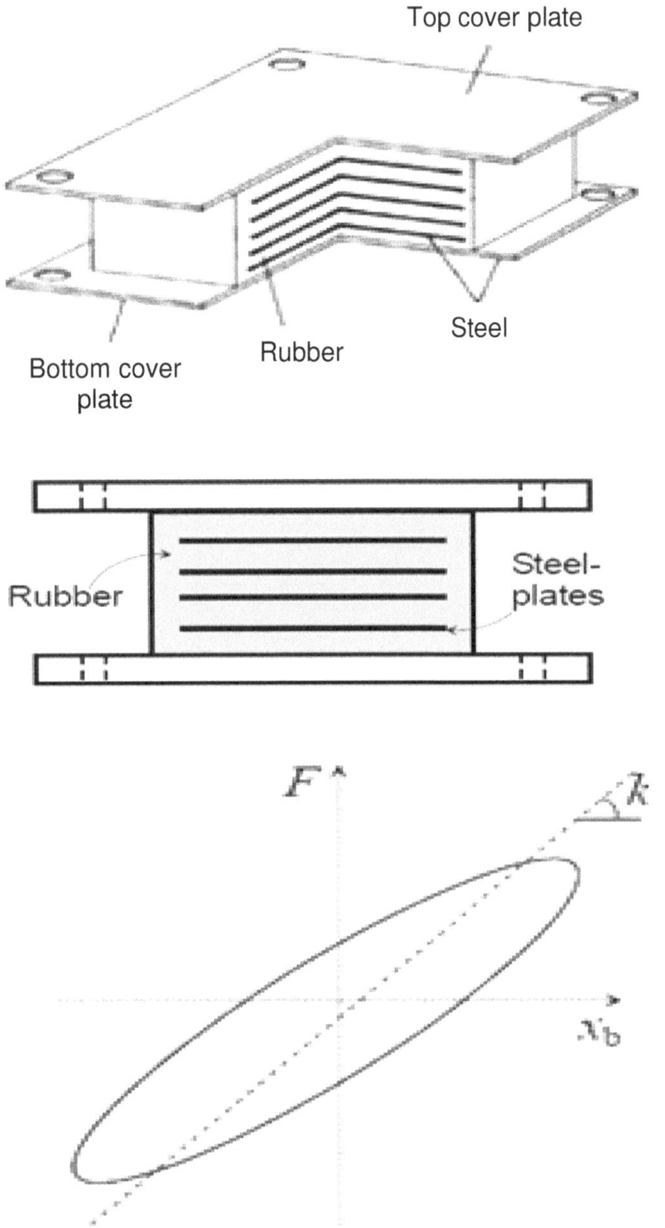

Figure 17. Geometry, schematic Diagram and ideal force – deformation behavior of HDRB.

(2) A pure sliding system will have unbounded displacements, with an upper limit equal to the maximum ground displacements for a coefficient of friction close to zero.

Two types of sliding systems are commonly used. A brief description with their basic functions, advantages and suitability is as follows.

Sliding support with rubber-pad (SSR)

When sliding base isolation system incorporates multilayer natural rubber pad then it is known as SSR.

Advantages of such bearings are:

(1) SSR can provide vibration isolation for light loads as well as large deformation performance like a large-scale isolation system.
(2) It provides protection against a wide range of tremors from small vibrations to major earthquakes.
(3) It can be used in conjunction with other isolation systems such as LRB, and HDRB.

Basic functions of SSR

For small vibrations, shear deformation of the rubber layers provides the same isolation effect as conventional multilayer rubber bearings.

For large vibrations, sliding materials slide to provide the same deformation performance as large-scale isolation systems.

Friction pendulum system (FPS)

Sliding friction pendulum isolation system (Figure 19) is one type of flexible isolation system suitable for small to large-scale buildings. It combines sliding a sliding action and a restoring force by geometry. Functions of FPS are same as SSR system.

Advantages of FPS include:

(1) It is possible to set the oscillation period of a building regardless of its weight.
(2) This system can reduce costs not only because of the low cost of its device but also due to the low cost of installation.
(3) The device is simple, works well and easy to install. Furthermore, it saves space and is practical for a seismic reinforcement.
(4) Performance of such device is stable due to the high durability of the device.
(5) As it requires only a simple visual check to maintain the device, maintenance is very easy.

Damping type

Damping provides sufficient resistance to structure against service loading. The effect of damping on dynamic response is beneficial. Generally all structural systems exhibit damping to various degrees. It is assumed that, structural damping is viscous by nature. Damping coefficient relates force to velocity. If damping coefficient is sufficiently large, it is possible to totally restrain the oscillatory motion. Damping that suppress totally the oscillatory motion is termed as critical damping. Damping is usually neglected in frequency and period calculations unless it exceeds about 20%.

Normally two types of damping are used in building. A

Figure 18. Installed sliding bearing.

brief description follows.

Elementary damping

This really means damping as a whole, i.e. the device (each and every element) itself acts as a purely damping device rather than an isolator. Purely damping devices can be used in low weight buildings to restrain the oscillating motion of the building. Another option may be using in conjunction with rubber bearing so-called as High Damping Rubber bearing.

In such devices amount of damping is significantly high, usually from 8 to 15% of critical damping. Lead plug damper is one of the forms of elementary damping.

The basic functions of such damper include:

(a) Vibration damping Junction: Lead plug damper absorbs large vibration of the building. As the layers of rubber are distorted, the lead plug is plastically deformed and at such stage it absorbs the earthquake energy and quickly damps the vibration.

(b) Trigger function: It also reduces vibration form source other than earthquake. For example, when vibration is generated by strong winds, the relative rigidity of the lead plug reduces the effect of such vibration.

Supplementary damping

There are some types of isolators (discussed in bearing portion), which are capable of providing flexibility but not significant damping, or resistance to service loads. In order to strengthen the damping phenomena, supplementary devices are included with general Isolators. Damping of such type can be termed as supplementary damping. One of the most popular types of supplementary damping is viscous damping. This device provides damping but not service load resistance. It does not have any elastic stiffness and for this reasons it adds less force to the system than other devices.

Other types

Apart from bearing and sliding type, there are some other types of isolators, which are also used in building but rarely. Springs, rollers, sleeved piles are some examples of such isolators. A brief description of them is also included here.

Springs

Spring isolators are devices whose working mechanism is based on steel springs. They are mostly used for machinery isolation. The main drawbacks of springs are two. Firstly, they are most flexible in both the vertical and horizontal directions. Secondly, springs alone have little damping and will move excessively under service loads.

Rollers

Cylindrical rollers and ball bearings are of this type. Like springs they are commonly used for machinery isolation. The resistance to movement and damping of rollers and ball bearings are sufficient under service loads.

Sleeved piles

The pin ended structural members, that is, piles inside a sleeve provide flexibility and allow movement of the soft first story in a building. This type of piles is known as sleeved piles. Sleeved piles provide flexibility but no

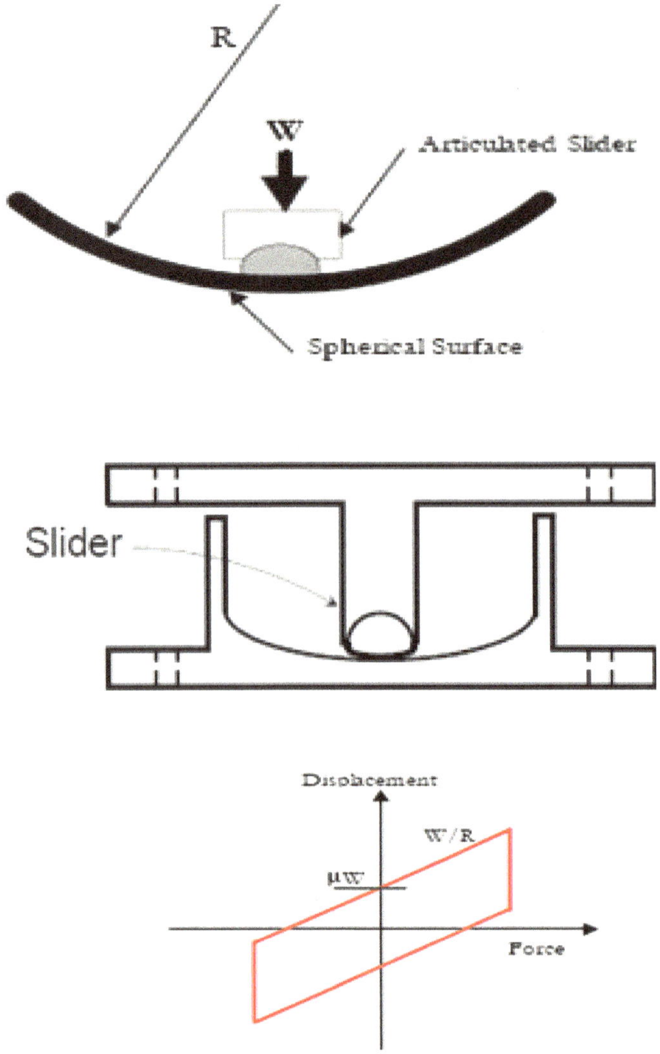

Figure 19. Geometry, schematic diagram and ideal force – deformation behavior of fps.

damping. Hence damping devices are required to use along with sleeved piles.

Design life of isolators

Most isolation systems are based on natural rubber bearings which have a long record of excellent in-service performance. As part of prototype testing, rubber tests including ozone testing and high temperature tests to simulate accelerated aging are performed to ensure the environmental resistance and longevity of the system. All steel components of the elastomeric based bearings are encased in a protective cover rubber except for the load plates. These plates are usually coated with a protective paint. The protective coating system adopted for the Museum of New Zealand, which had a specified 150 year design life, was a deposited metal paint system.

Fire resistance

Isolation bearings are generally required to achieve a fire rating equivalent to that required for the vertical load carrying assemblies.

One of two approaches can be used to provide acceptable fire rating with the method used decided on depending on specific project needs:

1. Design surrounding flexible protective "skirts" for the bearings as was done for Parliament Buildings.
2. Rate the resistance of rubber itself based on vertical load, dimensions and fire loading to determine whether it is more economical to provide the fire rating by providing extra cover rubber to the bearing.

Temperature range for installation

Base isolators are designed to resist the maximum seismic displacements plus the total R+S+T (creep, shrinkage and thermal) displacements. Therefore, the temperature of installation does not matter from a technical perspective. For aesthetic reasons it is desirable to install the bearings at as close to mean temperature as possible so that the bearings are not in a deformed configuration for most of their service life.

Availability of isolator

There are continual changes in the list of isolation system suppliers as new entrants commence supply and existing suppliers extend their product range. A large number of manufacturers of elastomeric bearings worldwide are also providing the requirements as these bearings are widely used. These manufacturers may offer to supply isolation systems such as lead-rubber and high damping rubber bearings, sliding bearings. However, standard bearings are designed to operate at relatively low strain levels of about 25%. Isolation bearings in high seismic zones may be required to operate at strain levels ten times this level, up to 250%. The manufacturing processes required to achieve the level of performance are much more stringent than for the lower strain levels. In particular, the bonding techniques are critical and the facilities must be of clean-room standard to ensure no contamination of components during assembly.

INSTALLATION OF ISOLATOR

The main requirement for installation (Figure 20) of a

Figure 20. Isolator installation in building.

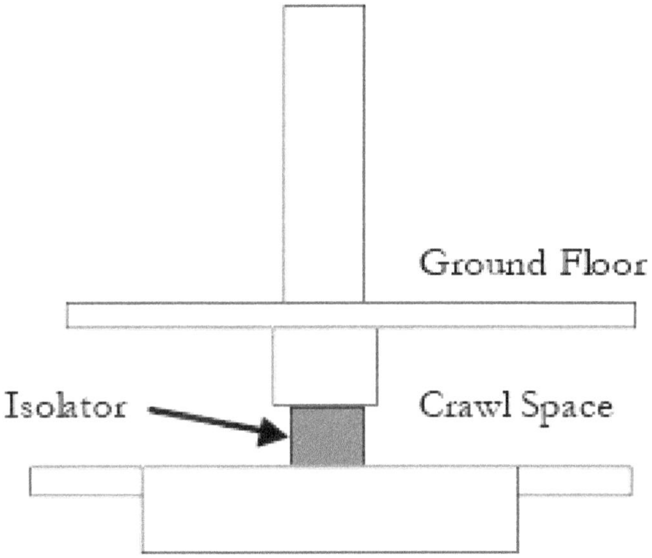

Figure 21. Isolator installation location in building with no basement.

base isolation system, that is, isolator is that the building be able to move horizontally relative to the ground, usually at least 100 mm and in some instance up to 1 m. A plane of separation must be selected to permit this movement. Final selection of the location of this plane depends on the structure.

Isolator installation techniques

The most common configuration is to install a diaphragm immediately above the isolators. This permits earthquake loads to be distributed to the isolators according to their stiffness. For a building without a basement, the isolators are mounted on foundation pads and the structure is constructed above them (Figure 21). If the building has a basement then the options are to install the isolators at the top, bottom or mid height of the basements, columns and walls (Figure 22).

In this study isolation technique is being discussed for a building without a basement as well as a building which has a basement presenting the practical example mentioning sequence in details.

Practical example of isolator installation in building

The installation technique for low to medium seismic region is of utmost importance. Here the gradual procedure about the installation of an isolator in an existing building is being introduced. The installation sequence of isolator into a reinforced concrete column is described below:

(a) Temporary steel columns on suitable foundations were installed to either side of the reinforced concrete column into which the bearing was to be installed (Figure 23). Hydraulic jacks were placed at the heads of the temporary columns, bearing onto the beams at first floor level, and stressed to a predetermined level, calculated as the gravity load in the permanent column from the analysis. The hydraulic fluid in the jacks is locked off.

(b) Bench marks were introduced on to the column just above and below the final position of the bearing, and measurements taken, to enable subsequent checks to be

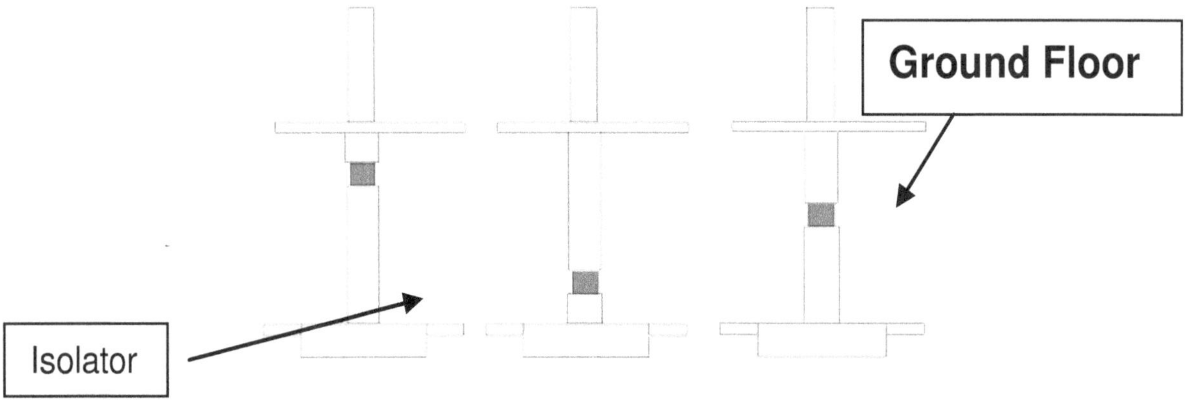

Figure 22. Isolator installation location in building having basement.

Figure 23. Installation of temporary steel column.

performed of possible movements of the column.

(c) Two horizontal cuts were made in the column using a diamond chain saw (Figure 24a). The block of concrete in between was removed (Figure 24b). The movements of the column above and below the cuts were then measured; in most cases this was small, but could reach as much as 6 mm (0.24 in.). This was considered acceptable.

A bed of epoxy mortar was placed on the lower half of the cut surface, and the LRB was then rolled into place on steel ball bearings. The gap above the bearing was then filled with epoxy mortar. The hydraulic jacks in the steel props were released and the props were removed after curing of the epoxy mortar.

(d) Steel jackets was welded into place above and below the bearing, and grouted to the column, to accommodate the stress concentrations at the cut surfaces of the column arising from the bearing and to replace the reinforcement that had been cut in Step c (Figure 25)

(e) The bearings were wrapped in fire insulation, and brackets introduced to support architectural finishes (Figure 26).

Conclusion

The obligations for practical isolation system to be incorporated in building structures are flexibility, Damping and resistance to service loads. Additional requirements such as durability, cost, ease of installation and specific project requirements influences device selection but all practical systems should contain these essential elements.

The entire superstructure is to be supported on discrete isolators whose dynamic characteristics are chosen to uncouple the ground motion. Displacement and yielding are concentrated at the level of the isolation devices, and the superstructure behaves very much like a rigid body. A through revise has been done performed in this research regarding the sequential development of seismic isolation systems.

This study also addressed the detail cram on isolation system, properties, characteristics of various device categories, recognition along with its effect on building structures. Meticulous schoolwork has also been accomplished about installation technique for various site stipulation at low to medium seismic region.

Rigorous reckonings illustrated the isolation system as very innovative and suitable in buildings to withstand the seismic lateral forces and also are contributed safety

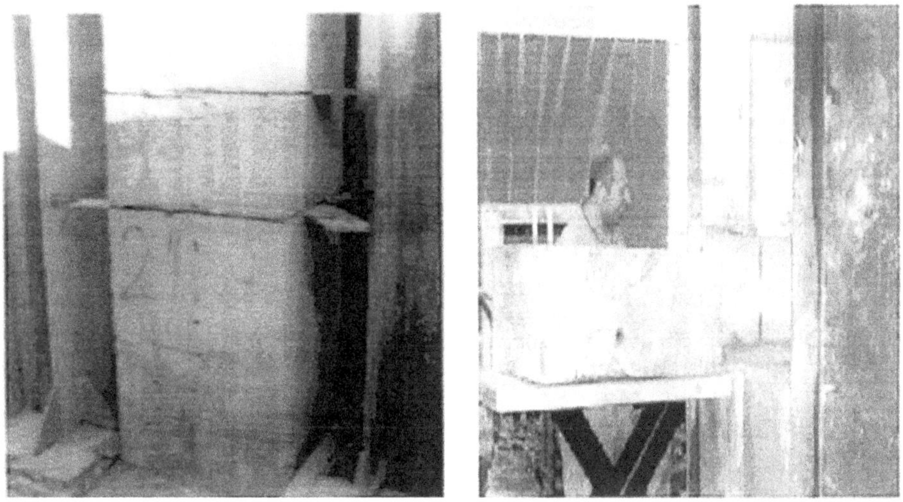

(a) (b)

Figure 24. Cutting and removal of concrete block. (a) Saw cuts through concrete Column, (b) removal of concrete Block.

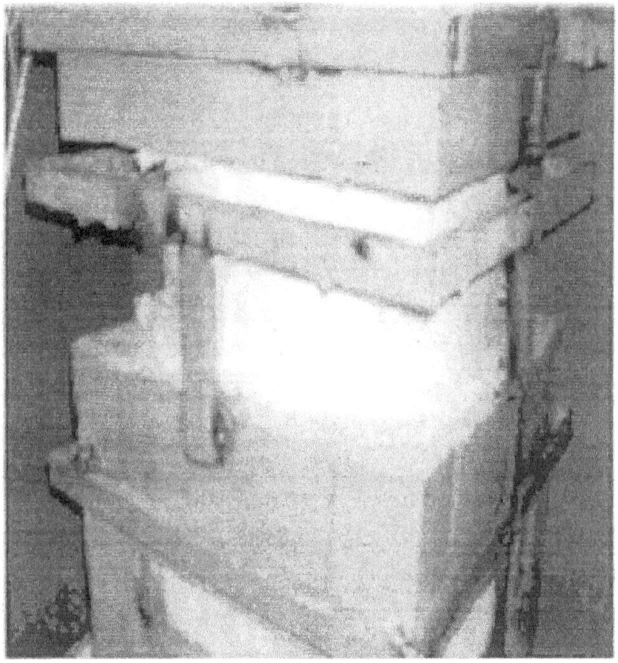

Figure 26. Installation of fire proofing elements.

Figure 25. Steel jackets replacing severed reinforcing bars.

ensuring flexibility of structures.

ACKNOWLEDGEMENTS

The authors gratefully acknowledge the effort by Bangladesh Earthquake Society as well as the grant provided from the University of Malaya to fund the research work.

REFERENCES

Ariga T, Kanno Y, Takewaki I (2006). Resonant behaviour of base-isolated high-rise buildings under long-period ground motions. Structural Design Tall Special Buildings, 15: 325-338.

Balkaya C, Kalkan E (2003). Nonlinear seismic response evaluation of tunnel form building structures. Comput. Struct., 81: 153-165.

Bhuiyan AR, Okui Y, Mitamura H, Imai T (2009). A rheology model of high damping rubber bearings for seismic analysis: Identification of nonlinear viscosity. Int. J. Solids Struct., 46: 1778-1792.

Casciati F, Hamdaoui K (2008). Modelling the uncertainty in the response of a base isolator. Probabilistic Eng. Mechanics. 23: 427-437.

Cheng FY, Jiang H, Lou K (2008). Smart Structures, Innovative Systems for Seismic Response Control: CRC Press. 672 p.

Dall'Asta A, Ragni L (2006). Experimental tests and analytical model of high damping rubber dissipating devices. Eng. Struct., 28: 1874-1884.

Dall'Asta A, Ragni L (2008). Nonlinear behavior of dynamic systems with high damping rubber devices. Eng. Struct., 30: 3610-3618.

Deb S (2004). Seismic base isolation-an overview. Curr. Sci., 87: 1426-1430.

Deb SK (2004). Seismic base isolation-an overview. Curr. Sci., 87(10): 1426-1430.

Di Egidio A, Contento A (2010). Seismic response of a non-symmetric rigid block on a constrained oscillating base. Eng. Struct., 32: 3028-3039.

Dicleli M, Buddaram S (2007). Comprehensive evaluation of equivalent linear analysis method for seismic-isolated structures represented by sdof systems. Eng. Struct., 29: 1653-1663.

Hong W, Kim H (2004). Performance of a multi-story structure with a resilient-friction base isolation system. Comput. Struct.. 82: 2271-2283.

Hussain RR, Islam ABMS, Ahmad SI (2010). Base Isolators as Earthquake Protection Devices in Buildings: VDM Publishing House Ltd. Benoit Novel, Simultaneously published in USA & U.K. 140 p.

ICBO (International Conference of Building Officials) (1997). Uniform Building Code. ICBO, Whittier, CA, USA.

Islam ABMS (2009). Evaluation of Structural and Economic Implications of Incorporating Base Isolator as Earthquake Protection Device in Buildings in Dhaka [Master's Thesis]. Dhaka, Bangladesh: Bangladesh University of Engineering and Technology (BUET). p. 118.

Islam ABMS, Ahmad SI (2010). Isolation System Design for Buildings in Dhaka: Its Feasibility and Economic Implication. Proceedings of the International Conference on Engineering Research, Innovation and Education. SUST. 11-13 January; Bangladesh, Sylhet. pp. 99-104.

Islam ABMS, Ahmad SI, Jameel M, Jumaat MZ (2010a). Seismic Base Isolation for Buildings in Regions of Low to Moderate Seismicity: A Practical Alternative Design. Practice Periodical on Structural Design and Construction, ASCE. [DOI: 10.1061/(ASCE)SC.1943-5576.0000093]

Islam ABMS, Ahmad SI, Al-Hussaini TM (2010b). Effect of Isolation on Buildings in Dhaka. 3rd International Earthquake Symposium. BES. 5-6 March; Bangladesh, Dhaka. Pp. 465-472.

Ismail M, Rodellar J, Ikhouane F (2010). An innovative isolation device for aseismic design. Engineering Structures. 32: 1168-1183.

Jangid RS (2007). Optimum lead-rubber isolation bearings for near-fault motions. Eng. Struct., 29: 2503-2513.

Kelly JM (1998). "Base Isolation: Origins and Development" A Paper submitted in National Information Services for Earthquake Engineering (NISEE), University of California, Berkeley.

Kelly TE (2001). Base isolation of structures: design guidelines. Auckland: Holmes Consulting Group Ltd. available from www.holmesgroup.com, 2001.

Kelly TE, Robinson WH, Skinner RI (2006). Seismic Isolation for Designers and Structural Engineers. Wellington: Robinson seismic Ltd. available from www.rslnz.com, 2006.

Kilar V, Koren D (2009). Seismic behaviour of asymmetric base isolated structures with various distributions of isolators. Eng Struct., 31: 910-921.

Komodromos P (2008). Simulation of the earthquake-induced pounding of seismically isolated buildings. Comput. Struct., 86: 618-626.

Komodromos P, Polycarpou P, Papaloizou L, Phocas M (2007). Response of seismically isolated buildings considering poundings. Earthquake Eng. Struct. Dynamics, 36: 1605-1622.

Lu L-Y, Lin G-L (2008). Predictive control of smart isolation system for precision equipment subjected to near-fault earthquakes. Eng. Struct., 30: 3045-3064.

Matsagar VA, Jangid RS (2004). Influence of isolator characteristics on the response of base-isolated structures. Eng. Struct., 26: 1735-1749.

Micheli I, Cardini S, Colaiuda A, Turroni P (2004). Investigation upon the dynamic structural response of a nuclear plant on aseismic isolating devices. Nuclear Eng. Design, 228: 319-343.

Olsen A, Aagaard B, Heaton T (2008). Long-period building response to earthquakes in the San Francisco Bay Area. Bull. Seismological Society Am., 98(2): 1047-1065.

Pocanschi A, Phocas MC (2007). Earthquake isolator with progressive nonlinear deformability. Eng. Struct., 29: 2586-2592.

Polycarpou PC, Komodromos P (2010). Earthquake-induced poundings of a seismically isolated building with adjacent structures. Eng. Struct., 32: 1937-1951.

Providakis CP (2008). Effect of LRB isolators and supplemental viscous dampers on seismic isolated buildings under near-fault excitations. Eng. Struct., 30: 1187-1198.

Skinner RI, Robinson WH, Mcverry GH (1993). An Introduction to Seismic Isolation. John Wiley and Sons Ltd.

Spyrakos CC, Koutromanos IA, Maniatakis CA (2009). Seismic response of base-isolated buildings including soil-structure interaction. Soil Dynamics Earthquake Eng., 29: 658-668.

Wibowo A, Wilson JL, Lam NTK, Gad EF (2010). Collapse modelling analysis of a precast soft storey building in Australia. Eng. Struct., 32: 1925-1936.

Permissions

All chapters in this book were first published in JETR, by Academic Journals; hereby published with permission under the Creative Commons Attribution License or equivalent. Every chapter published in this book has been scrutinized by our experts. Their significance has been extensively debated. The topics covered herein carry significant findings which will fuel the growth of the discipline. They may even be implemented as practical applications or may be referred to as a beginning point for another development.

The contributors of this book come from diverse backgrounds, making this book a truly international effort. This book will bring forth new frontiers with its revolutionizing research information and detailed analysis of the nascent developments around the world.

We would like to thank all the contributing authors for lending their expertise to make the book truly unique. They have played a crucial role in the development of this book. Without their invaluable contributions this book wouldn't have been possible. They have made vital efforts to compile up to date information on the varied aspects of this subject to make this book a valuable addition to the collection of many professionals and students.

This book was conceptualized with the vision of imparting up-to-date information and advanced data in this field. To ensure the same, a matchless editorial board was set up. Every individual on the board went through rigorous rounds of assessment to prove their worth. After which they invested a large part of their time researching and compiling the most relevant data for our readers.

The editorial board has been involved in producing this book since its inception. They have spent rigorous hours researching and exploring the diverse topics which have resulted in the successful publishing of this book. They have passed on their knowledge of decades through this book. To expedite this challenging task, the publisher supported the team at every step. A small team of assistant editors was also appointed to further simplify the editing procedure and attain best results for the readers.

Apart from the editorial board, the designing team has also invested a significant amount of their time in understanding the subject and creating the most relevant covers. They scrutinized every image to scout for the most suitable representation of the subject and create an appropriate cover for the book.

The publishing team has been an ardent support to the editorial, designing and production team. Their endless efforts to recruit the best for this project, has resulted in the accomplishment of this book. They are a veteran in the field of academics and their pool of knowledge is as vast as their experience in printing. Their expertise and guidance has proved useful at every step. Their uncompromising quality standards have made this book an exceptional effort. Their encouragement from time to time has been an inspiration for everyone.

The publisher and the editorial board hope that this book will prove to be a valuable piece of knowledge for researchers, students, practitioners and scholars across the globe.

List of Contributors

V. Ramamurti
Indian Institute of Technology Madras (IIT Madras), Chennai, India

S. Mithun
WABCO India Limited, Plot No. 3 (SP), III Main Road, Ambattur Industrial Estate, Chennai, Tamil Nadu 600058, India

N. Prabhakar
WABCO India Limited, Plot No. 3 (SP), III Main Road, Ambattur Industrial Estate, Chennai, Tamil Nadu 600058, India

T. Sukumar
WABCO India Limited, Plot No. 3 (SP), III Main Road, Ambattur Industrial Estate, Chennai, Tamil Nadu 600058, India

Samanpreet Kaur
Department of Soil and Water Engineering PAU, Ludhiana - 141004, India

Rajan Aggarwal
Department of Soil and Water Engineering PAU, Ludhiana - 141004, India

Satvinder Singh
Zonal Research Station for Kandi Area, Ballowal Saunkhri, Distt Nawanshaher, India

Harjit Singh Gulati
Department of Soil and Water Engineering PAU, Ludhiana - 141004, India

Kamal Karunananda
Department of Civil and Environmental Engineering, Ehime University, Bunkyo-cho 3, Matsuyama 790-8577, Japan

Mitao Ohga
Department of Civil and Environmental Engineering, Ehime University, Bunkyo-cho 3, Matsuyama 790-8577, Japan

Ranjith Dissanayake
Department of Civil Engineering, University of Peradeniya, Peradeniya 20400, Sri Lanka

Sudath Siriwardane
Department of Civil and Environmental Engineering, Ehime University, Bunkyo-cho 3, Matsuyama 790-8577, Japan

S. P. S. Matharu
Department of Mechanical Engineering, National Institute of Technology Raipur (C.G.) 492010, India

S. Sanyal
Department of Mechanical Engineering, National Institute of Technology Raipur (C.G.) 492010, India

D. S. Bal
Directorate of Technical Education, Government of Chattisgarh, Raipur (C.G.) 492001, India

P. R. P. Hoole
Department of Electrical Engineering, University of Malaya, 50603 Kuala Lumpur, Malaysia

S. R. H. Hoole
Department of Engineering and Science, Rensselaer Polytechnic Institute, USA

B. S. Shankar
Department of Civil Engineering, East Point College of Engineering and Technology, Bidarahalli, Virgonagar Post, Bangalore-560 049, Karnataka, India

S. A. Arabi
Centre for Energy Research and Training, Ahmadu Bello University, Zaria, Kaduna State, Nigeria

B. B. M. Dewu
Centre for Energy Research and Training, Ahmadu Bello University, Zaria, Kaduna State, Nigeria

A. M. Muhammad
Centre for Energy Research and Training, Ahmadu Bello University, Zaria, Kaduna State, Nigeria

M. B. Ibrahim
Centre for Energy Research and Training, Ahmadu Bello University, Zaria, Kaduna State, Nigeria

J. D. Abafoni
Department of Physics, University of Maiduguri, Maiduguri, Borno State, Nigeria

L. Devi
Department of Electrical Engineering, Intellectual Engineering College, Anantapur, AP, India

C. H. Sai Babu
Department of Electrical Engineering, Jawaharlal Nehru Technological University, Kakinada, AP, India

S. Siva Nagaraju
Department of Electrical Engineering, Jawaharlal Nehru Technological University, Kakinada, AP, India

M Gholizadeh
Chemistry Department, Ferdowsi University of Mashhad, Mashhad, Iran

H Arabshahi
Physics Department, Ferdowsi University of Mashhad, Mashhad, Iran

N. G.Goki
Department of Geology and Mining, Nasarawa State University, PMB 1022 Keffi, Nigeria

A. N. Amadi
Department of Geology, Federal University of Technology, PMB 65, Minna, Nigeria

P. I. Olasehinde
Department of Geology, Federal University of Technology, PMB 65, Minna, Nigeria

S. S. Dada
Department of Earth Sciences, Salem University, Lokoja, Nigeria

E. A. Ikpokonte
Department of Geology, Ahmadu Bello University, Zaria, Nigeria

J. I. D. Adekeye
Department of Geology and Mining, Nasarawa State University, PMB 1022 Keffi, Nigeria

S. K. Bhullar
Department of Mechanical Engineering, University of Victoria, Victoria, B.C. Canada V8W 3P6, Canada

J. L. Wegner
Department of Mechanical Engineering, University of Victoria, Victoria, B.C. Canada V8W 3P6, Canada

A. Mioduchowski
Department of Mechanical Engineering, University of Alberta, Edmonton, AB, Canada, T6G 2G8, Canada

Dagnachew Adugna Belete
Addis Ababa, Ethiopia

R. K. Tyagi
Department of Mechanical Engineering, Birla Institute of Technology, Mesra Ranchi, Jharkhand, India

K. K. Srivastava
Department of Mechanical Engineering, Birla Institute of Technology, Mesra Ranchi, Jharkhand, India

R. S. Pandey
Department of Applied Physics, Amity School of Engineering and Technology, Amity University, Uatter Pradesh, Noida, India

G. L. Oyekan
Department of Civil and Environmental Engineering, University of Lagos, Nigeria

O. M. Kamiyo
Department of Mechanical Engineering, University of Lagos, Nigeria

Dayal R. Parhi
Department of Mechanical Engineering, National Institute of Technology, Rourkela, India

Sasanka Choudhury
Department of Mechanical Engineering, National Institute of Technology, Rourkela, India

I. G. ADAMU
Department of Agricultural Engineering Technology, Adamawa State College of Agriculture, P. M. B. 2088 Ganye, Nigeria

H. U. KABRI
Department of Agricultural and Environmental Engineering, Modibbo Adama University of Technology, P. M. B. 2076 Yola, Nigeria

I. D. HUSSAINI
Department of Agricultural Engineering Technology, Adamawa State College of Agriculture, P. M. B. 2088 Ganye, Nigeria

A. D. MADA
Department of Agricultural Engineering Technology, Adamawa State College of Agriculture, P. M. B. 2088 Ganye, Nigeria

Dayal R. Parhi
Department of Mechanical Engineering, National Institute of Technology, Rourkela, India

Sasanka Choudhury
Department of Mechanical Engineering, National Institute of Technology, Rourkela, India

Y. Santosh Kumar Reddy
Department of Mechanical Engineering, Sri Venkateswara College of Engineering and Technology, Chittoor, Andhra Pradesh, India

C. Nadha Muni Reddy
Sri Venkatesa Perumal College of Engineering and Technology, Puttur, Chittoor (District) Andhra Pradesh, India

K. Hema Chandra Reddy
Academic Planning Unit, Jawaharlal Nehru Technological University, Anantapur Andhra Pradesh, India

H. M. A. Mahzuz
Department of Civil and Environmental Engineering, Shahjalal University of Science and Technology, Sylhet, Bangladesh

Mushtaq Ahmed
Department of Civil and Environmental Engineering, Shahjalal University of Science and Technology, Sylhet, Bangladesh

Md. Ashrafuzzaman
Department of Civil and Environmental Engineering, Shahjalal University of Science and Technology, Sylhet, Bangladesh

Rejaul Karim
Department of Civil and Environmental Engineering, Shahjalal University of Science and Technology, Sylhet, Bangladesh

Raju Ahmed
Department of Civil and Environmental Engineering, Shahjalal University of Science and Technology, Sylhet, Bangladesh

B. K. C. Ganesh
Department of Mechanical Engineering, College of Engineering (Autonomous), Andhra University, Andhra Pradesh India

N. Ramanaih
Department of Mechanical Engineering, College of Engineering (Autonomous), Andhra University, Vishakhapatnam, India

L. O. Mudasiru
Department of Mechanical Engineering, Ladoke Akintola University of Technology, Ogbomosho, Nigeria

I. A. Babatunde
Department of Mechanical Engineering, Ladoke Akintola University of Technology, Ogbomosho, Nigeria

W. A. Raheem
Department of Mechanical Engineering, Ladoke Akintola University of Technology, Ogbomosho, Nigeria

A. K. Lasisi
Department of Mechanical Engineering, Kwara State Polytechnic, Ilorin, Kwara State, Nigeria

S. Bali Reddy
Indian Institute of Technology Guwahati, India

A. V. Narashima Rao
Department of Civil Engineering, S. V. University, Tirupathi, India

A. Atta
Physics Department, National Center for Radiation Research and Technology (NCRRT), Atomic Energy Authority (AEA), Cairo Egypt

A. M. Abdel Reheem
Accelerators and Ion Sources Department, Nuclear Research Center, Atomic Energy Authority, P. O. Box 13759 Inchas, Cairo, Egypt

M. M. Abdel Rahman
Accelerators and Ion Sources Department, Nuclear Research Center, Atomic Energy Authority, P. O. Box 13759 Inchas, Cairo, Egypt

S. R. Ansari
Department of Mathematics, H. N. B. Garhwal University, Srinagar (Garhwal), India

P. K. Mittal
Department of Mathematics and R&D, Graphic Era University, Dehradun, Uttarakhand, India

Ritu Chandna
Department of Mathematics and R&D, Graphic Era University, Dehradun, Uttarakhand, India

Sameh S. Habib
Mechanical Engineering Department, Shoubra Faculty of Engineering, Benha University, Cairo, Egypt

Mohammed Mansour Kadhum Alkafaji
Civil Department, College of Engineering, Babylon University/ Babil, Iraq

O. R. Adetunji
Mechanical Engineering Department, College of Engineering, University of Agriculture, Abeokuta P. M. B. 2240, Abeokuta, Nigeria

P. O. Aiyedun
Mechanical Engineering Department, College of Engineering, University of Agriculture, Abeokuta P. M. B. 2240, Abeokuta, Nigeria

O. J. Alamu
Mechanical Engineering Department, Osun State University, Oshogbo, Osun State, Nigeria

A. S. Surakat
Mechanical Engineering Department, College of Engineering, University of Agriculture, Abeokuta P. M. B. 2240, Abeokuta, Nigeria

P. Sreeraj
Department of Mechanical Engineering, Younus College of Engineering and Technology, Kerala, 691010 India

T. Kannan
SVS College of Engineering, Coimbatore, Tamilnadu, 642109 India

S. Maji
Department of Mechanical Engineering, IGNOU, Delhi, 110068, India

A. B. M. Saiful Islam
Department of Civil Engineering, University of Malaya, Kuala Lumpur, Malaysia

Mohammed Jameel
Department of Civil Engineering, University of Malaya, Kuala Lumpur, Malaysia

Mohd Zamin Jumaat
Department of Civil Engineering, University of Malaya, Kuala Lumpur, Malaysia